HUMAN NATURE AND NATURAL KNOWLEDGE

BOSTON STUDIES IN THE PHILOSOPHY OF SCIENCE

EDITED BY ROBERT S. COHEN AND MARX W. WARTOFSKY

VOLUME 89

HUMAN NATURE AND NATURAL KNOWLEDGE

Essays Presented to Marjorie Grene on the Occasion of Her Seventy-Fifth Birthday

Edited by

ALAN DONAGAN
California Institute of Technology

ANTHONY N. PEROVICH, JR.
Hope College

and

MICHAEL V. WEDIN
University of California at Davis

D. REIDEL PUBLISHING COMPANY

A MEMBER OF THE KLUWER ACADEMIC PUBLISHERS GROUP

DORDRECHT / BOSTON / LANCASTER / TOKYO

Library of Congress Cataloging-in-Publication Data

Main entry under title:

Human nature and natural knowledge.

(Boston studies in the philosophy of science; v. 89)
Bibliography: p.
Includes indexes.
1. Philosophical anthropology—Addresses, essays, lectures. 2. Rationalism—Addresses, essays, lectures. 3. Philosophy—Addresses, essays, lectures. 4. Grene, Marjorie Glicksman, 1910– . I. Grene, Marjorie Glicksman, 1910– . II. Donagan, Alan. III. Perovich, Anthony N. (Anthony Novak), 1951– . IV. Wedin, Michael V. (Michael Vernon), 1943– . V. Series.
Q174.B67 vol. 89 [BD450] 128 85–24459
ISBN 90–277–1974–8

Published by D. Reidel Publishing Company,
P.O. Box 17, 3300 AA Dordrecht, Holland.

Sold and distributed in the U.S.A. and Canada
by Kluwer Academic Publishers,
190 Old Derby Street, Hingham, MA 02043, U.S.A.

In all other countries, sold and distributed
by Kluwer Academic Publishers Group,
P.O. Box 322, 3300 AH Dordrecht, Holland.

Printed in The Netherlands

FOR MARJORIE GRENE

MARJORIE GRENE

TABLE OF CONTENTS

PART IV: RENCONTRE

PART V: REFLECTIONS

EDITORIAL PREFACE

Everybody knows Marjorie Grene. In part, this is because she is a presence: her vividness, her energy, her acute intelligence, her critical edge, her quick humor, her love of talking, her passion for philosophy – all combine to make her inevitable. Marjorie Grene cannot be missed or overlooked or undervalued. She is there – *Dasein* personified. It is an honor to present a *Festschrift* to her. It honors philosophy to honor her.

Professor Grene has shaped American philosophy in her distinctive way (or, we should say, in distinctive *ways*). She was among the first to introduce Heidegger's thought . . . critically . . . to the American and English philosophical community, first in her early essay in the *Journal of Philosophy* (1938), and then in her book *Heidegger* (1957). She has written as well on Jaspers and Marcel, as in the *Kenyon Review* (1957). Grene's book *Dreadful Freedom* (1948) was one of the most important and influential introductions to Existentialism, and her works on Sartre have been among the most profound and insightful studies of his philosophy from the earliest to the later writings: her book *Sartre* (1973), and her papers 'L'Homme est une passion inutile: Sartre and Heidegger' in the *Kenyon Review* (1947), 'Sartre's Theory of the Emotions' in *Yale French Studies* (1948), 'Sartre: A Philosophical Study' in *Mind* (1969), 'The Aesthetic Dialogue of Sartre and Merleau-Ponty' in the initial volume of the *Journal of the British Society for Phenomenology* (1970), 'On First Reading *L'Idiot de la famille*' in the *California Quarterly* (1972), the splendid Presidential Address to the Pacific Division of the American Philosophical Association on 'Sartre and the Other' (1972), and 'Sartre and the Tradition' in the *Philosophical Quarterly* (1981). Few who were present at the Presidential Address or have read the published essay (*Proc. A.P.A.*, 1973) will forget her vivid treatment of Merleau-Ponty.

Throughout her philosophical life a deep student of Aristotle, in her essays and her book, *Portrait of Aristotle* (1963), Professor Grene offered an original understanding of his metaphysics, his logic, and his philosophy of biology. Her works range over much else in the historical development of Western philosophy, with informative and critical interpretive papers on Plato, Descartes, Leibniz, Spinoza, Hobbes, Hume, Kant, Kierkegaard. But far more, and beyond these historically situated studies, there is a central theme in Marjorie Grene's work that establishes her intellectual leadership. There is a whole field of philosophical thought which derives its principal character and importance within Anglo-American philosophy of science from her work. This must be seen as philosophy of biology *and* philosophical anthropology *and* theory of knowledge, but only if one understands the deep integration of these disparate fields in Professor Grene's approach. The key figures here include both leading working biologists in evolutionary theory (in which Grene's scientific competence is extraordinarily high) and philosophical thinkers: Plessner, Portmann, Mayr, Gould, and others. And as one focus, there is the profound teaching of Michael Polanyi, her colleague and friend of many years, in Grene's relentless and persuasive critique of reductionism, whether in epistemology or in biology proper.

Thus, in her books – *The Knower and the Known* (1966), *Approaches to a Philosophical Biology* (1969), *The Understanding of Nature: Essays in the Philosophy of Biology* (*Boston Studies* vol. 23, 1974), and the edited volumes, *The Anatomy of Knowledge* (1969), *Knowing and Being* (1969), *Interpretations of Life and Mind* (1971), and *Topics in the Philosophy of Biology* (with Everett Mendelsohn, *Boston Studies* vol. 27, 1975) – as well as in many essays, Professor Grene has set forth a complex, systematic and clarifying philosophy of man in nature. With Polanyi and against Aristotle, Grene places 'potency before actuality' (*The Knower and the Known*, 250). With Plessner, she develops the idea of the achievement of personhood in its ordered relation to "an artifact – the social world of a culture" ('People and Other Animals' in *The Understanding of Nature*, 357). And she goes on a page or two later to say what *being human* should mean: "Man's so-called rationality resides, not in his grasp of some transcendent truths, but in his power to doubt, to criticize, to ignore or deny the actual in favor of the barely possible, if not the impossible. . . . We *are* such stuff as dreams are made on." (*Ibid.*, 359). The last word?

Of course Grene is the honest realist: ". . . it is in the last analysis contingency, not necessity, in which we have our roots." ('Time and Teleology' in *The Knower and the Known*, 250). Nevertheless she continues "But if there is no end to our peril, neither is there any end, so long as life lasts, to our hope". Hopeful to the end, and without illusion. Our respect and love to Marjorie Grene.

October 1985

ROBERT S. COHEN
Boston University
MARX W. WARTOFSKY
Baruch College of CUNY

PREFACE

These essays are presented to Marjorie Grene with admiration and affection from a few of the many colleagues, students, and friends who wish her well on this occasion. The admiration derives in part from Marjorie's pursuit of no less than three careers, all with distinction. As an organizer and innovator in the profession and as a teacher, her record is second to none and continues to grow. At a time when most teachers would literally have retired, Professor Grene has accepted teaching invitations at Tulane, Temple, Rutgers, Yale, Boston University, Berkeley, University of Waterloo, Vassar College, Rochester Institute of Technology and Carleton College. Many of these have been distinguished appointments. Marjorie's service to the profession has been, to widely understate the point, important and extensive. But perhaps closest to her heart are her seminars and institutes: She has directed no less than four NEH seminars and two summer institutes for the Council of Philosophical Studies, including the important 1982 Institute at Cornell on the Philosophy of Biology. But it is as a scholar and philosopher that she has had the greatest impact. In nine books and over seventy-five articles Grene has addressed an extraordinary range of topics, always with imagination, incisiveness, and an unfailing ability to promote reflection. Her work spans traditions, combines disciplines, and balances science and letters with enviable ease. Even more is her career a model of that warmth to change and growth that marks the best of philosophical activity.

Most of the essays in this volume track themes that have engaged Professor Grene over the years and some are directly inspired by her own work. Five contributions focus on the rationalist tradition. Broughton isolates a central theme in Descartes, arguing that his deep commitment to the principle that a cause must contain what comes about in its effect explains a number of the problems that beset Descartes' treatment of the causal connections between mind and body and between bodies generally. Donagan and L. Cohen continue the focus on Descartes. While agreeing with Grene that Descartes errs in too narrowly conceiving the nature of animals, human and other, Donagan traces the defect to Descartes' concept of *cogitatio* rather than his dualism. Cohen argues for epistemological dualism without metaphysical complications on the grounds that the *cogito* can establish the first without entailing the second. Frankfurt and Marcus bring Spinoza into play. Frankfurt supports the interesting hypothesis that the divergence in Descartes and Spinoza's styles of rationalism reflects Descartes' preoccupation with the possibility of ultimate external disorder and Spinoza's preoccupation with the spectre of internal disarray. In a sense the essay by Marcus supports this, quite coincidentally, by arguing that Spinoza's so-called ontological proof differs crucially from Anselm's precisely by lacking epistemological or conceivability constraints in its premises.

Rorty, addressing Grene directly, argues that one ought to abandon the attempt to establish metaphysics on temporality based categories less because of the temporality than because of the metaphysics and that, in the manner of Heidegger, art and history rather than science are the revelators of what counts as being human. Harré takes up one side of this issue by sketching a metaphysical framework which gives times a central role and D. Frede assays prospects for realistic interpretations of the Heidegger of *Being and Time* in the course of examining his complaints against Kant's proof of the reality of objects of experience. Perovich gives the topic of conceptual frameworks an even more explicit historical setting in showing us how Kant can reconcile the stability of synthetic *a priori* principles with widespread theory change in science.

A theme of enduring interest to Grene, man as a social creature, is explicitly taken up by Berger, Hacking, and MacIntyre. Berger argues that moral theories such as utilitarianism can, after all, accommodate the moral relations that are intrinsic to love and friendship. Hacking uses the development of the concept of split personalities to illustrate the thesis that what individuals are is a function of culturally dominant categories of classification. MacIntyre neatly distinguishes types of suicide in tracing the instructive failure of the positivistic program to provide explanations that hold across social orders. All three authors underscore the necessity of non-reductionistic approaches in social theory. And Burian extends the range of social factors to scientific theories themselves, arguing that political considerations often are, and ought to be, relevant to acceptance and justification of one theory over another.

The theme of reductionism figures centrally in Cartwright's defense of a brand of teleological explanation that resists reduction to efficient causal explanation. Wimsatt pursues the topic of reduction by describing several varieties of emergence that involve reducibility but not of the logical positivistic sort. Gale sees in the recent emergence in physical theory of such strategies as the so-called anthropic argument a return to more classical, teleological modes of thought. The classical tradition is directly broached in the essays by M. Frede and Wedin. Focusing on the independent emergence in ancient medical science of a notion of explanation, Frede illuminatingly details, among others, early rationalist and empiricist approaches to explanation. Wedin returns to one of Grene's favorite philosophers in risking an interpretation of Aristotle's theory of receptive and productive *noûs.*

It is clear that each of the essays would be at home next to something that Grene has written. Even clearer, much would remain over. So, while this volume reflects what Professor Grene has given us, it cannot but fail to give the full measure. For all of us, Marjorie has been, as Henry James would have said, a quantity. We have been enriched and changed by her presence, both in word and in deed. It would fulfill our fondest hope on this occasion, should we succeed in returning a small part of this richness.

A. D.
A. N. P.
M. V. W.

ACKNOWLEDGEMENTS

We wish to extend our thanks to the editors of *Boston Studies in the Philosophy of Science* for their enthusiastic support of the volume and especially to Robert S. Cohen for his encouragement and guidance in bringing the project to completion.

Linda Carr-Lee of the University of California at Davis, Fred Musto and Caroline Smerk of the Indiana University Library, and Linda Visscher of the Hope College Library were of great help in tracking down bibliographical information.

Thanks are also due to Mrs. Jacqui Levy, Ms. Lesley Byrns, and John Leonard at the University of California at Davis and to Tom Andrews, Derek Emerson, Kathy Grindstaff, Connie Hoffman, and Sheila Young at Hope College for their able editorial assistance.

Actual production of the manuscript was made possible, in part, by a grant from the Committee on Research of the University of California at Davis.

The frontispiece photo of Marjorie Grene is by A. Sue Weisler of Rochester Institute of Technology.

PART I

SELF AND SOCIETY

FRED R. BERGER

LOVE, FRIENDSHIP, AND UTILITY: ON PRACTICAL REASON AND REDUCTIONISM*

> But love's a malady without a cure.
>
> — DRYDEN
>
> Friendship is constant in all other things
> Save in the office and affairs of love.
>
> — SHAKESPEARE
>
> To H[ayley]
> Thy friendship oft has made my heart to ache:
> Do be my enemy — for friendship's sake.
>
> — BLAKE
>
> You can't eat your friends and have them too.
>
> — BUDD SCHULBERG

Love and friendship relations are complex and play a multitude of roles in our lives. If the poets are to be believed (and here I think they are), love and friendship bring heavy burdens, even reducing us to unreason or to unprincipled behavior. But even if there is a sense in which "love knows no reasons," it plays an important role in *providing* us with reasons, and it is crucial in determining for us much of the behavior we owe to others, and the ways in which we can wrong them. Similarly, the details of the life of "the good person" are as much to be determined in real life by the requirements of friendship as by the promises we make. A moral theory that can explain how we are obligated by promises, but cannot explain the moral relations engendered in love and friendship is deficient in a fundamental way. And an account of human motivation and reason that ignores or distorts the roles of love and friendship in behavior cannot be wholly correct.

It has been contended that the chief moral theories held by professional philosophers cannot account for the moral relations associated with loving and friendship.[1] I want, in this paper, to explore this contention with respect to utilitarianism. I shall outline a possible line open to utilitarians that would rescue their view from the criticism. My concern, however, is not essentially with defending ultilitarianism. In the course of my discussion, I shall bring out how important features of love and friendship have certain implications for

A. Donagan, A. N. Perovich, Jr., and M. V. Wedin (eds.), Human Nature and Natural Knowledge, 3–21.

the nature of practical reasoning. I am as concerned to elucidate these as I am with utilitarianism. Finally, I believe the points I shall make concerning practical reasoning have implications for "reductivist" modes of explanation in the social sciences.

I

The problems I wish to consider are brought out by considering several imaginary stories. Suppose Smith visits her long-time, close friend Jones, asking to borrow $1,000. Smith has recently suffered the trauma of a bitter divorce and has been left with sizable legal debts and virtually no financial reserves. Moreover, Smith has just learned that she requires an expensive eye operation that is needed to keep her from going blind. The money requested is to go for that operation. Jones is financially comfortable, but certainly not wealthy. (He is a philosophy professor with no family responsibilities who has an inheritance sufficient for comfort, and he published a text in logic that received wide adoption and which appears in recurrent revisions.) The loan of the money will be felt, but would impose no hardship. Given no further special circumstances, we would expect Jones to lend Smith the money. Indeed, we would expect Jones to express concern, offer sympathetic support, seek out other ways to be of concrete help, and generally, to offer succor and assistance, as well as the requested financial grant.

Let us suppose, however, that Jones listens to the plea and responds: "You certainly have a genuine need for the money, but of course, being a good utilitarian, I must calculate the consequences for all who may be affected by my giving you the money." Jones then goes through the calculations, decides that the greatest good will be produced by giving Smith the sum requested, and grants the request. (In his head, Jones also calculates the consequences of showing care and concern, and decides that doing these too will be useful, and, along with the money, expresses concern and a desire to be additionally supportive.)

Consider another case. A man takes a woman out to dinner and a show. He brings her flowers, shows considerable attentiveness, displays interest in her background, her work, the sorts of activities she values. He shows considerable affection, and expresses an interest in seeing more of her. At the end of the evening, she thanks the man for a very enjoyable evening, to which he responds: "Well, I am very happily married to a woman whom I love dearly. But, when I met you, you seemed to be someone whom I might come to love even more. As a good utilitarian, I must always be concerned to maximize good consequences, so I am always on the lookout for a relationship that might be more rewarding for all concerned in the long-run."

So much strikes us as odd about these cases that it is hard to know where to begin the list. I shall concentrate on points central to a critique of utilitarianism. The first case makes a mockery of the friendship relationship because there are *certain* reactions to Smith's situation that we regard as part of what is involved in *being* Smith's friend. To be concerned with the other's ill-plight *as a matter of course* is integral to the friendship relationship. Barring especially burdensome consequences, special care, assistance, and aid *express* that relationship. These are provided as an aspect of standing in the friendship relation. To say *A* is a friend of *B means* that *A* is disposed to care for and help *B* in such situations *without* first having regard for self or for consequences. There is, of course, the situation in which one acts out of concern *for the friendship*, i.e., when one values the friendship, perceives act *X* is required to help preserve it, and performs *X* for these reasons. But, acting *for the sake of* a friendship is importantly different from acting *from* friendship. And both are different from acting in order to maximize good consequences. To the extent that the latter is one's reason to act one has *not* acted *from* friendship, and one need not have acted *for the sake* of friendship. Even were our fictional Jones a "good" utilitarian, he certainly is *not* a good friend.

The second sort of case is anomalous for a number of reasons also. Leaving aside the implications of the marriage relationship (it *is* incidental since a married couple could very well have an explicit agreement permitting such relations with others), some strain is put on the man's claim that he loves his wife "dearly," certainly the relationship with the woman in the story is undercut by the revelation that she is being pursued as part of a strategy for promoting maximally fulfilling relationships. It helps, of course, that the man thinks he could come to love her. But that is different from pursuing her because he likes her, enjoys being with her, takes pleasure in sharing her interests, enjoyments, achievements, and wants to be a source of comfort against disappointments, defeats, and hurts. Wanting to be with her because he cares for her is different from wanting to be with her to "try her out" as a potential substitute for his current relationship.

We should note also that the man shares a love *relationship* with his wife. One can be *in* love with someone without the complications of a relationship. But where there is a relationship, there is a reciprocity of feeling and concern, and there arise mutual needs *for* one another, together with concomitant emotional dependencies. (Again, the *marital* relationship is only a sign of such a complex of reciprocal relations. Many marriages are devoid of these relations, or involve *different* expectations, needs, and inter-dependencies than those typical of a love relationship.) The man's "utilitarian" strategy puts those needs and inter-dependencies at risk even if it is not absolutely inconsistent with them. Persons who share a love relation have need of one another — as companions, persons to share life's enjoyments and disappointments, and as persons *to be counted on* in life.[2] If he loves his wife, he *wants* to protect and further the interests she has that are invested in their relationship, and acting from this desire appears to conflict with always acting from a desire to maximize good consequences in the long run. Loving his wife, he will do or refrain from certain actions *as a matter of course*, without giving thought to maximizing good consequences.

There is a further dimension to friendship and love relationships that should be explicitly mentioned. The complex of desires, needs, and expectations that are involved in such relationships, gives rise to ways of wronging one another that are special to these relationships and norms come to be accepted as part of those relationships that serve as a basis for moral evaluation and criticism. The lover who is constantly testing others as possible replacements for his or her current love is not, for example, merely rude or boring or boorish. These are trivial, non-moral faults. Such a lover shows a disregard for central aspects of crucially important interpersonal relations and, in this way, a disregard for the very personhood of the other.[3] There is a moral dimension to friendship and love that is bound up with the nature of the relations in which they are embedded.

I can briefly summarize the problems utilitarianism may be thought to face:

1. Love and friendship relations involve a complex of beliefs, desires, needs, and expectations that characterize them as the relations

they are. As such, these relations have a "structure" that differentiates them from others.

2. The friendship and love relations people have engender norms of behavior and special commitment. Thus people often act from desires that are integral to those relationships, and the reasons and justifications for their actions then make reference to aspects of those relationships. (That one is disposed to act from *those* desires, or for *those* reasons is also part of the "structure" of the relationships, as are the norms and commitments that arise.)

3. Utilitarianism requires that people always act from the desire to maximize good consequences, or for the reason that one's act will maximize good consequences. But one cannot consistently act in this way and still maintain full love and friendship relations, since such relations provide motives and reasons internal to those relationships.

4. Finally, the actions that express and spring from friendship and love are valued for themselves. Our friend's act is of value to us *because* it is an act of concern. If it cheers us up, it will do so because of this. Hence, not all the value of an act is determined by its consequences.

In the next section, I shall assess the overall force of this critique of utilitarianism, and I shall outline a utilitarian approach to the problems that I believe is promising as a solution. I am not entirely convinced, however, and I shall indicate the grounds of my hesitation.

II

The critique of utilitarianism that is generated by consideration of love and friendship relations assumes utilitarianism to be a theory about correct motives, or correct reasons for action or about how to choose what acts to do, i.e., it identifies utilitarianism with maintaining one or all of the following theses:

1. One ought always and only to act from the desire to maximize good consequences from one's actions.

2. The only proper reason for an agent to perform an act is that it will produce the best possible consequences among the alternatives open to the agent.

3. One ought always to calculate the consequences of a contemplated act and perform that act that is likely to produce the best consequences.

The criticisms of utilitarianism given in the last section all invoke at least one of these claims as central to utilitarianism and many other criticisms of utilitarians appeal to such a characterization of utilitarian theory.[4] It is possible, of course, that there have been utilitarians who have held one or more of these theses, and someone who held any of them would almost surely be a utilitarian. Some utilitarians, e.g., John Stuart Mill, have explicitly rejected all of them.

Mill, for example, described the basic principle of utilitarianism as follows: "The creed which accepts as the foundation of morals, Utility, or the Greatest Happiness Principle, holds that actions are right in proportion as they tend to promote happiness, wrong as they tend to produce the reverse of happiness."[5] Though this proposition may seem like a variant of 1–3 above, especially 3, there is a significant difference. The principle Mill gave is a *criterion* of right and wrong action. It is *neither* a prescription concerning the motives or reasons for action nor an injunction concerning how one is to make practical decisions. Given a criterion of what properties an act must have to be right, there can remain the further question as to how best to make practical decisions that will yield actions that meet the test. Inculcating *certain* motives rather than others as bases for behavior or encouraging use of *certain* reasons rather than others may, in the long run, be more productive of right actions than always acting from the desire to maximize good consequences, or for the reason that one's act will do so. And the attempt to decide how to act by first calculating consequences might very well be a *poor* procedure for maximizing good consequences, since employment of the calculative process *itself* has bad consequences. It is sometimes difficult to assess consequences, takes valuable time that interferes with leading a normal life, and may very well defeat expectations others legitimately have about our behavior.[6]

In fact, Mill explicitly rejected the view that calculation is required for each act. He held that virtuous states of character (ones which it is useful to have and encourage) involve having "fixed purposes" or habits, so that one acts as a matter of course, not for the sake of the consequences of action.[7] Furthermore, he chastized critics for confusing "the rule of action with the motive of it." Ethics should tell us what our duties are, "or by what test we may know them; but no system of ethics requires that the sole motive of all we do shall be a feeling of duty."[8]

A utilitarian such as Mill could still hold that there is a sense in which happiness is the *ultimate* end of all action. Pursuit of things other than happiness itself can be explained and justified by their connection with happiness. Either they *promote* happiness, or their pursuit is an *aspect* of what we understand human well-being to consist in. With the special motives and reasons they provide for behavior, friendship and love relations could then be accepted *because* of this connection with happiness.

There is an apparent stumbling block to this sort of utilitarian approach to friendship and love that should be explored in some detail because the discussion sheds light on utilitarianism as well as on aspects of love and friendship.

The problem I have in mind can be brought out in the following way. The process I described above, whereby rules and fixed states of character come to provide motives and reasons for actions, appeals throughout to what might be termed "strategic" considerations. It is *in order* to adopt the best *means* of achieving well-being that the rules and character states are developed. It would appear, then, that alternative motives and reasons enter into the nexus of determinants of our behavior *via* an assessment of what strategic devices will be most productive of happiness in the long run. But this is *not* how we acquire friends and lovers. Indeed, the person who is always calculating in developing friendships, who picks and chooses according to the ultimate fruits of those relationships, is suspect as a friend. Friendships arise in spontaneous interaction. They satisfy deep-seated needs for social interaction and the resulting relationships are themselves *expressions* of the "caring-for" and "cared for" feelings that naturally occur. The "strategic" stance of the utilitarian appears at odds with some of the most crucial aspects of these relations.

I do not think the utilitarian is forced to accept this conclusion. Only if happiness is conceived of as something separable from the activities and relationships that give rise to it, would such a conclusion seem to follow. A utilitarian could hold that we are by nature social creatures who cannot be happy without friendship and love relationships. Indeed, these relationships are *expressive* of deep-seated needs and they reflect natural desires, likings, dislikings, etc. Nor does the spontaneous origin of these relationships undermine the utilitarian approach. No doubt a music lover need not "try out" music because he or she thinks it might be desirable. But he or she may come to discover a liking for it merely in experiencing it. That the person is happier with music in his or her life might still be good reason to keep it a part of that life. The utilitarian could be thus concede that a happy human life requires friendship and love relations that are expressive of feelings, beliefs, and attitudes that arise in interaction without planning and calculation. Having such relations is *part* of what it is to live well as a human being.[9]

We should also note that this sort of response does not make the "strategy" conception completely irrelevant to love and friendship relationships. This is because there is more to such relationships than merely having feelings or beliefs that have arisen in (usually) spontaneous interaction. There is much that we can and do do *about* these feelings and beliefs. For example, we can try to avoid encounters in which they arise or positively seek them out; we can foster, develop, and maintain the relationships that are associated with them, and so on. *Certain* friendships *should* be avoided, e.g., when one party is a public official, the other a member of the Mafia. And if we are already deeply involved in a love relationship, perhaps we ought to avoid situations that can give rise to competing ones. Given the nature of love and friendship relationships, there are grounds (*in utility*) for avoiding systematic appeals to strategy considerations, but there remain special circumstances where it is appropriate to bring them in. Some have already been cited. Consider also the case of someone who has a tendency to enter into infelicitous love relationships — a woman who is attracted to married men but is often left broken-hearted. Such a person might do well to take deliberate steps to avoid such relationships in the future. Similarly, when relationships have become strained, when one's "friend" has been too inconstant or unreliable, it is surely appropriate to raise the question of whether to make a break or seek accommodation and repair. And a married woman who finds she is very much in love with a man other than her husband would surely be foolish not to ask herself hard questions about continuing her prior relationship.

Because friendship and love relations *are* satisfying in themselves, and *because* having them is partly constitutive of a happy human life, there is strong utilitarian justification for having them, even though this entails acting from motives and reasons that do not themselves look to the production of good consequences. That it is natural for us to enter into and maintain such relationships means that we do not, normally, *decide* to have them. But that consequential considerations become relevant "at the edges" so to speak, i.e., in various special circumstances, shows (from the utilitarian's point of view) that such considerations are involved in the ultimate explanation and justification of the behavior associated with these relationships.

In addition, it should be pointed out that there would be no inconsistency for a utilitarian to hold that the actions that spring from and express love and friendship have value in themselves. It is of utmost importance to stress that utilitarianism, in the form for example, expressed in Mill's formulation, is a theory of *right and wrong conduct.* That criterion cannot, of course, be applied without a criterion of value. But there appears to be no one theory of value that one *must* hold in order to be rightly classified as a utilitarian. If acts of friendship or love are pleasing in themselves, desirable as such, and partly constitutive of our happiness both when we receive and give them, then they may well have intrinsic value. There are *further* questions to be answered in any particular case, e.g., whether we should go on and *do* them or whether it was *right* to perform such acts. To these questions the utilitarian has a characteristic answer. But that answer is not inconsistent with ascribing value to the act itself.[10]

I believe this outline of a utilitarian account of friendship and love goes a long way toward resolving the problems posed earlier for utilitarian theory. It is not conclusive however. There are three reasons why not. First, though the argument shows that *reflection* and *reassessment* on those relations is relevant, that they are not self-justifying, there has been no argument to show that *only* consequentialist considerations are relevant. These relationships involve relevant promises, commitments, expectations and these are past events not future consequences. Second, even if only those considerations are relevant that can be given a consequentialist rationale, it is not clear that justification consists in showing that a course of action will *maximize* good consequences. An action that produces the death of someone might be objectionable even if it produces pleasure for an enormous number of others. One *could*, of course, say that the death *outweighs* the other goods. But it is not clear if *that* is some further value judgment or a necessary consequence of having done the "arithmetic." In other words, the concept of "maximizing good consequences" is not so clear as one would need to establish the utilitarian conclusion.[11]

Finally, there are hard problems concerning how to resolve conflicts, e.g., when friendship requires one act, while maximizing good consequences would require another. Perhaps there is no problem in the extreme cases. Where small gains only are to be had from violating the demands of friendship, strategy considerations alone can justify ignoring them. The risk of misjudgment together with the weight of what is at stake dictate overlooking these utilities. At the other end, where disaster will occur by observing friendship or love demands scrupulously, the utilitarian also seems vindicated. At least it is nowhere near so obvious that consequentialist considerations do not win out. But there are myriad cases in between, where life is really lived, and it is not so obvious that there is a consequentialist answer or strategy which is defensible. This problem is, of course, one that can be generalized. So long as *any* other guides to conduct are employed – intermediate rules, appeal to character motives, social role requirements, requirements of law or justice – when these conflict with what will produce the greatest good, it is not obvious there is a defensible utilitarian solution. These objections, however, are standard ones raised against utilitarianism. They do not, for the most part, spring from the special nature of friendship and love. *If* the utilitarian can resolve the more general problems of utilitarian theory, I see no further incompatibility in being a utilitarian *with* friends and lovers.

In the final section, I want to briefly indicate some implications of my analysis for practical reasoning and reductivist explanation in the social disciplines. Though I cannot argue in detail for the claims I shall make, I do want to sketch the outlines of the argument.

III

Consider practical reasoning first. It is sometimes alleged that utilitarians are committed to an account of practical reason according to which people always do or always should reason concerning actions by determining what act would produce the best consequences. All practical reasoning is, or should be of one form. Indeed, insofar as critics have assimilated utilitarianism to a monistic theory of human motivation, they tend to perceive it as giving a monistic account of *all* decision-making.

The account of friendship and love in the previous sections are inconsistent with a monistic account of practical reason. The special "structure" of the interpersonal relationships we enter into provides motives as well as reasons for our actions that are enormously diverse. My belief that my friend is in trouble or in danger or is ill, provides me reasons to act in certain ways. My desire to share my lover's success, or to share physical joy in sex, or to reassure of my constancy, also provide diverse reasons for widely different actions.[12] Similarly, my belief that it is *wrong* to let down a friend in need, or to be insufficiently sensitive in criticizing a loved one in public, are reasons for my behavior. Furthermore, all the sorts of reasons I have cited can function both as *explanatory* and *justificatory* reasons. No account of practical reason that fails to give a role to reasons such as these can be adequate.

Furthermore, among the sorts of reasons that love and friendship provide for behavior are ones that do not make reference to the *ends* our conduct serves. When I act out of friendship upon perceiving my friend's need for reassurance, my act is chosen because it *is* an act of reassurance. It may, of course, be chosen also for its further consequences, e.g., because it will restore my friend's self-confidence. But providing reassurance at such times is as much an *expression* of my friendship as it is an attempt to act in a way that will aid my friend. *Being* a friend, I provide reassurance as a matter of course.

This point can be made stronger. Only when we deliberate do we entertain and weigh reasons. Here is where a notion of practical *reasoning* gets its clearest setting. But even here not all action is produced by consideration of the ends to be served by action. In a case where assistance to a friend may be burdensome, or where allegiance to our friend puts a strain on relations with other friends or lovers, or where such allegiance strains our standards of morality, the reasoning involved can be quite complex. Consider the sorts of questions it is appropriate to raise: Just how strong and committed *is* our relationship? *Is* this the sort of act he or she can demand of me? Is what is required outside of the limits of friendship entirely? Are there *other* acts possible that sufficiently express my friendship but which are also consistent with my own needs, those of other friends, or my moral standards? These are questions about the *nature* of the relationship between these individuals, or about the limits of friendship itself. The reasoning needed to provide answers involves understanding what has transpired between the parties and the obligations that are thereby entailed. It involves questions about the nature and limits of friendship obligations in general and understanding how *this* friendship relates to others and how they are to be ordered in cases of conflict. None of these are explicitly questions about the *ends* the contemplated acts serve. They are more like conceptual questions about how the acts "fit in" with this relationship and the nexus of obligations and expectations generated from it, how this relationship is limited by others or by general moral considerations.

In such situations one can and should go on to ask explicitly consequentialist questions: What will my failure to act do to the friendship? How will I feel about it afterward, e.g., will I be plagued with guilt feelings? How will my friend, or others, feel? Furthermore, there would be no inconsistency in holding that there is some "ultimate" teleology that explains and justifies our incorporating such reasoning into our everyday reasoning. But no teleological view of either the explanation or the justification of behavior can be correct that fails to give a role to reasoning about action that is not of the means/end variety. At least, no such view can be correct if it also is to give a significant role to love and friendship relations.[13]

A few, final remarks are in order concerning so-called "reductivist" programs for the explanation of human behavior. There is a kind of reductivist program that I think the analysis of love and friendship relations gives reason to think is conceptually misconceived. Such accounts of explanation may be termed "eliminative explanatory reductionism." They hold that human behavior can be explained solely by reference to physical, chemical, or biological entities or events. Such a view "eliminates" a role for explanations referring to intentions, beliefs, social roles, rules, and so on (except insofar as *these* can be eliminated in favor of reference to the preferred entities or processes). Moreover, the claim is about "explanation" and need entail no further claim about ontological reductionism.

One example of such a view is B. F. Skinner's behaviorist conception of social science.[14] Behavior, according to Skinner, is the final link in a three-link causal chain. First is an external "operation" performed on an organism, second an "inner condition" and, finally behavior. Water deprivation produces a feeling of thirst or some physiological process, which, in turn produces drinking. Since the process is conceived of as "causal," each link is lawfully connected with the next. This entails that there will be lawful connections between the first and third. Skinner conceives of the first and third "links" as fully describable in physical terms, hence, as "observable" events. Further, since he also conceives it as integral to the scientific enterprise that it permit prediction and control of events, a science of behavior can ignore entirely the inner conditions in the causal chain. Behavior can be subsumed under laws connecting external physical events and subsequent movements of the organism and these will suffice for prediction and control. Although, as is well known, Skinner *does* go on to deny there *are* non-reducible mental events,[15] no ontological claims follow from the explanatory model.

Skinner's account may be thought somewhat out of date but it has the virtue of being clear, whereas many contemporary advocates of, for example, a biological or genetic or evolutionary account of human behavior are far less clear concerning the scientific program they adopt. There is, I think, a perfectly acceptable way of conceiving "sociobiology", to cite one example, that does *not* entail any form of "eliminative explanatory reductionism." But some of its advocates have made claims for it that seem to me intelligible only as claims for an eliminative explanatory program. E. O. Wilson has argued for ethics to be "biologicized" and "to be removed temporarily from the hands of the philosophers."[16] When philosophers argue for an ethical position, according to Wilson, they take their emotional responses to alternatives to be a "hidden oracle." In his view:

> That oracle resides in the deep emotional centers of the brain most probably within the limbic system ... Human emotional responses and the more general ethical practices based on them have been programmed to a substantial degree by natural selection over thousands of generations. The challenge to science is to measure the tightness of the constraints caused by the programming, to find their source in the brain, *and to decode their significance through the reconstruction of the evolutionary history of the mind.*[17]

As he put the point elsewhere, philosophers "intuit" ethical principles by "consulting the emotive centers of the hypothalmic-limbic system ... only by interpreting the activity of the emotive centers as a biological adaptation can the meaning of the canons be deciphered.[18]

These statements seem to imply the "eliminative explanatory reductivist" model of behavior explanation. As in the case of Skinner, reference to the social and cultural factors involved, or the norms or standards relied on, is eliminated in favor of reference to prior genetic history, together with present brain functions.

Friendship and love relations are inconsistent with these views. The Skinnerian model gets part of its support from his claim that it is prediction and control that is central to science. Though this seems to me mistaken (they are *tools* of science, not the ultimate objectives of science as such), even if one accepts this view, it is clear they do not exhaust the scientific enterprise. For, surely, science is as much concerned with *understanding*, in particular with *how* events come about. And if some of our behavior comes about via our having love and friendship relations, i.e., as a result of our concerns and fears for others, or because we feel we *owe* them certain things, or as an expression of our caring, and so on, then any explanation of such behavior that omits reference to these will fail to provide a full account of *how* the physical, chemical, or biological factors produce the behavior to be explained. Even *if* there are causal, lawful connections between (say) external events, brain physiology, and behavior, recitation of those alone will always fail to cite the internal features of friendship and love that provide the reasons and motives from which we act.

Proponents of such views hold, of course, that *these* elements can in turn be "explained" on their model. But there is considerable evidence that they do not really believe this. Wilson, for example, who apparently believes that Rawls "consulted" a part of his brain in deciding that "justice as fairness" is correct, does not tell us what part of his brain *he* "consulted" in concluding that sociobiology can provide a true account of social behavior. Nor did he cite any lawlike connections between prior genetic evolution, parts of his brain, and his subsequent conclusion that sociobiology is adequate. What he did was to recite and interpret evidence, answer critical points, draw inductive and deductive consequences, etc., to explain *why* and *how* he came to conclude sociobiology is true. He had, in other words, to draw attention to what he took to be internal connections between evidence and conclusion, reason and consequence, fact and its implications, to explain why and how he arrived at his conclusions. Any *other* sort of explanation would be *irrelevant* to the explanatory activity he was engaged in. And, no doubt, he is as aware of this as I. Of course, other sorts of explanation might enlighten our understanding in other dimensions. This, however, could hardly justify elimination of the explanation that refers to reasons and evidence, implications and consequences.

But, I have argued, social relations have an internal dimension as well, the understanding of which is crucial to the explanation of behavior that arises from them.

There is a further reason why eliminative reductionism must fail.

When I reciprocate my friend's act of friendship, what I am responding to is *not* a mere physical movement. Nor is my response a mere physical movement. At the very least, I must perceive my friend's act as done with a certain intention, or as arising from certain beliefs, desires, etc. The reassuring caress I give my lover is *that* action only if *I* have certain intentions expectations, and so on. It would not *be* that act absent such further features.[19] Similarly, some actions are conceptually dependent on a background of rules, roles, and institutions. Casting a vote, paying someone $5, fulfilling one's promise, all require reference to such cultural and social matters to identify them as such. The eliminative model of explanation, if it omits reference to such matters, cannot adequately *identify* either the items that produce behavior or the behavior to be explained.[20] It has been noted by a number of critics, for example, that sociobiological "explanations" of altrusim and selfishness systematically misinterpret these notions, hence, fail to "explain" them.[21] We should note further that Wilson's interpretation of ethical claims as expressions of emotions, though a somewhat respectable philosophical claim, is just *that* — *an* interpretation of the nature of ethics. Furthermore, even *if* this bit of philosophizing were correct, philosophers and others have accepted a *range* of ethical theories which they have thought imply certain rather than other consequences, and which provide reasons and motives for certain rather than other behaviors. So, the acceptance of such a theory is itself an event that must be identified by reference to a range of factors that are not merely physical, chemical, or biological.

Before closing I should make it clear that I do not believe my arguments show that there cannot be genuine sciences of behavior, or even that the causal model of explanation is irrelevant to explaining behavior. There undoubtedly *is* an underlying biology to love and friendship that figures in important ways in producing these behaviors. And it is undoubtedly true that for *some* explanatory purposes, leaving these out of account would be a crucial omission. Furthermore, I see no reason *in principle* why there could not be lawful regularities between prior events and conditioning and the behaviors involved in love and friendship relations. Insofar as nothing I have said seems to imply a denial of such a possibility, I do not think the arguments resolve such larger issues as the free-will *vs.* causal determinism issue. I have argued only that we cannot fully understand love and friendship relations and the behaviors that express them without understanding their "internal" features — the feelings, beliefs, values, norms, that comprise them. Thomas Hardy once wrote: "Smile out; but still suffer:/ The paths of love are rougher/Than thoroughfares of stones." To understand why this is true for many people will require more than the biologist or technologist of behavior can tell us.

University of California at Davis

NOTES

* I first began thinking about the matters involved in this paper while considering the moral dimensions of gratitude. That resulted in a paper, Berger (1975), which anticipates some points made here. Further development arose out of research for a series of papers and a book I have done on Mill's moral theory, and from teaching courses in the philosophy of the social sciences. Furthermore, though she is unlikely to remember the conversations, a number of the points made in this paper were discussed with Marjorie Grene while shlepping back and forth together as commuters between Berkeley-Richmond and Davis, California, over the course of several years. I am grateful for her discussion, as well as for her love and friendship.

[1] This appears to be the claim of Stocker (1976). Since I wrote this paper, however (with the exception of the remainder of this footnote), I have read his (1981). The more recent paper emphasizes and argues for a central claim of this paper — that acting from friendship is not to be fully understood as a case of acting to achieve some end, and that it has value *as* the act it is. There is also some reason to think that he would find the defense of utilitarianism I give acceptable insofar as he recognizes the distinction I rely on between a theory of motivation — either psychological or evaluative — and a theory of right action that appeals to consequences. (See, especially, his footnote 19, p. 758.)

[2] I want to make it clear that I think one can love and share love relationships with more than one person. The notions of "fidelity" and "loyalty" that love relations involve are not easily accounted for, and are certainly *not* to be explicated with reference to exclusionary sexual relations. Nor is my argument meant to rule out all relationships that could put existing ones at risk. There are limits to what lovers and friends have a right to demand of us.

[3] I hesitate, however, to speak of violations of duties or obligations. This language seems inappropriate because it suggests strict rules that are broken. While this may sometimes be the case, it seems more apt to speak of *commitments* that are undermined, where these are conceived in some rather loose way. Moral philosophy, which has concentrated on such things as promise-keeping and justice, has tended to develop a standard vocabulary that is not entirely apt for discussing the morality of interpersonal relations. For more on this point, see my(1975), pp. 305—307.

[4] John Rawls, for example, notes that John Stuart Mill held that people desire to be in harmony with others and (at least when enlightened) do not desire to benefit at the expense of others. The principle that correlates with this desire is akin to one of Rawls's own principles of justice, labeled by Rawls "the difference principle." Of Mill's

position, he remarks: "Now the desire Mill characterizes here is the desire to act upon the difference principle (or some similar criterion), and not the desire to act on the principle of utility. Mill does not notice the discrepancy." (Rawls (1971), p. 502) It is not clear that the desire to act on the principle of utility is quite the same as the desire to act so as to maximize good consequences, but it *is* clear, as I shall argue below, that Mill did *not* hold that one does or should always act from such a desire.

5 John Stuart Mill (1963–present), vol. 10, p. 210 (*Utilitarianism*). The sort of theory I attribute to Mill is explored in detail in my (1983).

6 John Stuart Mill (1963–present), vol. 10, pp. 224–225 (*Utilitarianism*).

7 *Ibid.*, p. 238. See also *ibid.*, pp. 12–13 ("Remarks on Bentham's Philosophy") and *ibid.*, vol. 8, p. 842 (*A System of Logic*).

8 *Ibid.*, vol. 10, p. 219 (*Utilitarianism*). See also *ibid.*, vol. 8, pp. 839–840, (*A System of Logic*), and *Ibid.*, vol 10, p. 335 ("Auguste Comte and Positivism").

9 It should be noted that Mill held sociality to be "natural" to humans as such, and that the highest happiness requires the development and fulfillment of our social natures. See, for example, the concluding paragraphs of the much neglected chapter 3 of *Utilitarianism.* Many criticisms of utilitarianism take the theory to be committed to some form of psychological egoism. Mill not only *rejected* that view, but held social relations to be *central* to human well-being. For elaboration of these points, see Berger (1978), and also chapter 2 of Berger (1984).

10 The only explicit consideration of these points that I know of in the classical utilitarian literature is in James Mill (1870). In one place he wrote that "acts are performed, not for their own sake, but for the sake of their consequences." (p. 310) But this is misleading, as in the sentence preceding he referred to "the act itself, the muscular contraction," whereas his view throughout the book was that an act is a set of movements *with* an intention. Moreover, elsewhere in the book, he wrote that there are acts that we like for their own sakes, and he added: "That is to say, the acts; for the outward act, detached from the mental part, is not an object of liking or disliking." (p. 85)

11 In Berger (1984), chapter 6, I argue that Mill's concept of happiness is logically incompatible with a strictly additive conception of "the greatest happiness."

12 I am not saying that lists of beliefs and desires *by themselves* give a full account of the reasons for an act, only that *in a suitable context,* citation of one of these sufficiently gives our reason. Both desires or wants *and* beliefs may be requisite to produce behavior.

13 It is somewhat unclear to me how "ultimate" ultimate is. Even the consequentialist questions we can ask in these cases raise "utimately" such questions as "What sort of person should I be (or do I *want* to be)?" and "What really *is* the 'good life'?" And it is unclear that these are questions that themselves can be answered *solely* by asking about the ends certain actions serve.

14 I am drawing largely from Skinner (1953).

15 See, for example, Skinner (1971).

16 Wilson (1975), p. 562.

17 Wilson (1978), p. 6. Italics added.

18 Wilson (1975), p. 563.

19 I note that an example that Skinner uses in one place of an action-event is: "She slammed the door and walked off without a word." While it is hard to imagine what

physical movements alone, without reference to intentions or states of mind, could constitute "slammed" the door, it is clear enough that "walked off without a word" is *not* a mere physical movement. In this context, it is clearly an intended *omission* that is described, and there appear to be insuperable difficulties in reducing these to movements.

[20] Though there is much in his account that I do not accept, Peter Winch has shown persuasively that many acts are conceptually tied to background rules and institutions. See Winch (1958). Winch goes too far, I think, in holding that *all* identifications and reidentifications of things must take place *via* socially learned rules. This would, I think, make a mystery of how anyone *could* learn a language in the first place since there could be no identification and reidentification of anything to begin with. *Recognizing* anything would be literally impossible.

[21] Richard M. Burian has a very helpful discussion of the "description" problem, and especially of Wilson's identification of "altruistic" behavior in Burian (1978). See also Gunther S. Stent's "Introduction" in Stent (1980), pp. 1–18. Similar points are made in Singer (1981), pp. 126 ff.

BIBLIOGRAPHY

Berger, F.: (1975), 'Gratitude', *Ethics* **85**, pp. 298–309.

Berger, F.: (1978), 'Mill's Concept of Happiness', *Interpretation* **7**, pp. 95–117.

Berger, F.: (1984), *Happiness, Justice, and Freedom: The Moral and Political Philosophy of John Stuart Mill*, University of California Press, Berkeley.

Burian, R.: (1978), 'A Methodological Critique of Sociobiology', in Caplan, 1978, pp. 376–395.

Caplan, A.: (1978), *The Sociobiology Debate: Readings on the Ethical and Scientific Issues Concerning Sociobiology*, Harper and Row, New York.

Mill, James: (1870), *A Fragment on Mackintosh*, Longmans, Green, Reader, and Dyer, London.

Mill, John Stuart: (1963–present), *Collected Works*, 21 vols. to date, University of Toronto Press, Toronto.

Rawls, J.: (1971), *A Theory of Justice*, Harvard University Press, Cambridge, Massachusetts.

Singer, P.: (1981), *The Expanding Circle: Ethics and Sociobiology*, New American Library, New York.

Skinner, B.: (1953), *Science and Human Behavior*, The Macmillan Company, New York.

Skinner, B.: (1971), *Beyond Freedom and Dignity*, Alfred A. Knopf, New York.

Stent, G. (ed.): (1980), *Morality As a Biological Phenomenon: The Presuppositions of Sociobiological Research*, rev. ed., University of California Press, Berkeley.

Stocker, M.: (1976), 'The Schizophrenia of Modern Ethical Theories', *The Journal of Philosophy* **73**, pp. 453–466.

Stocker, M. : (1981), 'Values and Purposes: The Limits of Teleology and the Ends of Friendship', *The Journal of Philosophy* **78**, pp. 747–765.

Wilson, E.: (1975), *Sociobiology: The New Synthesis*, Harvard University Press, Cambridge, Massachusetts.

Wilson, E.: (1978), *On Human Nature*, Harvard University Press, Cambridge, Massachusetts.

Winch, P.: (1958), *The Idea of a Social Science and Its Relation to Philosophy*, Routledge and Kegan Paul, London.

RICHARD M. BURIAN

THE "INTERNAL POLITICS" OF BIOLOGY AND THE JUSTIFICATION OF BIOLOGICAL THEORIES*

FOREWORD

Marjorie Grene has played an extraordinary role in the philosophy of biology. As an academic outsider, working her way back into philosophy after fifteen years' absence from the profession, she devoted a considerable amount of energy to what was then an unpopular non-discipline. At the time, most of the philosophers who wrote on biological topics were driven to take absurd postures because they accepted a general account of science deriving from the philosophy of physics. Against this orthodoxy, Grene's was the voice of a dissenter.[1] Further compounding the difficulty of gaining acceptance for her work was the fact that many of her early papers in philosophy of biology accorded unorthodox evolutionary theories greater importance than mainstream biologists and philosophers were, at the time at least, prepared to grant to them,[2] as well as the fact that she worked within a (very roughly) Aristotelian tradition unfamiliar to the philosophers and biologists interested in what we now know as philosophy of biology.[3] Given all of these handicaps, it is marvelous that her voice has been heard at all.

Yet, in spite of her penchant for swimming against the stream and in spite of the obstacles that she had to overcome, Marjorie Grene has done as much as any individual (with the possible exception of David Hull) in forming and reforming the newly-flourishing discipline of philosophy of biology. And at an age when most thinkers are content to rest on their laurels or to mop up the loose ends left in their earlier work, she has revised many of her positions in philosophy of biology on the basis of substantial new knowledge.[4] In reviving some of her old interests while proceeding in new philosophical directions, she has done an immense amount to shape the thinking of the new generation of philosophers of biology. This is so not only thanks to her writing and the symposia that she has organized,[5] but also because of the impact of the summer institute that she directed at Cornell University under the auspices of the Council for Philosophical Studies in the

A. Donagan, A. N. Perovich, Jr., and M. V. Wedin (eds.), Human Nature and Natural Knowledge, 23–45.

summer of 1982. That institute, devoted to the teaching of philosophy of biology, has helped create a community of philosophers, biologists, and historians whose work will surely play an important role in reshaping the philosophy of biology in the next decade or two. The many indirect products of that institute[6] will go a long way toward fostering serious and valuable interactions between biology and philosophy of biology.

Her influence is, of course, due to a great variety of factors. These include the strength of her personality and her ability to charm and irritate those who come into contact with her into thinking hard about the difficulties she so forcibly brings to their attention. Most important of all, I think, is her stubborn persistence in raising problems that matter, in worrying at them from a variety of angles, in insisting that they be set into the context of real science, and in insisting on a concrete understanding of any proposed solution to those problems. Especially noteworthy is her unwillingness to settle for anything short of genuine understanding. This quality is admirable, as is the person in whom it is embodied. It is a great pleasure to be able to present this paper to her and to join so many of her friends in celebrating her accomplishments.

SETTING THE EPISTEMOLOGICAL STAGE: ON THE DIFFICULTY OF DRAWING SHARP BOUNDARIES

In this paper I am concerned to examine the epistemological situation of scientists and interested bystanders during periods of theory competition. I have two main targets: first, to show that, at least in biology (though the point is entirely general), when one deals with larger theoretical problems from the perspectives of those who are in the midst of the fray,[7] one cannot usefully draw sharp boundaries among ideological, political, and more narrowly scientific issues and, second, that knowledge of the internal politics of the science plays a legitimate role in estimating the relative support for competing positions. I had also hoped to argue that this important truth, shattering as it is to most traditional epistemologies of science, nonetheless does not rule out the possibility of justifying a decision between theories *over the long term* on grounds of their relative likelihood of being true or of their relative success in picking out and describing so-called theoretical entities. But, though it is central to my own views, there will not be room to deal with this issue in the present paper.

Assuming for the nonce, then, that scientific realism is not dismissed out of hand, the position set forth below amounts to the first step in an extended project. The hope is to reconcile long-term scientific realism with the recognition that a great variety of processes traditionally dissociated from the search for truth play a dominant role in the epistemological assessment of workaday science. Indeed, I shall argue in the present paper that political processes not only play an immensely important role at the surface of scientific life (hardly a startling claim), but also that they should do so. An understanding of the details of these processes is of considerable importance in reaching a proper *epistemic* or *cognitive* evaluation of the theories which scientists propound during periods of theory construction and theory competition. The argument supporting these claims is intended to show the legitimacy and relevance of considering the internal politics of science in weighing the merits of competing theories and scientific points of view. In the course of the argument, I shall attempt to categorize the sorts of considerations which might be taken to be epistemologically relevant to theory assessment in the short term and try to convince you to come part way with me toward the views here advocated.

Although the cases dealt with below are exclusively biological, there is nothing peculiar about them with respect to the principal concerns of this paper except, perhaps, for the clarity with which they illustrate some of the points at stake. The cases may be arranged, for present purposes, on a crude spectrum from those which are most ideologically laden to those which appear, at least at first glance, to turn on purely scientific considerations. At one extreme are the debates over Lysenkoism and creationism, at the other extreme are the debates (usually characterized as purely scientific) over the technical genetics of Barbara McClintock, Carl Correns, and Richard Goldschmidt. Scattered about the middle are various debates over Darwinism, eugenics, intelligence testing, the evolutionary importance of retroviruses, the risks involved in research into recombinant DNA, sociobiology, the so-called punctuated equilibrium theory of speciation, and so on. Since all of these debates are immensely complex and exhibit a wide variety of ramifications, one will be tempted to locate them at quite various positions along this spectrum according to the particular aspects with which one happens to be concerned. Although there is not room to argue the matter in detail, I maintain that a case-by-case analysis shows that there is no boundary which one can draw such that, across *this* line, political and ideological considerations are irrelevant to the evaluation of a particular project or of a particular scientist's work. And among the sorts of evaluation for which this is true, I explicitly include epistemic evaluation.

SOME EPISTEMOLOGICAL ISSUES

Before turning to cases, let me indicate the sorts of epistemological matters which, I believe, require consideration of more-or-less political processes. A number of traditional epistemological categories are not considered seriously in most contemporary theories of knowledge. Two of these — authority and responsibility — are crucial to our topic. I have in mind, specifically, the *authority* of an individual (or, for that matter of a lab or a discipline) to speak to an issue (to speak authoritatively, as we say) and the *responsibility* of an individual (or a lab or a discipline) to account for certain kinds or classes of evidence. Individuals, labs, theories, disciplines, and so on are engaged in an immensely complex political process, one in which they struggle for authority in a variety of ways. (They also struggle, of course, for a variety of other goods, including funding.)

When scientific inquiry works properly, one finds that a great deal of the "political" struggle concerns the acknowledgement of expertise and knowledge, and that this, in turn, is connected to issues regarding the proper discharging of responsibilities regarding evidence. These responsibilities include, most prominently, responsibility to create new and germane evidence bearing on the evaluation of the theories being investigated and responsibility to account properly for evidence already in hand — or at least for that portion of it for which the relevant party legitimately owes an account. (One of the deep issues regarding theory competition, one deserving a paper of its own, concerns the determination of the limits of the evidential domain for which a given party should be held accountable.) To the extent that I remain a scientific realist, it is in virtue of the belief (not to be justified on this occasion) that over a very long time scale, involving many scientific generations, it is possible to develop evidence and to handle it in such a way that the outcome of the complex political processes to which I have just alluded is substantial knowledge of new sorts of entities — neutrons, DNA molecules, black holes, and so on — entities which genuinely exist and whose behavior is well enough characterized and understood to justify certain classes of claims to knowledge about them.

Before turning, finally, to some cases it will prove useful to provide a preliminary classification of the sorts of matters which philosophers have traditionally considered to be extra-scientific, but which influence the evaluation and acceptance of theories in ways that bear on our present concerns. As a first approximation there are four sorts of traditionally "extra-scientific" considerations which are often held to be relevant to the evaluation of theories, hypotheses, experimental methodologies, and empirical results. These concern (1) the internal politics of science, (2) the external politics of science, (3) the socially significant consequences of the acceptance of particular scientific views (or the execution of particular experiments),[8] and (4) the interplay of scientific and ideological considerations. All four of these are alleged to have epistemic import at least some of the time, i.e., to affect the degree to which particular scientific claims or positions are justified or the extent of justification required before they ought to be pursued or accepted. It is possible to make at least a plausible case, that, in some instances at least, factors of each of these sorts legitimately enter into the epistemic evaluation of scientific positions.

The case for the legitimate epistemic relevance of the internal politics of science, to which this paper is devoted, is perhaps the strongest of the lot. I should explain, therefore, that I do not intend anything fancy by the distinction between the internal and external politics of science. Internal politics is what occurs within professional associations, in deciding which professional colleagues or arguments to take seriously, in choosing who will edit journals, in allocating resources within a laboratory, in tenure decisions, in peer review panels, and (to some extent at least) in the competition between laboratories or departments for authority and intellectual standing within a discipline or group of disciplines.

The rest of this paper deals chiefly with case studies of science-in-progress. It may be helpful to remember that the point of the cases is to show the legitimacy of taking the internal politics of science into account when conducting epistemic evaluations of developing scientific positions. While theories are still under construction or still in the midst of competition with serious alternatives, one is typically not able to muster a knock-down, drag-out argument to the effect that the theory in question can handle all the relevant phenomena adequately or as well as its competitors, much less that the theory is true. Part of the job of evaluation, then, must be to establish some sort of balance sheet on which one not only tots up the evidence in hand, but also assigns some sorts of weights to the individual pieces of evidence and to the various approaches to the class of problems (or theories) in question. This is an onerous task, normally requiring a long term process of repatterning, probing, testing, and debate. Such debates, typically, cannot be resolved in the short term by use of the traditional sorts of "objective" evidentially-based grounds preferred by classical epistemology. (One reason for this is that the weighting of evidence is discipline and theory *specific*.) Since such objective resolution is not feasible in the short term, it is often necessary to resort to other considerations, including, I shall argue, considerations of the standing and authority of the relevant parties and of the assignment of responsibility to them in dealing with the available evidence with the limited means at hand.

THE AGE OF THE EARTH

Consider, for example, the issue between Darwin and Kelvin regarding the age of earth.[9] As things stood in, say, 1880, Kelvin had produced, among other arguments, a rigorous thermodynamical argument, based on rather uncertain premises, to show that when one took account of all the sources of heat (the combustion of fuel on the sun, gravitational pressure on the earth, tidal friction, and so on), the total period of time which it could have taken for the earth to cool to its present temperature was of the order of twenty million to no more than 400 million years. (Kelvin's later estimates, some of which were already published, only served to reduce the upper limit below 100 million years.)

Darwin was keenly aware that the gradual process of evolution which he had described, like the gradual processes of uniformitarian geology to which he also subscribed, required far more time than Kelvin's thermodynamic calculations allowed. His rough and ready calculation of the length of time required for the erosion of a single limestone formation (the Weald) down to sea level in England, withdrawn from the fourth and subsequent editions of the *Origin*, had been severely criticized by geologists independently of Kelvin's arguments. Though Darwin's calculations were not nearly as refined as Kelvin's, they clearly required for the erosion of the Weald alone nearly the same amount of time as the maximum Kelvin allowed for the existence of a crust on the earth. Forced to abandon his sloppy argument regarding the age of the Weald, Darwin recognized that he had no adequate answer to Kelvin's arguments against the reliability of geological estimates of the age of various formations and the immense time required by uniformitarian geology and evolution by natural selection (especially in Darwin's gradualist version).

In the fifth edition of the *Origin*, Darwin admitted the thorniness of the problem, though he drew some consolation from the fact that physical (and hence biological) change was probably more rapid at an early phase of the earth's history.[10] Beyond this, he never took an explicit public stand regarding Kelvin's argument. In his private correspondence he agonized over the problem. Among his letters to Alfred Russel Wallace on the topic is one in which he says, "But then comes Sir W. Thomson like an odious spectre."[11] Nonetheless, in the correspondence he presented himself as prepared to argue from a massive base of stubborn "facts" — facts, however, whose value in setting an absolute time scale remained, at the time, quite uncertain. He also did everything he could to shorten the time required for biological evolution. One of the many ways in which he tried to speed up the course of evolution was to allow a far greater role for the inheritance of acquired characteristics than he had in earlier editions of the *Origin*. The need to cope with this problem helps crucially in explaining Darwin's weakening of the role of natural selection between the first and sixth editions of the *Origin*.

Here is a case where the contemporaneous evaluation of the competing arguments could not be satisfactorily completed on the basis of the available knowledge. As a stopgap, evaluation of the competing positions required some sort of cross-field assessment. This is no simple matter. Involved, at the very least, are comparisons of the general standing of quantitative arguments in physics with the more qualitative arguments in biology and geology, assessment of the relative strength of the competing premises employed by Kelvin and Darwin, and an estimate of the security of the facts and theories in question as a means of projecting over an immense time scale. It is important to insist here that one not read subsequent knowledge or results into the dispute, but think it through from the perspective of an individual who has available only the information of the day and who truly does not know how to resolve the conflict in a definitive way.

From this perspective, the relative standing of the disputants within their respective disciplines and of the disciplines as wholes both count as relevant information which a contemporary ought to take into account in getting a fix on the relative standing of the arguments. Now the standing of geology in England at the time was very much greater than a twentieth century reader might expect; at the time geology was, in certain respects, the queen of the inductive sciences. Nonetheless, the reactions of many of Darwin's contemporaries suggest that they felt it necessary to defer to an argument from physics. Wallace, for example, simply tried to reconcile the course of evolution with a time scale approximating Kelvin's. The major exception to this is Huxley who offered no counterargument to Kelvin, but maintained, in effect, that biologists should take their time scale from geology and that if they waited the matter out they would be vindicated since, at least roughly, the massive evidence of the geologists outweighed the arguments of physics.[12] Darwin, to repeat, did what he could to shorten the time required for evolution, grumbled about the matter privately, and never took a strong public stance or presented a strong counterargument to Kelvin or Jenkin on this issue.

As a further point, kindly suggested by Rachel Laudan, it is seldom appreciated that, in spite of the success of Lyell's *Principles*, uniformitarianism, with its commitment to immensely long intervals of geological time, was a minority position among the geoglogists of the day. Accordingly, most geologists were not greatly concerned with Kelvin's arguments about the total length of geological time, especially after some physicists began to debate Kelvin's premises. Nor did they engage his much stronger, if unoriginal, argument (from the conservation of energy and the dissipation of energy by the sun and the earth) to show that the rates of geological processes had been decreasing throughout terrestrial history. In sum, although neither side had a decisive argument (which helps account for the fact that the dispute stayed lively), various *prima facie* considerations lent great authority to Kelvin's argument. Darwin seems to have thought that this was one issue on which authority ran against him, and in this he was probably right. No matter how sure Darwin was of his facts, no matter whether he thought Kelvin’s argument good or not, he had to take this “odious spectre” seriously, precisely because of the degree to which Kelvin spoke from authority.

GOLDSCHMIDT AND MCCLINTOCK

How seriously should this sort of argument be taken? I have deliberately couched it in a fairly weak form. It can now be strengthened by demonstrating the need, on occasion, to fall back on second order considerations of a political character. Consider the difficulty that a biologist whose central research interests did not lie in genetics would have had circa 1950 in evaluating the status of Richard Goldschmidt’s arguments against the classical theory of the gene. Goldschmidt was an eminent biologist, most of whose career had been spent in Germany, who had done considerable work on development, physiological genetics, genetic systems of sex determination, the disturbances of those systems which yielded intersexes, and so on. Yet his views departed from those of mainstream evolutionists and geneticists, at least in the Anglo-American countries. He maintained steadfastly that Mendelian mechanisms, while utterly essential to the adjustments and adaptions exhibited by the members of local populations to locally prevalent ecological conditions, simply could not account for speciation, for the crossing of what he called the “bridgeless gaps” between species. He believed that what he called “systemic mutations” occurred extremely rarely, but that these rare events, by altering the development of the organism at an early stage, brought about in one fell swoop a major reorganization of its structure, creating a “hopeful monster,” a member of what was potentially a new species. These views which he had argued since publishing *The Material Basis of Evolution* in 1940,[13] were not only the subject of considerable criticism, but also the butt of an immense amount of ridicule and animus, for these were the days of the hardening of the synthetic theory of evolution, which held that all evolution is Mendelian microevolution writ large.[14]

Goldschmidt’s views in genetics proper were equally controversial and equally far from the mainstream.[15] After World War II he had done a considerable amount of work on *Drosophila*, the favored organism of geneticists for exploring the issues with which he was dealing. He had digested the technical literature in *Drosophila* genetics thoroughly enough to base his criticisms of that literature on knowledge comparable to that of an insider, though never achieving inside status in that field. He argued that the classical theory of the gene rested, in many respects, on a tissue of technical mistakes. Among other things, he argued that there was no experimentally adequate way of telling whether what was identified as a single gene by the usual techniques was in any useful sense a single entity and whether what was standardly identified as a mutation was a change in that entity or rather a miniscule rearrangement of the material on the chromosome creating what was technically called a position effect, i.e., a change of gene function caused by a rearrangement of the genetic material. Goldschmidt strongly favored the latter view, though the logic of his argument is peculiar: The very strong considerations he offered to show the impossibility of discriminating between the presence of discrete genes and the occurrence of miniscule rearrangements undermined his own positive argument for the latter.

This rough sketch of the issues that Goldschmidt raised shows how hard it would be to evaluate the merits of his case *vis a vis* that of such proponents of the classical theory of the gene as H. J. Muller. For those whose knowledge of the technical literature or whose competence to evaluate the evidence in hand was not up to the task, the best strategy was to obtain an estimate of the standing and authority of the various disputants, the reliability of their experimental technique and techniques, the thoroughness with which they had met the objections of their opponents, the relative standing of their disciplines, their following — or lack of a following — by the experts in the field, and so on. One can see this strategy being followed on all sides in the private correspondence of those interested biologists who had little expertise in *Drosophilia*, and, to a lesser extent, in the literature of the day. But to follow such a strategy is to employ what I am calling the internal politics of science as a means of supporting a rough and ready estimate of the relative *epistemic* standing of Goldschmidt's theory versus the classical theory of the gene, of the strength of his case against the orthodox theory.

This point cuts far deeper than is yet apparent. Thus far it has been claimed only that *if one is out of one's depth* one ought to use the best available estimate of authority and expertise (together with an account of the views of the experts in the field) as a *secondary* way of judging the epistemic status of work which one is not competent to judge firsthand for oneself. This claim, though interesting in other respects, has no fundamental epistemological interest since it could be correct while the primary means of settling epistemological disputes remained unchanged. Thus far, then, the point has only limited value in clarifying the epistemological relevance of knowledge of the actual sociology of scientific disciplines as an ancillary tool in evaluating theories when, *faute de mieux*, one is not able to execute a "first order" adjudication. The deeper point is that there are occasions on which even the experts can do no better than to fall back on this strategy. I shall offer two arguments, based on the present case, to support this claim.

First, if (as I am prepared to argue in detail on another occasion) Goldschmidt's technical arguments about the limitations of the then available experimental techniques are correct,[16] the experts themselves could not resolve the dispute in an epistemically satisfactory way on the basis of the materials then available to them. Nor could they dodge the issue: *Since the very techniques for identifying and localizing genes were the subject of the debate*, the proper interpretation of experimental results and the proper planning of genetic experiments bearing on the issues were at the heart of the dispute.[17] The only evidence available (and the only evidence cited by Goldschmidt or by the orthodox) was *indirect* in speaking to matters below the threshold of direct experimental resolution. Thus, until new techniques became available, the experts too had to have recourse to arguments that were unsatisfactory by the primary epistemological standards traditionally required within the philosophy of science.

The second argument to show the necessity of falling back on an epistemologically secondary strategy turns on the fact that different scientific fields differ legitimately about the criteria of adequacy of such theories as the theory of the gene. In precisely this respect, there is more to the story about Goldschmidt than has yet been indicated. Both developmental and physiological genetics — Goldschmidt's primary fields of genetic expertise — envisage the task of genetics differently than the classical Morgan school did. The primary concern of the Morgan school was what has since come to be labelled "transmission genetics," i.e., the study of the patterns of transmission of the traits of organisms to their offspring and later descendants and the factors in the egg responsible for the transmission of various traits or the tendencies to manifest them. Thus the choice of problems of the Morgan school biased their work — and their criteria for the evaluation of genetic work in general — toward acceptance of a particulate theory of the gene. How, after all, were they to identify genes? In the first instance (though the full story is much more complex than this suggests) any trait that was transmitted discretely must be brought about in consequence of the presence of a discrete genetic unit. And the penchant to interpret things in this manner was reinforced by the actual outcome of experiments in which various "complex" traits, eye color, for example, were analyzed as the products of a series of discrete simple traits — in this instance, the presence or absence of the normal or abnormal forms of a whole series of catalysts and pigments and the precursor substances out of which they were built. Success in localizing *regions* of chromosomes which, given the normal genetic context, affected particular traits was readily extrapolated into specific localization of *the* genes for those traits.[18]

In contrast, the internal problem-settings of developmental and physiological genetics both worked to devalue particular solutions to genetic problems and to emphasize the radical incompleteness of the classical theory. If one is interested in how the stuff inside a fertilized egg brings about the development of a complete organism (and remember that in interesting cases like that of *Drosophila* an organism may metamorphose into two or more seemingly finished phenotypes), one must ask how whatever it is that is inherited guides the processes which increase the differentiation of the organism and guide it to develop in certain ways rather than other ones. And, as had been obvious to all developmental biologists concerned with these issues since the rediscovery of Mendel in 1900, any solution to the problem of what it is that is inherited which only specifies and/or localizes, however precisely, a material particle or string of material particles, simply has not dealt adequately with the problem of inheritance, *for what is inherited must be a system or means for guiding development.* Even as late as 1950 it certainly could not be reasonably maintained that a particle or molecule, of whatever constitution, had the capacity to guide development. Molecules are not, after all, homunculi, and although they can serve in a limited way as templates for the formation of other molecules, for example in the process of crystallization, there simply was no way to get from a molecule or complex of molecules of the sorts found in the egg to anything so complicated as an organism. The processes involved required some sort of dynamic account — whatever the proper story of inheritance might turn out to be, it had to make intelligible how the genes could guide development.

(Indeed, I hold that in spite of the success of molecular genetics over the last thirty years in answering how genes or, better, genetic material guides the formation of particular proteins, it remains true even now that most molecular geneticists do not recognize the depth of the problem of development and that a general solution to that problem remains far out of sight.)[19]

Thus, as of roughly 1950, different subdsciplines of genetics, although not nearly as independent as Kelvin's physics and Darwin's theory of evolution, nonetheless led to incompatible expectations about the character of genes, much as Darwin's and Kelvin's theories led to incompatible expectations regarding the age of the earth. The conflict between these incompatible positions had to be resolved sooner or later yet the resources in hand simply were not adequate to allow a definitive resolution of the disagreement. This is not a case of conflicting Kuhnian paradigms (*as these are usually interpreted*) because, in spite of the fundamental differences in the theoretical approach and problem definition, the differing subdisciplines were constrained to accomplish an eventual reconciliation of their differing accounts of the units of inheritance or to abandon their claims to offer an adequate theory. And the issues between developmental and transmission genetics are not properly described as being resolved (or to be resolved) "by negotiation," for it is the working up of decisive evidence, *not* a political resolution which is at stake in the end. Yet the impasse is one which has continued over a good many scientific generations. Worse yet, while the impasse continues, one cannot afford to be neutral as to the relative merits of the differing theories, for the central details of one's approach to serious experimental questions turns on whether or not the classical school's claims to have located and properly identified genes are correct, at least to a first approximation. The position one takes cannot wait on the ultimate resolution of the disputed matters, but must turn (at least in part) on other considerations.

To reinforce the claim that there is a kind of politics involved here, briefly contrast the case of Goldschmidt with that of Barbara McClintock, recent Nobel laureate for her 1950's work on mobile, non-Mendelian, genetic elements in maize.[20] McClintock's path-breaking research had relatively little influence in the 1950's and 1960's, indeed well into the 1970's. Yet there was a qualitative difference between the reception of her work and that of Goldschmidt.

In McClintock's case, unlike Goldschmidt's, there was no question of the reliability of her experimental technique and no question that she was obtaining experimental results not readily interpreted along orthodox lines. Nearly all of the concerned parties admitted that if McClintock claimed that identifiable bits of the chromosome were mobile and that, by moving, they altered the expression of neighboring genes and of the traits of the plant, then these claims were almost certainly true. (Unlike Goldschmidt, McClintock provided substantial experimental evidence of the phenomena described in her theory.) Yet, such non-Mendelian behavior of the genetic material stood in fundamental conflict with the findings of orthodox transmission genetics and with the supposed stable linear ordering of chromosomal DNA. A variety of factors, inextricably interconnected, played a role in delaying the general recognition of McClintock's work on mobile elements and of its fundamental importance. First of all, she was working on corn. Plant genetics was, on the whole, far behind animal genetics, and not considered to be nearly as reliable. Furthermore, although maize was in certain respects a moderately well understood and characterized organism, it was also far more intractable a subject for genetic experimentation, especially on the problems which had recently moved to the center of genetic interest, than *Drosophila* or *Escherichia coli* or various yeast and fungi.

From the standpoint of an expermental transmission geneticist, the sheer fact that maize allowed only one generation (or at most two generations) per year was a tremendous disadvantage. Transmission geneticists traditionally ensured the reliability of their results by taking advantage of the statistical powers accrued by dealing with a large number of generations and many organisms with the same genotypes or with specific known similarities of genotype. All of this required rapid turnover of generations and the ability to reconstruct organisms with identical genotypes, or, at the very least, with identical chromosomes or chromosome segments among their chromosome complement. None of these things could be reliably done with maize. Furthermore, with an annual plant, there was always the risk of confusing environmental effects or effects of subtle alterations of gene expression with the primary effects of genes. For all of these reasons, work on corn was thought by many not to carry great weight in debates over fundamentals. Finally, McClintock's work used classical techniques just as the cutting edge in genetic research turned to molecular techniques. The facts that her techniques were not up to date, that they employed old-fashioned transmission-genetic and cytological analyses in a way which could not be meshed with the new molecular techniques developed for bacteria, and that she worked on an organism which at first proved intractable when approached with the new molecular techniques for analyzing DNAs all impaired the reception of her work. If one were to take results like those of McClintock seriously — i.e., if one were to consider them as offering evidence against the stability of the genome and against the correctness of the classical account of gene transmission — one would simply have to have a far stronger body of evidence than work on corn alone could afford. Perhaps if one found similar phenomena in *Drosophila, E.-coli*, or phage,[21] or meshed the cytological with a molecular analysis, it would be worth pursuing the recondite and confusing theoretical arguments embedded in McClintock's papers. But in the absence of such results (and no one had yet found such phenomena), and until she could show that her results had some generality, it was better to treat them as well attested and, perhaps, important oddities that should not be allowed to undermine the flourishing progress of the classical theory of the gene.

SOME PRELIMINARY CONCLUSIONS . . .

I do not pretend that this simple-minded sketch of an historically complex situation is accurate in every detail. Those who are interested in further details without undue complexity can find them, together with a somewhat different interpretation, in Evelyn Fox Keller's excellent biography of McClintock. Nor do I mean to suggest that the account I just offered will serve unmodified as a correct account of the epistemological situation. Rather, the point is that, considered together with the related case of the reception of Goldschmidt's work, the reception of McClintock's work provides a fairly representative and suggestive account of the extremely mixed sorts of evaluative considerations employed when important scientists take a stand in opposition to the entrenched orthodoxy. In situations like this, scientists must make some rough estimate, more or less well informed, of the authority with which the protagonist or protagonists speak to the issues at hand. To make such an estimate, one needs to evaluate the competence and reliability of the protagonists, the strengths and suitability of the disciplines on which they draw, the appropriateness of the techniques employed, the adequacy of the contact with the orthodox position, and so on. In addition to such measures of the standing of the protagonists, one also needs some sense of the distance between the orthodoxy and the stance taken by its critic, for this provides a decisive measure of the responsibilities of the critic. The greater the distance between the two positions, the greater the responsibility of the critic. In certain ways, the measure of distance in question opens a can of worms, for it may include, crudely speaking, ideological distance. The more "foreign" the point of view *vis à vis* the orthodoxy, the greater its burden to produce decisive new evidence, the more carefully it must handle the evidence already in place, and the less reasonable it is to accept its "negative" arguments against the orthodoxy without the additional support of powerful positive arguments. A great variety of considerations are involved here, but some of them, at least, belong to what I have been calling the internal politics of science. The short way of putting it is that one's place in the social system of science has consequences for the epistemic weight which one's contemporaries attach, and ought to attach, to one's work.

It is both inevitable and legitimate that a scientist's postion in the social system of science should influence the initial evaluation of that scientist's claims and results. It is also inevitable that the place that a scientist occupies in the social system of science turns in part on the field in which the scientist works, on the kind and quality of the technical work that the scientist does and on the actual results the scientist obtains. But it is also true that other factors, institutional, ideological, personal (McClintock was, for example, in many ways a loner), and fiscal, influence a scientist's place within the social systems of science. For science-in-progress, the weighting of results and arguments is almost always a mixed lot, turning on an immense diversity of considerations and circumstances. It is only over the long haul that the influence of adventitious considerations and circumstances can be recognized and removed. And because, when work is in progress, there is normally no satisfactory way to disentangle the complex factors entering into the evaluation of competing positions, one usually cannot extricate more narrowly cognitive or epistemic factors from others which bear on the place of the scientist in the social system of science. For those of us impressed with the progress of science, fitful as it is, the remarkable thing is how often this complex social system tends to place authority in the hands of those who make cognitive progress.

... AND A MORAL

It is time to draw a moral, however tentative, from the cases discussed and to add a few words about the larger picture into which this discussion should be fit. First the moral.

Cases like the ones I have been discussing support the premises and the conclusion of the schematic argument displayed below. Although the argument needs to be made more rigorous, I am confident that it points in the right direction.

P1. Sciences which are "in contact" with one another on a fundamental issue may disagree deeply. (Cf. Goldschmidt and other developmental geneticists vs. the classical school on the nature of the gene.)

P2. Such disagreements may be irreconcilable on the basis of present (and presently obtainable) evidence.

P3. Yet disagreement may be unavoidable in that
(a) it stems, in part, from the different problems considered in different disciplines,
(b) it deeply affects experimental methodology, interpretation of experimental results, and degree of justification for conclusions drawn from experiments, and
(c) the alternative theories may employ seriously conflicting ontologies or entail central conflicts over factual claims.

P4. Among the legitimate clues, as strong as any available, about which way to turn are the following:
(a) the relative authority assigned to the relevant (sub)disciplines regarding the question at hand,
(b) the relative authority and standing of the leading figures in the controversy, and
(c) the assessment of the larger community (e.g. of leading geneticists in our primary example) regarding what evidence weighs most heavily and who has best discharged the responsibilities to the evidence.

P5. One cannot extricate the assessments required in P4 from the "internal politics" of science.

Conclusion. For science-in-progress, epistemic evaluation of competing theories may legitimately turn, at least in part, on consideration of the "internal politics" of science.

One way of looking at this argument is the following: there is an authority structure in and among the various sciences, a structure which can be (and to some extent has been) made the subject of straightforward sociological and anthropological study. Traditional epistemology and epistemologists have thought either that that authority structure is a *consequence* of the cognitive virtues of the individuals, institutions and disciplines in question or that it is *irrelevant* to epistemological analysis of those cognitive virtues. I have been arguing that knowledge of the politics of science and its authority structures has a positive contribution to make to the epistemic appraisal of science in progress. If the considerations dealt with in premise four legitimately alter the balance of evidence, the traditional epistemological assumption that one can work up an epistemic appraisal of factual claims, hypotheses, theories, and so on with respect to the available evidence and irrespective of the social and institutional positions of the advocates of those claims, hypotheses, and theories is simply wrong.

When it comes to science-in-progress, the very fact that one is at the cutting edge means that *all* theories are faced with such powerful epistemic liabilities, that a "pure" epistemic appraisal, taken strictly, will land one in a paralyzing scepticism, will force one to say "a plague on all your theories." One can avoid this disastrous impasse by broadening the class of considerations with which one is allowed to work in making cognitive appraisals to include those which are actually employed by the members of the relevant disciplines at the time of the dispute. And, as has by now been amply demonstrated, these include judgments of the relative authority and reliability of the individuals and techniques involved and, given the unfortunate imperfections of knowledge and technique, an assessment of what issues the relevant parties are responsible to address or resolve. For example, the (justifiable) claims that McClintock was a far more responsible critic than Goldschmidt and that she was more responsible both in generating and interpreting data count as legitimate grounds for taking her worries more seriously than his. For another example, compare the difference in the reception of the rigorous negative arguments which Kelvin offered against evolution by means of natural selection with those which Goldschmidt offered against localization of the gene. In both instances there was considerable rigor in the argumentation (though it was probably greater in Kelvin's case). But the respect accorded to the arguments and their influence on their intended audiences are markedly different. A partial explanation of these differences turns on the fact that Kelvin spoke with the full authority of 19th century thermodynamics behind him, while Goldschmidt spoke with only the more limited authority of mid-twentieth century physiological and developmental genetics. Together with the more personal factors of the sort discussed above, the comparison provides a partial and I think, legitimate, account of the greater weight attached to Kelvin's arguments against evolution than Goldschmidt's against the classical theory of the gene.

An issue that all this suggests concerns the special features (if any) of the politics of science. On the surface, that politics is rather like the politics of any profession. Yet, on the whole, it *seems* to have done remarkably well in insuring that those outsiders and outcasts whose work had some sort of handhold or foothold in fundamental truths would be heard and, if things panned out, would have their work sooner or later assimilated into the mainstream. (For that matter, it seems usually to be outsiders who manage to divert the mainstream.) Are these appearances correct? Are there features of the social or institutional structure of science, are there long-term forces, which tend to bring scientific and cognitive authority into line? How good is the politics of science in this respect? How fragile in this respect are the institutions of science? And how would incorporation of some consideration of the politics of science into cognitive appraisal transform the epistemology of science? I am keenly aware that I have done nothing to resolve these larger questions in this paper, but they are eminently worth raising and, if the position I have been exploring is correct, the need to deal with them is pressing.

Virginia Polytechnic Institute and State University

NOTES

* A preliminary version of this paper was read at the 20th Annual Philosophy Colloquium of the University of Cincinnati in 1982. I have benefitted greatly from the helpful discussion there as well as from the incisive criticisms (far too many of which I have ignored) of the following friends: M. Grene, E. F. Keller, L. Laudan, R. Laudan, E. Lloyd, J. Pitt, R. Richardson, and B. Wallace. The research for this paper was conducted in part with the support from a National Science Foundation Scholar's Award (Grant Number SES-8309214). I am very grateful for this support. The paper

is, of course, only a small token of my thanks to Marjorie Grene for her role in bringing me into philosophy of biology, for her work, and for the inspiration that she has been to me and to so many others.

[1] Cf., for example, Parts One and Two of Grene (1966).

[2] Cf., e.g., Grene (1958), reprinted (among other places) in Grene (1974a), and her essays on Portmann and Plessner in (1974a) and in Grene (1968).

[3] Grene's own reading of Aristotle is set forth in Grene (1963). Cf. also Grene (1972), reprinted, together with her (1974b), in Grene (1974a).

[4] Cf. Grene (1981), Grene (forthcoming), and Grene (1985).

[5] The principal volumes, in addition to those already mentioned, are Grene (1971), Grene and Mendelsohn (1976), and Grene (1983).

[6] Among these are two further summer institutes in History, Philosophy, and Sociology of Biology (held at Denison University, Granville, OH, in 1983 and at St. Mary's College, South Bend, IN, in 1985); a major conference at California State University, Fullerton, the proceedings of which constitute the major part of Depew and Weber (1985); a further anthology, Brandon and Burian (1984); a special issue of *Philosophy of Science* (July 1984) devoted to philosophy of biology; and a large number of ongoing research projects. In addition, a large number of courses in philosophy of biology were inspired and shaped by the institute and its curricular booklet, *Philosophy of Biology in the Philosophy Curriculum* (1983).

[7] On the importance of the perspective from which one addresses such matters, cf. my (1977).

[8] An illustration will make clear the sort of thing that I have in mind. Shortly before the Trinity test of the first atomic bomb, one of the physicists involved in the project calculated that there was a small but finite risk that the test explosion might ignite atmospheric nitrogen, thereby causing an unprecedented and uncontrollable catastrophe. Some quick but agonizing calculations were undertaken to examine the matter, with the result that the risk estimate was significantly downgraded to something like 10^{-12}. The suggestion under consideration is that in cases where the risks are extraordinarily high, the degree of epistemic support that a theory, calculation, or methodology (etc.) should enjoy should also be extraordinarily high. More generally, the suggestion is that epistemic hurdles should be scaled to certain non-epistemic features of the issue in hand, proceeding along the lines of a cost-benefit analysis that takes into account the consequences of "false negatives" and "false positives".

[9] The best secondary source on this topic is Burchfield (1975). Two of the principal salvos that set off the debate are W. Thomson (Lord Kelvin), "On Geological Time," Presidential Address before the Geological Society of Glasgow, 1868, reprinted in Thomson (1892), pp. 10–64, and T. H. Huxley, "Geological Reform," Presidential Address before the Geological Society of Glasgow, 1869, reprinted in Huxley (1909), pp. 308–342. My treatment of this case has been greatly improved by discussions with R. Laudan and R. C. Richardson.

[10] Cf. Darwin (1959), pp. 513–514 and 728.

[11] Marchant (1916), p. 268, letter from Darwin to Wallace, dated 12 July, 1871, as quoted in Burchfield (1975), pp. 78–79.

[12] Huxley (1909), see note 9.

[13] Goldschmidt (1940).

[14] Cf. S. Gould, "Introduction", in the 1982 reprint of Goldschmidt (1940) and Gould (1983).

15 A sample of the relevant works includes Goldschmidt (1955), (1946), (1950a), (1950b), (1952), and (1954).
16 One piece of evidence in support of this contention, available now but not at the time, is the fact that, on molecular analysis, the mutations affecting classically identified genes include a great variety of insertions, deletions, alterations of regulation, and minute alterations of linear order in the genome as well as straightforward mutations of a coding sequence within exons, i.e., strings of DNA that actually code for protein.
17 Stadler (1954) illustrates the difficulty here. Stadler attempts to separate operational and theoretical concepts of the gene with the consequence that it becomes utterly impossible to give a first order characterization of genes (in terms of their location or their intrinsic properties). Such an account risks turning genetics into an impoverished bookeeping of patterns of trait transmission, intolerant of any theory of underlying mechanisms. Neither Goldschmidt nor the classical school was willing to accept Stadler's way out.
18 As early as the revised edition of Morgan *et al.* (1922), one finds this attitude clearly expressed:

> It is sometimes said that our theories of heredity must remain superficial until we know something of the reactions that transform the egg into the adult. There can be no question of the paramount importance of finding out what takes place during development. The efforts of all students of experimental embryology have been directed for several years toward this goal. It may even be true that this information, when gained, may help us to a better understanding of the factorial [i.e., Mendelian] theory — we cannot tell; for a knowledge of the chemistry of all of the pigments in an animal or plant might still be very far removed from an understanding of the chemical constitution of the hereditary factors by whose activity the pigments are ultimately produced. However this may be, the far-reaching significance of Mendel's principles remains, and gives us a numerical basis for the study of heredity. Although Mendel's law does not explain the phenomena of development, and does not pretend to explain the phenomena of development, it stands as a scientific explanation of heredity, because it fulfills all the requirements of any causal explanation.

19 Cf. Maynard Smith *et al.* (in press).
20 A sample of the relevant articles includes McClintock (1950), (1951), (1952), (1953a), (1953b), (1956), and (1961). My reading of this material has been considerably helped by Evelyn Fox Keller's valuable (1983) biography of McClintock.
21 In fact, it was the discovery of "transposable elements" in bacteria that helped to rekindle widespread interest in McClintock's work.

BIBLIOGRAPHY

Brandon, R. and Burian, R. (eds.): (1984), *Genes, Organisms, Populations: Controversies Over the Units of Selection*, Bradford Books/MIT Press, Cambridge, Massachusetts.

Burchfield, J.: (1975), *Lord Kelvin and the Age of the Earth* Science History Publications, New York.

Burian, R.: (1977), 'More Than a Marriage of Convenience: On the Inextricability of the History and Philosophy of Science', *Philosophy of Science* **44**, pp. 1–42.

Darwin, C.: (1959), *The Origin of Species, by Charles Darwin: A Variorum Text*, ed. M. Peckham, University of Pennsylvania Press, Philadelphia.

Depew, D. and Weber, B. (eds.): (1985), *Evolution at a Crossroads: The New Biology and the New Philosophy of Science*, Bradford Books/MIT Press, Cambridge, Massachusetts.

Goldschmidt, R.: (1940), *The Material Basis of Evolution*, Yale University Press, New Haven. (Reprinted (1982) with a new introduction by S. Gould.)

Goldschmidt, R.: (1946), 'Position Effect and the Theory of the Corpuscular Gene', *Experientia* **2**, pp. 197–203 and 250–256.

Goldschmidt, (1950a), '"Repeats" and the Modern Theory of the Gene', *Proceedings of the National Academy of Science* **36**, pp. 365–368.

Goldschmidt, (1950b), 'Marginalia to McClintock's Work on Mutable Loci in Maize', *American Naturalist* **84**, pp. 437–455.

Goldschmidt, R.: (1952), 'The Theory of the Gene', *Cold Spring Harbor Symposia in Quantitative Biology* **16**, pp. 1–11.

Goldschmidt, R.: (1954), 'Different Philosophies of Genetics', *Science* **119**, pp. 703–719.

Goldschmidt, R.: (1955), *Theoretical Genetics*, University of California Press, Berkeley.

Gould, S.: (1983), 'The Hardening of the Modern Synthesis', in Grene, 1983, pp. 71–93.

Grene, M.: (1958), 'Two Evolutionary Theories', *British Journal for the Philosophy of Science* **9**, pp. 110–127 and 185–193.

Grene, M.: (1963), *A Portrait of Aristotle*, Faber and Faber, London.

Grene, M.: (1966), *The Knower and the Known*, Basic Books, New York. (Reprinted (1974), University of California Press, Berkeley.)

Grene, M.: (1968), *Approaches to a Philosophical Biology*, Basic Books, New York.

Grene, M. (ed.): (1971), *Interpretations of Life and Mind: Essays Around the Problem of Reduction*, Humanities Press, New York.

Grene, M.: (1972), 'Aristotle and Modern Biology', *Journal of the History of Ideas* **33**, pp. 395–424.

Grene, M.: (1974a), *The Understanding of Human Nature*, D. Reidel, Dordrecht, Holland.

Grene, M.: (1974b), 'Is Genus to Species as Matter to Form? Aristotle and Taxonomy', in Grene, 1974a.

Grene, M: (1981), 'Changing Concepts of Darwinian Evolution', *The Monist* **64**, pp. 195–213.

Grene, M. (ed.): (1983), *Dimensions of Darwinism: Themes and Counterthemes in Twentieth-Century Evolutionary Theory*, Cambridge University Press, Cambridge, England.

Grene, M.: (1985), 'Perception and Interpretation: Toward a New Philosophy of Science', in Depew and Weber, 1985, pp. 1–20.

Grene, M.: (forthcoming), 'Philosophy of Biology, 1983: Problems and Prospects', *Proceedings of the Seventh International Congress for Logic, Methodology, and Philosophy of Science.*

Grene, M. and Mendelsohn, E. (eds.): (1976), *Topics in the Philosophy of Biology*, D. Reidel, Dordrecht, Holland.

Huxley, T.: (1909), *Discourses, Biological and Geological Essays*, Appleton, New York.

Keller, E. F.: (1983), *A Feeling for the Organism: The Life and Work of Barbara McClintock*, W. H. Freeman, San Francisco.

Marchant, J. (ed.): (1916), *Alfred Russel Wallace: Letters and Reminiscences*, vol. 1, Cassell, London.

Maynard Smith, J. et al.: (in press), 'Developmental Constraints and Evolution: Conference on Development and Evolution', *Quarterly Review of Biology*. (Copies available from R. Burian.)

McClintock, B.: (1950), 'The Origin and Behavior of Mutable Loci in Maize', *Proceedings of the National Academy of Science* **36**, pp. 344–355.

McClintock, B.: (1951), 'Chromosome Organization and Genic Expression', *Cold Spring Harbor Symposia in Quantitative Biology* **16**, pp. 13–47.

McClintock, B.: (1952), 'Mutable Loci in Maize', *Carnegie Institution of Washington Yearbook* **51**, pp. 212–219.

McClintock, B.: (1953a), 'Mutation in Maize', *Carnegie Institution of Washington Yearbook* **51**, pp. 227–237.

McClintock, B.: (1953b), 'Induction of Instability of Selected Loci in Maize', *Genetics* **38**, pp. 579–599.

McClintock, B.: (1956), 'Controlling Elements and the Gene', *Cold Spring Harbor Symposia in Quantitative Biology* **21**, pp. 197–216.

McClintock, B.: (1961), 'Some Parallels Between Gene Control Systems in Maize and Bacteria', *American Naturalist* **95**, pp. 269–277.

Morgan, T., Sturtevant, A., Muller, H., and Bridges, C.: (1922), *The Mechanism of Mendelian Heredity*, Henry Holt, New York.

Philosophy of Biology in the Philosophy Curriculum (1983), The Council for Philosophical Studies, San Francisco.

Stadler, L.: (1954), 'The Gene', *Science* **120**, pp. 811–819.

Thomson, W. (Lord Kelvin): (1892), *Popular lectures and Addresses*, vol. 2, Macmillan, London.

HARRY G. FRANKFURT

TWO MOTIVATIONS FOR RATIONALISM: DESCARTES AND SPINOZA

I

This essay deals with certain relatively neglected features of the theories of knowledge of Descartes and Spinoza. My aim is not to criticize or to evaluate the views of these philosophers, but to bring out certain parallels and contrasts in the motives which guided their epistemological preoccupations. In the case of Descartes, the pertinent materials appear almost exclusively in his letters. Even though they hardly figure at all in his systematic philosophical and scientific writings, they are highly theoretical and unmistakably germane to his central doctrines. With respect to Spinoza, the situation is somewhat the reverse. While the considerations in question appear in one of his major philosophical treatises, they are of a quite concrete and indeed personal nature. It is customary to classify both Descartes and Spinoza as rationalists, although the differences between them are far from peripheral or insignificant. So far as concerns the underlying motivations of their inquiries, the divergence between them is particularly sharp and unequivocal.

What they did share was an ambition to construct a body of philosophical and scientific knowledge grounded in absolutely certain fundamental principles. Neither was willing to assent to any proposition whose acceptability could not be thoroughly guaranteed by a rigorous demonstration from these principles. Nothing is more characteristic of Descartes and Spinoza than the importance ascribed to *order* and to unshakeable conviction: that is, to starting with the most indubitable beliefs, and to proceeding by carefully planned and fully justified steps to develop an unequivocally reliable understanding of everything within the scope of human intelligence.

My primary interest here is, with respect both to Descartes and to Spinoza, in the *other* side of their uncompromising demands for rationality and logical order. I want to call attention to the particular way in which each of them was fundamentally preoccupied with *disorder*, and how for each of them his commitment to reason was

A. Donagan, A. N. Perovich, Jr., and M. V. Wedin (eds.), Human Nature and Natural Knowledge, 47–61.

conceived as the solution to a disturbing problem concerning irrationality. I shall delineate the disorder which Descartes contemplated, identify the quite different source of Spinoza's compelling need for order, and suggest how the philosophical position of each man resolved or allayed the pre-theoretical concern — with a sort of chaos — by which his craving for the rationalist ideal was mobilized.

II

Descartes begins his effort to understand the world with the rather discouraging concession that, for all he knows or can hope to prove, the world he wishes to understand may be totally and inherently unintelligible to a rational creature. The concept of unintelligibility pertinent here is a severely logical one. What troubles Descartes is that the canons of rational discourse or thought — in other words, the fundamental principles and standards of logic — may not coincide at all with the structure of reality. His anxiety is that the inconceivable may be actual, that what is logically impossible (what has a self-contradictory description) may nonetheless be a fact, that propositions which are internally incoherent may nevertheless be true.

This appears not to be a very promising starting-point for an inquiry devoted to discovering how a comprehensive knowledge of the natural world can be deduced with unshakable certainty from self-evident first principles. It is a commonplace that human reason is, even at its best, disappointingly limited in power and scope. But Descartes's initial lack of confidence in rationality is of a far more radical and inhibiting sort. It derives from a belief that the possession of reason may actually be an insurmountable *obstacle* to the acquisition of any genuine knowledge at all: our rationality may be not merely too little of a good thing but in fact a crippling defect, sufficient to guarantee the failure of any inquiry into how things really are.

Descartes's anxiety concerning the authority of reason has its source in his distinctive view that the so-called eternal truths are created by God and that God's power is therefore in no way constrained by them.[1] Eternal truths are propositions concerning the essences or essential natures of things. They specify characteristics which a thing necessarily has just in virtue of being of a certain type — as, for instance, the Pythagorean Theorem specifies a characteristic which any right triangle will inescapably possess. Truths or facts of this kind can be established entirely on the basis of logical or conceptual considerations; and, by the same token, denying any of them produces a contradiction. But Descartes insists nonetheless that their truth is wholly subject to God's will. The alternative, he claims, is that the eternal truths are *independent* of God. This alternative would fatally compromise divine omnipotence: for if the eternal truths are independent of God, then what God can do is limited by truths or facts over which He has no control.

The orthodox view has always been that God is truly and sufficiently omnipotent as long as He can do anything *conceivable.* Since it is not conceivable that a self-contradictory proposition should be true, most philosophers and theologians have felt that acknowledging that He cannot make contradictions true implies no defect in His power. Descartes does not really provide any substantial argument for his claim that God's power is unconditioned by the laws of logic or, to put the same point in another way, that God is not necessarily rational. Perhaps this is because he construes the claim as serving only a heuristic purpose, rather than as a doctrine to whose truth he definitively commits himself. Descartes's general philosophical project is to consider how (if at all) a reasonable person can attain absolutely certain knowledge of the natural world. In order to carry out this project in the most convincing way, he begins by supposing that the conditions under which he must proceed are as unfavorable as can be imagined. He approaches his problem, in other words, by considering the worst possible case. Let me elaborate his assumption concerning divine omnipotence and explain just how unfavorable to the prospects for knowledge it really is.

Descartes's way of construing divine omnipotence means that when God creates things He is not limited by any antecedent knowledge of what they must be like, how it would be best for them to be, or whether they need or ought to exist at all. The standards of what must be and of what should be are themselves creatures of God; so God is free to create or to alter them entirely as He likes, in any way whatever. His creative will cannot be determined or even influenced by any antecedent considerations of value or of rationality, because there *are* no truths before He creates them. Thus the divine will is entirely arbitrary. Here is how Descartes describes the situation:

He did not will that the three angles of a triangle equal two right angles because He knew that it could not be otherwise. On the contrary, it is because He willed that the three angles of a triangle equal two right angles that it is now true that this is so and that it cannot be otherwise; and similarly for everything else.[2]

In Descartes's view, then, God has no reasons for His decrees. His choices are submissive to no moral or rational constraints at all.

Descartes quite openly embraces the remarkable implication that a proposition might be true even if it is self-contradictory: "God cannot have been required to make it true that contradictories cannot be true together, and therefore . . . He could have done the opposite."[3] One can see, of course, what Descartes has in mind. After all, if God is all-powerful then nothing can *require* Him to do anything; so there is nothing that He *must* do or that He *cannot* do. On the other hand, it is not so easy to see what sort of rationalistic program can possibly be feasible if the world which is to be explained has no rational explanation. The eternal truths are the very foundation and substance of rationality. Yet on Descartes's account they are inherently as contingent and in need of explanation (which they cannot be given) as any other proposition.

The principles of logic delineate the boundaries of what it is possible for us to understand, but Descartes maintains that it is only a psychological fact that they do so. Propositions that violate logic — that is, self-contradictions — are unintelligible to us; we simply cannot conceive their truth. But our inability to conceive the truth of a contradiction is merely a contingent characteristic of our minds. God simply happened to make us that way: "He has given me such a mind," Descartes says, "that I cannot conceive . . . an aggregate of one and two which is not three."[4] This is just a fact about *me*, about what I can and cannot do, and not about the inherent relationship of one and two. My inability to conceive something, Descartes holds, signifies nothing beyond itself except that God freely chose to create me like that. It does not signify that He might not have created me with a mind of another sort, or that the world outside of me must conform to the characteristics of my intelligence.

We cannot legitimately presume that the world is subject to the necessities of logic and reason. We only know that, as it happens, *we* cannot conceive a world that does not conform to them. Here is what Descartes says:

I do not think we should ever say of anything that it cannot be brought about by God. For since everything involved in truth and goodness depends on His omnipotence, I would not dare to say that God cannot make a mountain without a valley, or that one and two should not make three. I merely say that He has given me such a mind that I cannot conceive of one and two which is not three, and that such things involve a contradiction in my conception.[5]

The upshot is that we cannot arrive at any certainty concerning how things are by relying upon the standards of rationality set by the human mind. In other words, we cannot dare to assume that rational inquiry leads to the truth as God created it. That truth may be inaccessible to us, because reason may not be a suitable instrument for discovering it.

On the other hand, however, there is no need for us to be concerned about our inability to discover what is absolutely true. Since any absolute truth that conflicts with the beliefs reason requires us to adopt is *ipso facto* self-contradictory and unintelligible to us, we could not make use of it even if we were somehow able to identify it as absolutely true. The possibility of a discrepancy between rationally warranted belief and absolute truth therefore does not prevent Descartes from relying confidently upon reason. In fact, he says, "it would be entirely contrary to reason to doubt what we understand very well because of something else which we do not understand and which we see no need to understand."[6] We can reasonably be satisfied with the products of reason, since no *comprehensible* truth — no truth comprehensible by us — can conflict with them. Descartes's account leads him, then, to a conception of rational inquiry as an exploration by reason of its own limits and necessities, whose goal is to determine what it is reasonable for us to believe, or what it would be irrational for us to doubt, rather than what is true in the absolute sense of grasping the inherent nature of things as they are in themselves and as they appear to the eyes of God.

This way of understanding the point of rational inquiry liberates reason from the uncertain and dangerous competition in which, during the early seventeenth century, it found itself engaged with religious faith. By acknowledging that reality may be in itself unintelligible, Descartes concedes that reason cannot claim authoritative access to the inherent nature of things. In order to know how God made the world, we must rely upon whatever truth divine revelation provides; for rational methods of inquiry are incommensurate with the range of alternatives available to God in creation. By the same token, however, Descartes frees reason from the need to acknowledge any dependence upon what is revealed to faith because he denies that it is the proper aim of reason to discover absolute truth. Thus science becomes autonomous, albeit at the cost of renouncing any claim to precedence over revelation as a description of the inherent nature of things. By assigning different objectives to reason and to revelation, Descartes seals them off from each other and eliminates the possibility of competition or conflict between them.

In more personal terms, the movement of Descartes's thought constitutes a withdrawal, from a universe whose unconstrained and unintelligible power might well have seemed uninviting, into the secure coherence and familiarity of his own mind. Contemplating the onmipotence of God, Descartes recognizes that there may be an irreconcilable discontinuity between his own needs and capacities and the objective character of the universe. So he proceeds to find within himself, in his own rational faculty, the stability and intelligibility which he requires and which he cannot be confident of finding in a world created without concern for his requirements and tastes. The possibility that the real is inescapably alien to him becomes irrelevant, as he undertakes to be satisfied with the rational whether or not it coincides with the real.

Descartes sacrifices the grandiose ambition to become intellectually the master and possessor of the world, in other words, and resolves to content himself with the opportunity to develop and to complete his own thought. He turns away from the world, which he cannot presume to be even tolerably congenial to his intelligence, and seeks only to become harmoniously attuned to the regularities and standards of his own nature. As I hope to make apparent, the movement of Spinoza's thought is just the opposite of this.

III

Spinoza's *Ethics* is written in a forbiddingly austere and impersonal way, in the form of a geometric treatise set out with definitions, axioms, postulates, theorems, corollaries, lemmas, and scholia, linked together by elaborate and often obscure trains of deductive argumentation. Unlike Descartes's *Discourse on Method* or *Meditations*, it is not an engaging piece of work. It gives the impression that Spinoza wishes to expunge from his writing every trace of his own individuality and personality, in order that the universal voice of reason — which is, presumably, identical in all men, in quality if not in quantity — should be more plainly heard and recognized as such.

In his excellent book on Spinoza, Stuart Hampshire accepts this impression. "Spinoza thought it right," he suggests, "that a philosopher should remain impassively concealed behind his philosophy." Perhaps Spinoza did actually think that. But Hampshire goes on to declare further that Spinoza has, in fact, "effectively concealed himself behind his work."[7] It may well be that Spinoza attempted to do this, and that his attempt to do it accounts for the peculiar style and form of his writings, but the concealment is by no means entirely complete. For one thing, the very attempt at concealment, and the devices employed in it, are themselves revealing. The degree of austerity and rigidity which Spinoza attempted to impose upon the expression of his thought strongly convey that he may have felt an especially powerful tendency in the contrary direction.

But there is another source of insight into Spinoza's more intimate and personal characteristics, which does not depend upon supposing that people who present themselves in exaggerated or extreme ways are striving to conceal characteristics which they consider to require special restraint or cosmetic treatment. The first few pages of one of Spinoza's more important philosophical books — his *Treatise on the Improvement of the Understanding* — are quasi-autobiographical. They do not tell us about Spinoza's inner life in unqualified or explicit terms. Nonetheless, reading them attentively does bring to light far more of what motivated Spinoza's philosophical work — how he was quite personally involved in his philosophical project — than Hampshire's characterization of his selflessness might lead us to expect.

The opening section of the *Treatise* (its first 16 or 17 paragraphs) has been regarded by many readers as one of the most inspiring pieces of writing in our philosophical literature. One commentator speaks of the "sweep and grandeur of the thoughts it embodies," and declares that it "undoubtedly merits the universal admiration it has received."[8] My own opinion of it is somewhat less enthusiastic. I find its "grandeur" a trifle conventional and its moral stance perhaps even somewhat banal. But it is certainly an interesting text, if not for the novelty or the inherent value of its ideas then for the rather surprising illumination of Spinoza's intimate psychic processes and concerns which it provides.

Here is the paragraph with which the *Treatise* begins:

> After experience had taught me that all the usual surroundings of social life are vain and futile; seeing that none of the objects of my fears contained in themselves anything either good or bad, except insofar as the mind is affected by them, I finally resolved to inquire whether there might be some real good having power to communicate itself, which would affect the mind singly, to the exclusion of all else: whether in fact there might be anything of which the discovery and attainment would enable me to enjoy continuous, supreme and unending happiness.[9]

In the early parts of his *Discourse on Method* Descartes had complained of how empty and unsatisfying he had found the science and other learning to which his early education had been devoted. Spinoza starts by expressing a more general complaint, against the vanity and futility not merely of education and of standard learning, but of *all* the usual surroundings of social life. And the goal to whose pursuit he commits himself is correspondingly general and uncompromising: not merely a reliable and productive method of scientific inquiry, but nothing less than the secret of "continuous, supreme and unending happiness."

This is extremely ambitious. It makes Spinoza's project rather unreal, a bit preposterous, and perhaps somewhat intimidating. In any case it is notable that Spinoza makes a close link between the improvement of the understanding or intellect — which is, after all, the primary theme of the *Treatise* — and the attainment of happiness, which is the subject to which he first turns. Of course, it is natural enough to suppose that increasing our capacity for reliable knowledge would facilitate the improvement of our lives. But Spinoza connects perfecting our rational capacities and perfecting our lives in a particularly intimate way.

After stating the goal to which his *Treatise* is to be devoted, Spinoza explains that he cannot conduct his proposed inquiry effectively without abandoning, or at least substantially modifying, the customary preoccupations of life. Most people, he says, assume that the highest goods are money and material things, fame, and sensual pleasure; and they devote themselves to pursuing these. But he has found in his own experience that pursuing them is too absorbing and too distracting. Since it does not permit a serious and fruitful effort to discover the nature of what is ultimately valuable, he must turn away from it.

It is quite commonplace for moralists to warn people against supposing that sensual pleasure, fame, and money are genuine or unequivocal goods. It is worth looking in some detail, however, at the particular critique of these attractions that Spinoza offers. Of sensual pleasure he says this: "By sensual pleasure the mind is enthralled to the extent of quiescence, as if the supreme good were actually attained, so that it is quite incapable of thinking of any other object; when such pleasure has been gratified it is followed be extreme melancholy, whereby the mind, though not enthralled, is disturbed and dulled".[10] Spinoza is clearly not talking about the sort of physical pleasure one gets from a brisk walk in the country. That is not the sort of activity which enthralls us to the point of quiescence or whose appeal drives out all thoughts of other things. Nor is it common, or even understandable, that an activity of that sort should be followed by melancholy, much less by "extreme melancholy." A bit further on Spinoza asserts that sensual pleasure is followed not merely by melancholy but by *repentance.* This only confirms the strong impression that, whether or not he is clear about this himself, Spinoza is describing sexual pleasure, which he evidently finds extremely enticing but which he also finds to be mixed characteristically with unpleasant experiences of depression and guilt.

Spinoza's initial point about money and fame is a somewhat different one. It is that there is no limit to how much of those things people who care about them desire; indeed, the more people have of them the more they tend to want. The pursuit of sensual pleasure is limited by our susceptibility to fatigue or to the exhaustion of our capacity for engaging in, or for enjoying, whatever activity is in question. But no limitation of this kind is inherent in the pursuit of money or the pursuit of fame, which are by nature endless and uncompleteable and which will continue as long as they are not limited by something outside themselves — some goal which defines how much money or how much fame is enough. Spinoza finds that these goods cannot of themselves bring satisfaction, because there is no particular amount of them which is naturally or inherently satisfying. Moreover, pursuing either of them is inevitably competitive and leads inescapably to undesirable experiences such as those of envy, hatred, fear, and disappointment. People who are committed to the pursuit of conventional goods, Spinoza warns, expose themselves to extremes of contradictory emotion. They swing from intense pleasure to feelings of melancholy and guilt; and their satisfactions are often mixed with frustration when they discover that attaining what they desire serves only to arouse a further desire for more than they already have.

As Spinoza elaborates these points, the quality and tone of his account undergo a conspicuous change. The conventional goods, he says, are not only unreliable and unsatisfying. They are actually *evil* and extremely *dangerous* to us; and anyone who devotes himself to them is "in a state of great peril." As he goes on, Spinoza seems to be more and more carried away. The peril to which he refers turns out to be not just a danger of moral corruption or of misery or of some sort of deterioration of the soul. He insists that it is literally a peril of *death.* And then it becomes not only a peril but even a *certainty* of death! Conventional goods *often* cause the deaths of those who possess them, Spinoza declares, and they *always* cause the deaths of those who are possessed by them. The context makes it umistakably clear, by the way, that these references to death are not metaphorical. Spinoza's statements really are just as wild as they seem.[11]

Spinoza never completed the *Treatise on the Improvement of the Understanding.* He clearly had some notion of completing it, however, since he made a number of notes in the margin of his manuscript concerning changes to be made in a subsequent version which he never actually wrote. One of those notes occurs at the point where he makes the remarkable — indeed, incredible — claim that giving in to or allowing oneself to be possessed by desires for sensual pleasure, money, or fame brings certain death. His note to himself reads: "These considerations should be set forth more precisely."[12] Indeed! Evidently he became aware of the exaggeration into which he had permitted himself to be swept, caught himself up short in his marginal admonition to himself, and intended one day to give a more measured and less fantasy-ridden account of his subject.

As I read Spinoza's text, my impression is that he is expressing very considerable anxiety concerning urges and desires within himself which he recognizes as having no inherent limit and which he therefore fears he may be unable to control. His desires for sensual pleasure *are* self-limiting in a way, since satisfaction exhausts them. But they are difficult to resist, and they and their satisfactions are naturally linked in his experience to melancholy and to guilt. When he imagines himself giving in to the other desires with which he is concerned, his thought of doing so is closely associated in his mind with a fantasy of being severely punished and even, if the urges are permitted to careen altogether out of control, of being killed.

In discussing the desires for money and fame, Spinoza attributes much of the unsatisfactory quality of these objects as ends in themselves to the fact that they are necessarily *scarce*. Since there cannot be enough of them to satisfy all possible desires, the satisfaction of one person's desires diminishes the chances of satisfaction for others. Those who desire money or fame are therefore inevitably in competition with others who are also ambitious to acquire them. It is the inevitability of competition which leads Spinoza to regard the value of attaining these goods as inextricably compromised by the evils of hatred, envy, fear of loss, and other disturbances of the mind.

The general impression given by Spinoza's examination of conventional goods as a whole is that he is concerned — only slightly beneath the surface and with so much intensity that it bubbles up at times into more or less open view — about powerful sexual desires and competitive urges, by which he is strongly moved but which he believes he must control and limit because of intense feelings of guilt and fear of punishment which are closely associated with them in his mind. I think it very natural and plausible to identify these competitive and sexual urges, the anxiety that they will get out of control, and the exaggerated fear of punishment which Spinoza associates with them, as comprising elements in an Oedipal syndrome. But I will not insist upon this characterization of his underlying preoccuption. It will suffice, for my purposes, to notice that he is concerned with desires which lead to trouble because they do not naturally limit themselves, because they bring those who give in to them into conflict with others, and because for a variety of reasons they entail an array of undesirable disturbances of the mind.

What is the cure for all this? How are we to avoid these disturbances of the mind, which obstruct and interrupt the sustained serenity and joy which Spinoza seeks? The secret, he declares, is to care deeply only for what is eternal and infinite — in other words, for what is neither transitory nor scarce. The enjoyment of something of that kind will be, he assures us, unmingled with any sadness; no contrary pain will be inherent in the pleasure it brings. Now it is pretty clear that being eternal and infinite are not really enough. After all, the number six is eternal; and even if we add all the other positive integers, so that we get an object that is not only eternal but infinite as well, this hardly solves the problem of how to achieve perfect happiness.

Of course Spinoza has something more particular in mind, which comes out when he begins to describe the ideal condition of human life. Here is the paragraph in which his central claim emerges:

All things which come to pass, come to pass according to the eternal order and fixed laws of Nature. However, human weakness cannot attain to this order in its own thought, but meanwhile man conceives a human character much more stable than his own, and sees that there is no reason why he should not himself acquire such a character. . . . What that character is we shall show in due time, namely that it is the knowledge of the union existing between the mind and the whole of nature.[13]

It is clear enough what eternal and infinite object it is that Spinoza identifies in this passage as capable of playing a fundamental role in the achievement of human happiness. It is "the eternal order and fixed laws of Nature." But the observation Spinoza makes next is sometimes misunderstood. After referring to the eternal order and fixed laws of Nature, he observes that we are too weak to attain this order in our own thoughts. Now this has been construed by some readers as a lament over our inability to arrive at a totally comprehensive knowledge of Nature. That is how Joachim, for example, appears to read the passage. He evidently supposes that the human weakness to which Spinoza calls attention is an intellectual or scientific inadequacy; we are simply not intelligent enough to grasp the eternal order in all its details.[14]

Spinoza himself makes it rather clear, however, that his attention is not focused primarily on our intellectual limitations. After alluding to human weakness and to the incapacity it entails, he declares that "man conceives a human character much more stable than his own," and that this conception of a more stable character provides the ideal goal towards which human endeavor must strive. The ideal is not formulated in terms of intelligence or of knowledge or of understanding, but as a matter of *stability*. In other words, what we conceive as the ideal character for ourselves is not one distinguished primarily by greater knowledge than we possess, but one which emulates the characteristics which Spinoza has just been ascribing to Nature — namely, order and fixity or, to use his word, stability. Our aim is to be rid of the disturbances which unbalance our condition, interrupt the evenness of our thoughts and feelings, and make us suffer passively the effects upon us of forces with which we do not identify and which we experience as alien to ourselves.

How does Spinoza imagine this stability can be achieved? There, of course, is where knowledge comes in. We achieve stability by understanding "the union existing between the mind and the whole of nature" — that is, by recognizing ourselves as products of forces which are generated systematically and in a lawful manner according to the fixed nature of the world, and by understanding that what goes on within us is by no means random or unintelligible but that it is (like everything else that happens) a necessary consequence of the fundamental substance and structure of the universe. The more we come to see the events of our own lives — and especially the events of our minds — as manifestations of an eternal and fixed order of natural law and natural necessity, the more intelligible they become to us and the less we are beset by emotions which breach and undermine the order of our nature and the stability of our existence. This reduces our sense that the power of the universe is alien to us and that we are merely passive with respect to it.

IV

Both Spinoza and Descartes are fundamentally preoccupied with what each takes to be — or fears may be — a crippling discrepancy between his own nature and the general nature of reality. The problem is in each case one of order and intelligibility. Descartes's concern is that the world around him may be literally a chaos, altogether lacking in any coherence or intelligible structure which he might grasp. Spinoza, on the other hand, is attentive primarily to the experience of disorder within himself, which represents an estrangement between himself and the rational stability of the order of Nature.

Descartes's response to the problem posed by his realization that he may not "belong" in the world — that his intellectual standards may not be satisfiable by the inherent nature of things — is to decide that he does not and need not care, because he can find sufficient satisfaction within the processes of his own thought without establishing that these processes reflect those of the world as it is in itself. Spinoza's solution to the problem posed by his inner disorder is to turn away from himself and to have his life ordered and structured through being embraced within the comprehensive rational order forming the intelligible structure of the world as a whole. Descartes seeks to escape the irrationality of the universe by discovering and cultivating an orderliness and coherence within his own mind. Spinoza attempts to escape the irrationality and violent disorder he finds within his own psychic experience by discovering and connecting himself to the coherence and necessity of the world outside himself.

This difference between Descartes and Spinoza reflects a different sort of confidence in the two men: the former believes in himself and is willing to rely upon his own inner resources and character without caring whether or not they can be validated by an authority external to themselves; on the other hand, the latter despairs of satisfying his own inner needs without turning for support to a source of strength and stability outside himself. Descartes substitutes the order in himself for the chaos in the world. Spinoza substitutes the order of the universe for the chaos within himself.

Each of the two philosophers wishes to avoid the implications of a fundamental conflict between his own nature and the nature of the world. But Descartes's notion is to achieve harmony by insisting that the world be describable (whatever its inherent nature) in terms congenial to the requirements of his own mind, while Spinoza's idea is to achieve the harmony for which he recognizes a need by submitting himself to an order which he gladly acknowledges to be superior to his own. On the one hand there is the rebel, who defies opposition to the satisfaction of his own requirements; on the other there is the mystic, who copes with conflict and opposition by denying himself and by seekng to be absorbed into the whole from which he finds himself separated. The rebel and the mystic share a desire to overcome the discrepany between themselves and what is other than themselves, one by changing the world to suit himself and the other by merging himself into the external order.

I do not know what moral to draw from these comparisons and contrasts, or whether there is in fact any significant or useful moral to be drawn from them. No doubt the differences between Spinoza and Descartes refect to some considerable extent the difference between two fundamentally divergent types of individual character structure. Not much serious work has been done to illuminate the relationships between individual character and psychic structure on the one hand and types of philosophical thought on the other. I believe that these relationships might prove to be very interesting, but I am uncertain whether they can reliably be pursued in satisfactory depth.

Yale University

NOTES

[1] The following discussion draws heavily on Frankfurt (1977).
[2] Descartes (1897–1910), vol. IX, p. 233; Descartes (1972), vol. II, p. 248. (Sixth "Replies to Objections".)
[3] Descartes (1897–1910), vol IV, p. 118; Descartes (1970), p. 151. (Letter to Mesland, 2 May, 1644.)
[4] Descartes (1897–1910), vol V, p. 224; Descartes (1970), p. 236. (Letter to Arnauld, 29 July, 1648.)
[5] Descartes (1897–1910), vol V, p. 223–224; Descartes (1970), pp. 236–237. (Letter to Arnauld, 29 July, 1648.)
[6] Descartes (1897–1910), vol IX, p. 236; Descartes (1972), vol. II, p. 251. (Sixth "Replies to Objections".)
[7] Hampshire (1951), pp. 234–235.
[8] Joachim (1958), pp. 14–15.
[9] Spinoza (1913), vol. I, p. 3; Spinoza (1951), vol. II, p.3.
[10] Spinoza (1913), vol. I, p. 3; Spinoza (1951), vol. II, p. 4.
[11] Spinoza (1913), vol. I, p. 4; Spinoza (1951), vol. II, p. 4.
[12] Spinoza (1913), vol. I, p. 4; Spinoza (1951), vol. II, p. 5.
[13] Spinoza (1913), vol. I, pp. 5–6; Spinoza (1951), vol. II, p. 6.
[14] Joachim (1958), pp. 22–23.

BIBLIOGRAPHY

Descartes, R.: (1897–1910), *Oeuvres*, ed. C. Adam and P. Tannery 11 vols., Cerf, Paris.

Descartes, R.: (1970), *Philosophical Letters*, trans. A. Kenny, Clarendon Press, Oxford.

Descartes, R.: (1972), *Philosophical Works*, trans. E. Haldane and G. Ross, 2 vols., Cambridge University Press, Cambridge, England.

Frankfurt, H.: (1977), 'Descartes on the Creation of Eternal Truths', *Philosophical Review* **86**, pp. 36–57.

Hampshire, S.: (1951), *Spinoza*, Penguin Books, Harmondsworth, Middlesex, England.

Joachim, H.: (1958), *Spinoza's Tractatus de Intellectus Emendatione*, Clarendon Press, Oxford.

Spinoza, B.: (1913), *Opera*, ed. J. Van Vloten and J. P. N. Land, 3rd ed., 4 vols. in 2, Martinus Nijhoff, the Hague.

Spinoza, B.: (1951), *Chief Works*, trans. R. Elwes, 2 vols., Dover Publications, Inc., New York.

IAN HACKING

THE INVENTION OF SPLIT PERSONALITIES

(An illustration of Michel Foucault's doctrine of the constitution of the subject)*

In one of his essays about power, Michel Foucault writes:

> Rather than ask ourselves how the sovereign appears to us in lofty isolation, we should try to discover how it is that subjects are gradually, progressively really and materially constituted through a multiplicity of organisms, forces, energies, materials, desires, thoughts, etc. We should try to grasp subjection in its material instance as a constitution of subjects. This would be an exact opposite of Hobbes' project in *Leviathan.*

In the individualistic world of which Hobbes is our greatest spokesman, people come together to constitute the sovereign in order to save their goods and their skins from their own predatory instincts. The individual human beings are given, but the sovereign is not. He must be created, and it matters little whether the Sovereign is Queen or Republic, State or Despot, Pope or Parliament. That has proven to be an extraordinarily fertile fable. It starts with the same assumption made in all seventeenth century philosophy: the person, the individual, is given. The sovereign is an artefact of the subjects. Foucault's *exact opposite* says that the subjects are an artefact of the power relations into which they unwittingly enter. This too is a fable, but it may turn out to be as instructive a fable as that which we owe to Hobbes.

I shall not defend the new fable in general terms, but instead seek for examples. In Foucault's recent work there is an intense concern for the ways in which new kinds of knowledge about people are part of a system of power relations. I would especially emphasize the following. What it is to be a person is in part determined by the possible categories which you and others may use to describe you. This is not only a static matter, for the options that you see open to you are put, in your inmost thoughts, in terms of possible descriptions of your future self. So a knowledge that brought into being new systems of classification and destroyed old ones would be a most potent vehicle. It would truly constitute you, not only in the eyes of others, but also in terms of the possible courses of action which you, as agent, might choose to make your own.

Now there is at least some trite truth in this. I cannot, now, choose

A. Donagan, A. N. Perovich, Jr., and M. V. Wedin (eds.), Human Nature and Natural Knowledge, 63–85.

to be exorcist, for I have no idea what it is to think and act as an exorcist. Nor does anyone else with whom I communicate. Those categories and intentional practices are perhaps lost forever. They may be pathetically imitated by various satanic cults that have no true sense of a dead past, but it is not open to any of us to be the sort of person embodied in a standard exorcist of the European sixteenth century. Very well, but in detail how can the constitution of more familiar concepts of the person occur in the systems of power relations imagined by Foucault?

One sketch of his answer lies in two ways in which knowledge has interacted with "Life" itself. You will recall that Foucault contends, in *The Order of Things*, that Life, Labor, and Language are new objects of thought emerging towards the end of the eighteenth century. One way in which power relations bear on Life concerns the sheer size of populations, while the other concerns individual modes of procreation. Late in the eighteenth century, power over these two constituted "two poles of development linked together by a whole intermediary cluster of relations." One pole centered on the body as a speaking, working, procreating machine, subject to personal discipline. It was organized by procedures that Foucault calls an *anatamo-politics of the human body.* Key words in Foucault's own expositions can be taken from the titles of some of his books: the prison, madness, clinical medicine, sexuality. The second pole "focussed on the species body" that serves as the "basis of the biological processes: propagation, births and mortality, the level of health, life expectancy and longevity. " This is a *bio-politics of the population* that "gave rise to comprehensive measures, statistical assessments, and interventions aimed at the entire social body or at groups as a whole."

My field of expertise is statistical assessment. In a companion paper — 'Biopower and the Avalanche of Printed Numbers' — I direct attention at a fetishism for counting that snowballs in the period 1820–1840. It brings into being not only new enumerations and new bureaucracies, but also, in a more subtle way, new ways in which to classify people. The new categories are specifically designed and defined in terms of ways in which it would make it easier to count. Examples are not merely the obvious reorganizations of the family structure for purposes of counting, but also the redefinitions of people in terms of enumerable classes within the work force. As the reader of Foucault will expect, statistics had the counting of deviancies as its main concern, and in the early days, the statistics of people was actually called *moral science*! I claim that many of the ways in which we now think of being a person are artefacts of the fetishism for numbers.

By dedicated reading of dusty archives we can witness new conceptualizations of the person coming into existence. They are not the creations of novelists nor the inventions of preachers; they are not the analyses of philosophers but the devices of forgotten bureaucrats. Thus at the pole of bio-politics we do indeed see one of the endless ways in which the "constitution of the subject" has occured. This is of course a thorough *nominalist* theme. It says that systems of classifications, at least of people, are artefacts of society. But ordinary old-fashioned nominalism is pretty unintelligible because it gives no account of how man-made systems of naming could actually latch on to the world. It is "wishful thinking nominalism." Somehow thinking and naming is supposed to create a natural class. Foucault's is a dynamic nominalism about personal classifications, which shows, through an historical event, how an invented classification is adopted by a populace so that the glove, in the end, fits.

Having written about some bio-political examples at length it might seem natural for me to attend here to anatomo-politics. I think that would be a mistake, at least for someone who, like myself, tries part of the time to be a junior collaborator in the exercise. Foucault mentions the statistical assessments but needed someone else to write about them. It needs no one else to write about anatomo-politics, for that is his own on-going project. But there are many elements in the "inter mediary cluster of relations" to discuss, especially for a philosopher. At least since Locke there has been a standard problem of "personal identity." Can this teach us anything about the constitution of the person?

The Integrated Personality

Suppose we take seriously the idea of the person being constituted by relations of power. We may wish to consider the idea of "the whole person," the healthy person, the intergrated personality. That is the most general level at which, we may suppose, the whole person *ought* to be constituted. Now of course Foucault from the very beginning has approached any given "norm" obliquely. To consider the age of reason, reflect on what it excludes, namely madness. Let us try that here. To reflect on the integrated person, let us consider what that idea excludes. It excludes many things, but here I shall consider only the most simple minded. It excludes the split personality.

I shall urge that the concept of multiple personality was invented by doctors and lawyers, with an assist from a host of charlatans and politically engaged philosophers.

Now that story is rather easy to tell. It sits on the surface of facts, waiting to be read by the industrious. It is especially easy because although the "split personality" has become a permanent part of our folklore, the majority of psychiatric practitioners is very dubious if the idea is useful at all, today. Some will welcome what I say as an extreme version of what they have always believed. I can even assert this: there were no multiple personalities until a quite definite moment, after which there were a great many. It is as if disturbed people found a new way to be disturbed. The new way was largely invented by doctors, but later became a spontaneous way in which to express unhappiness. Thus I don't deny that the multiple personalities of the late nineteenth century were ill at ease. I assert only that multiple personality is a culturally acquired form of madness. That will sound pretty extreme, but because I have chosen the easy case of multiple personality, it will not be all that shocking, especially after one has read a few of the tests that support this reading.

A more ambitious undertaking would try to locate the level at which the "whole personality" becomes an ideal. It has always been a possible ideal, an ideal of celebrated Greeks and Asiatic seers. Yet the doctrine that people ought to integrate different aspects of themselves is not a human universal. If there is a universal method in the "treatment" of individuals troubled by incompatible and unbearable fracturing of their souls, it is surely not integration but exorcism.

The idea of exorcism has not quite left us. I do not refer to unspeakable movies and the odd bizarre cult, but rather to behavioral modification. That therapy tries to eliminate some kinds of behavior with which a person is uncomfortable or even finds life-destroying. Instead of integrating the unwanted acts into a large person, the behavior is to be driven out, often by procedures of which wizards would be ashamed. Having made this admission, we may freely note that the last wave of successful, fashionable, highbrow exorcism went through Europe in the 1770's. The last great exorcist was Johann

Joseph Gassner (1727–1779). Perhaps there are still strands of exorcism in the new animal magnetism of F. A. Mesmer (1734–1815) whom we now, nervously, call a hypnotist. He was at his apogee when the French Royal Commission of 1784–85 sceptically examined his claims to cures. The commissioners, who included Lavoisier and Benjamin Franklin, were not impressed, although magnetism went underground, and became popular rather than fashionable, for more complex social reasons that have been studied by Robert Darnton.

Kant

The end of exorcism is, I suspect, the beginning of the European idea of the whole person. About this time it even appears cast on to a flickering philosophical screen. What is the transcendental unity of apperception but the doctrine of the integrated personality in philosophical garb? The "I think" which must accompany all my perceptions as a precondition for objective knowledge is indeed a wonderful integrator of the personality. What of Kant's noumenal self that makes human freedom possible? Insofar as I am a free agent, my choices must originate in that primary *one.*

When we turn to Kant's elderly and cheerfully informal writings, we find numerous remarks that show that, in real life, he quite favored the integration doctrine to which his philosophy might have led him by more abstract paths. This is especially true of his *Anthropology* and his *Struggle with the Medical Faculty,* both appearing in 1798. The latter was a specific reply to a curious book by Hufeland called *Macrobiotics, or the Art of Prolonging life.* Here is the origin of the word "macrobiotic." Hufeland's word as his book was much translated into languages, including Chinese and hence back into Californian. Among the numerous bits of Kant's reply to this essay on dietetic and healthy living, is a chapter called, "On the Power of the Mind to be Master of One's Morbid Feelings through Simple Decision." It is something of an essay on the unified control of the diverse and less agreeable aspects of oneself. There is even a careful distinction between the use of plural "we" for an address by a preacher, as opposed to the personal "I" when the patient comes to the doctor "as if to the confessional." Kant has, at any rate, fitting attitudes to be the first philosopher of the integrated personality. He is properly scornful of Gassner, Mesmer, and other quacks. He admires J. C. Reil (1759–1813). Reil conceived a wholly new kind of lunatic asylum, complete with diet therapies, role-acting in the "therapeutic theatre," occupational therapy including art therapy, and, in general, psychic cures based on a precise "practical empirical psychology." Notable is his claim that there is mental disease that results from a disintegratoin of the sense of self, or centricity (*Gemeingefühl*). The old Kant is contemporary with a new emphasis upon the whole person, and his philosophy and his personal attitudes positively foster it.

Such observations are merely suggestive. There is a far more precise connection between philosophers and the invention of split personality. It will be significant that the philosphers in question are fighting Kant's metaphysical, unified self by creating split personalities that furnish living counterexamples to it. I divide what follows in two parts: (I) Doctors and (II) Philosophers. The first describes the cases, and the second states their philosophical utility.

PART I: DOCTORS

Classic Multiple Personality

Possession (by demons or whatever) was the old model of something like multiple personality, and exorcism is the cure. In the new model (of, say, 1875) we have one individual with a number of alternating personalities. This would take the form of an individual who has two characteristic states of mind, and passes from one to the other by a trance, fit, or abnormally deep sleep. In State 1 the individual is depressed, and exhibits standard symptoms of "hysteria." In State 2 the person is light-hearted, excitable, unreliable. An important feature is, as Littré put it in the *Revue de philosophie positive* for 1875, "the two existences have no memory of each other, and each is ignorant of the other." Naturally all the obvious variations were "found": State 2 knows State 1 but not vice-versa, there are more than two states, there are "simultaneous" states and so forth. Incidentally, the conceptual emphasis on one-way or two-way amnesia, and the passage through an intervening state of "somnambulism" help to distinguish the new multiple personality from the old concept of "possession."

On the surface level, it is easy to tell the tale of multiple personality. In the last quarter of the eighteenth century, Erasmus Darwin has a two sentence mention of a village girl who, with hindsight, sounds like a "split personality." In the first quarter of the nineteenth century, there is one remarkable case whose adventures I describe in detail. In the second quarter there was another case. In the third quarter there came a third case, also of great interest. In the fourth quarter there were ever so many multiple personalities. 1875 is the great divide.

The parties interested in multiple personality were a motely lot. Because of the deep sleep separating the two states, nineteenth century students of somnambulism were gravely concerned. The transitions between states seemed to resemble the results of Mesmer and the magnetizers. Hence they, and then their successors, the hypnotizers, had a hand. The great wave of spiritualism is surely pertinent. It will be recalled that spiritualism began at the home of Mr. and Mrs. J. D. Fox in upstate New York in 1848, and soon swept over America and Europe. In one of its guises the medium took on the personality of somebody dead. Hypnosis and spiritualism became closely connected, and indeed much of the best work on both of these and on multiple personality is to be read in the English *Proceedings of the Society for Psychical Research.* There is the related phenomenon of automatic writing, in which a person writes, for example, answers to spoken questions while at the same time saying she does not know the answers, or else is not aware that the question is asked. Automatic writing was used in spiritualism, hypnotism, and in the investigation of hysteria.

Then, on a different tack, are the legal concerns: what if an individual in State 2 commits a crime, and then reverts to State 1, with no knowledge of the crime? Is the individual responsible? That leads us to the fictional aspects, epitomized by Stevenson's 1886 *Dr. Jekyll and Mr. Hyde,* but also represented by dozens of other tales published in that same period. Finally there are the positivist philosophers who sought to use the phenomena of multiple personality in order to insist on their own conception of personal identity, rejecting transcendental or metaphysical ideas of the self.

Thus the space of multiple personality includes the following vectors: somnambulism, hypnosis, hysteria, spiritualism, automatic writing, forensic science, macabre fiction, and positivism. In one essay I can but attend to the main vector, the medical histories and the medical history.

The Lady of Mac-Nish

The first divided person to be described in more than a few lines was written up in the *Medical Repository* for 1816. A Major Ellicot, who taught mathematics at West Point, had come across a young woman in the "Wilds of Pennsylvania." The good Major wrote about her to his friend Dr. J. K. Mitchell, a co-editor of the *Repository*, who included it in his "Intelligence" column of current curiosities. I shall soon describe the trajectory of this case. Henri Ellenberger writes in his splendid book, *The Discovery of the Unconscious*, that neither he nor the able staff of the New York Public Library could locate the legendary communication by Major Ellicot. Luckily I looked for the report before I read Ellenberger, or I might have been discouraged. After a few hours reading of the *Repository*, I did find the case. It really is there. But it is relevant that it is hard to find. Despite the subsequent fame of this lady, I suspect that hardly anyone read this report.

Ellicot's story concerns a young woman, not named. She was reported as being perfectly healthy, sensible, and sound. One day she fell into a profound and lengthy sleep. When she awoke she had lost every trace of "acquired knowledge. Her memory was a tabula rasa — all vestiges, both of words and things, were obliterated and gone." She gradually relearned words, things, and people. Then it happened again. When she awoke from a second "fit of somnolency" she was restored to her former state but knew nothing of what passed in between. And so, back and forth. She came to speak of her "Old State" and her "New State." She had splendid penmanship in the Old State, but poor writing in the New State for (Ellicot said) she had not had time to perfect her New State hand. When she was introduced to a new acquaintance, she had to be introduced to the person in both states in order to avoid an obvious embarrassment. Notable are the concluding lines of the report:

> Both the lady and her family are now capable of conducting the affair without embarrassments. By simply knowing whether she is in the old or the new state, they regulate the intercourse and govern themselves accordingly.

This first report was published when Mary Reynolds was 23; her infirmity had begun in 1811 when she was 18. In the first instance it was likened to extreme cases of sleep-walking, or rather somnambulism. People were known who talked, walked, and so forth in a fairly coherent way when "asleep" yet could recall none of it when awake. There is something of a continuum, for consider a story told in the *Philosophical Transactions of the Royal Society of Edinburgh.* Dr. Dyce reports that a young domestic was well known to have such fits. A wicked fellow servant brought a young man into the house who, during the fit, "treated Maria in the most brutal and treacherous manner" "stopping her mouth with bedclothes" and overcoming "the vigorous resistance she made to their villainy, even in her somnolent state." Naturally she remembered nothing, but on the next occasion of somnambulism "the events came flooding back and she related to her mother all the revolting particulars."

This story is repeated by the Glasgow physician Robert MacNish in his little book called *The Philosophy of Sleep*, 1830. He says of Maria "that the state of mind in this case was perfectly analogous to that which is said to occur in the magnetic sleep." Thus he already made the connection to Mesmerism, magnetism, and what came to be called hypnotism. In a footnote he then quotes much of Ellicot's story verbatim, calling it "in some respects similar," to the tale of Dr. Dyce. Thanks to this footnote the young lady then became famous, although at second hand. The French doctors began to refer to her as "The Lady of Mac-Nish." In this role she entered a slightly arcane medical literature.

Meanwhile, back in the Wilds of Pennsylvania, the selfsame lady lived until the age of 61, dying in 1854. In 1860 a family friend, Rev. W. S. Plumer wrote up the case as a popular anecdote for *Harper's Magazine.* Finally when double personality was making its sensational appearance after 1875, Dr. W Mitchell described it from family papers in a more solemn source, the *Transactions of the College of Physicians of Philadelphia*, April 4, 1888. The young woman was Mary Reynolds, but the two cases were not seen to be identical. The French "Lady of Mac-Nish" medical specimen was not at first recognized as the same Mary Reynolds medical specimen. Third hand sources regularly describe the woman as *two* rare cases of double personality.

It must seem too good to be true that the first recorded double personality also had two personalities in the medical profession, not seen as the same. Ellenberger (who as noted had not read the original account of 1816) conjectures that this is due to the fact that the first account lacked details. That contributed, perhaps, but I would say she also had two lives in the medical literature. In the one she was a medical curiosity, an anecdote for Harper's magazine. In the other, *after 1875*, she was assuming the role of an important proof of the existence of double personality.

I have noted that the classic double personality alternates between a gay, almost abandoned personality and one which is severely depressed. Curiously the fact that Mary Reynolds was of this type came out in public only after the type had itself been defined, and then it was written up in 1888 with a power of description worthy of the finest novelist. After a state-switch we learn "Instead of being melancholy she was now cheerful to extremity. Formerly taciturn and retiring, she was now merry and jocose." She became enamoured of nature, which apparently was no wise thing out there in the Wilds of Pennsylvania:

> She knew no fear, and as bears and panthers were numerous in the woods, and rattlesnakes and copperheads abounded everywhere, her friends told her of the danger to which she exposed herself, but it produced no other effect than to draw forth a contemptuous laugh, as she said, "I know you only want to see your bears and I am perfectly convinced that they are nothing more than black hogs."
>
> One evening, after her return from her daily excursion, she told the following incident: "As I was riding to-day along a narrow path a great black hog came out of the woods and stopped before me. I never saw such an impudent black hog before. It stood up on its hind feet and grinned and gnashed its teeth at me. I could not make the horse go on. I told him he was a fool to be frightened at a hog, and tried to whip him past, but he would not go and wanted to turn back. I told the hog to get out of the way, but he did not mind me. 'Well,' said I, 'if you won't for words, I'll try blows;' so I got off and took a stick, and walked up toward it. When I got pretty close by, it got down on all fours and walked away slowly and sullenly, stopping every few steps and looking back and grinning and growling. Then I got on my horse and rode on...."

You will agree that this is a remarkable description, given that the events occurred about 72 years earlier. It is clear that the period described is that ancient for we are also told that the contrast between "the gay hysterical mischievous woman" to one "sobered down to levels of practical usefulness" gradually became muted, even to the extent that her family began to describe her in a "third state," characterized by none of the above, but simply by swings in mood. Indeed here she is "rational, industrious, and very cheerful, yet reasonably serious, and not having the slightest indication of an injured or disturbed mind." She taught school and at age 61 died almost instantly, complaining of a dreadful headache.

Mary Reynolds has passed away, but I am scarcely the first to keep her memory alive. It is all reported again with some care in *The International Journal of Clinical and Experimental Hyponosis* in 1962. Charles Marks, in his excellent monograph, *Commissurotomy, Consciousness & Unity of Mind*, uses that source. He argues, with the admirable caution that characterizes his essay, that he can't be quite sure how to describe this case, but that it is at any rate different from those cases of commissurotomy which recent doctors use to tell us that we are a right brain person and a left brain person in one body. If there were to be sensible talk of being two persons, says Marks, it would be Mary Reynolds, rather than the patients produced by modern surgeons who mechanically separate the two brain lobes.

Lucky Mary Reynolds, the Lady of Mac-Nish! She had a sensible family, and an unpretentious medical man who did not think her "mad." She lived a pretty good life, subject, in youth, to some restraints attempted by parents and family. She also lives in history as an example for us all to use for our own purposes. My purpose is to show that in the first quarter of the nineteenth century there was one candidate for double personality, who was the basis of no theory. She was compared to the somnambulist or the victim of hypnotism and she was a curiosity. She probably did have the mood changes associated with the later classic multiple personalities, but the publication that fits her into that pattern appears only half a century after she had the changes described. It is published just when those are the classic symptoms which the patient is "supposed" to display.

Félida X

We owe the next couple of generations to the hypnotists. I shall skip one generation omitting the 1837 case due to A. Despine. It is possible to claim that he manufactured his example, as it resulted from use of animal magnetism. Less savory hypnotists certainly deliberately created nonstandard states of a person, which then could not be remembered in the waking state. They would assign names to the person in the non-standard state. But double personality is supposed to be "spontaneous." It is a little hard to sort out a spontaneous state of a previously magnetized patient, given that post-hypnotic suggestion was the rage.

I pass at once to Félida X, discovered in 1858 by an otherwise obscure Bordeaux surgeon named E. E. Azam. Azam's own account is clear and modest, but its reception was something else. For the background I quote from some Harvard lectures given in 1906 by Pierre Janet. The lectures are written in a light-hearted style but entirely conform to the author's intensely serious work in the subject. His is the study of the unconscious which, before Freud, I find more original and probably more faithful to the "facts" than that of anyone else of stature. You will note that he begins, even in 1906, by treating Reynolds and the Lady of MacNish as separate people.

> If America can boast of having presented in the person of the "Lady of MacNish" and in that of Mary Reynolds the finest examples of the first form, the history of Félida X gives now to France an unquestionable superiority. Allow me to make you acquainted with Félida. She is a very remarkable personage who has played a rather important part in the history of ideas. Do not forget that this humble person was the educator of Taine and Ribot. Her history was the great argument of which the positivist psychologists made use at the time of the heroic stuggles against the spiritualistic dogmatism of Cousin's school. But for Félida, it is not certain that there would be a professorship of psychology at the College de France, and that I should be here, speaking to you of the mental state of hystericals. It is a physician of Bordeaux who has attached his name to the history of Félida: Azam reported this astonishing history first at the "Society of Surgery," then at the "Academy of Medicine," in January, 1860. He entitled his communication, "Note on Nervous Sleep or Hypnotism," and spoke of this case in connection with the discussion of the existence of an abnormal sleep during which it would be possible to operate without pain. And this communication, thus incidentally made, was to revolutionize psychology in fifty years. Subsequent to that time, Azam understood better the interest and success of his observation; he published various memoirs, and even books on this subject, in 1866, 1876, 1877, 1883, 1890. As I told you, first Taine, in his book on "Intelligence," then Ribot, in his "Diseases of Memory," took possession of this history, which has gone round the world, and to-day there is a whole library written about this poor woman.

Janet's account is entirely just, but we must add a few footnotes to see quite what was going on. The long string of Azam's publication dates is misleading: we get from Janet the impression of regular observation of the Lady. On the contrary, Azam saw her only in 1858 and 1859, after which she largely got better and was lost from sight. Azam chased her up only after he had read of the idea of double personality invoked in an 1875 medical report on a sensational Belgian case, Louise Lateau, the "stigmatisée de Bois d'Haine." That report, he writes, set him in search of *his* old case. His case had now become a phenomenon. His chief previous medical interest had been, as Janet says, in the possibility of painless surgery.

Azam saw the girl when she was 15 and 16; at that time he was attached to an asylum for madwomen, as an adjunct. As a teenager Félida would almost every day pass from a "first state" through a brief trance to a "second state." Later in the day she would revert. In the first state she had the hysterical symptoms held to be standard at the time — defective tactile sensations, no sense of taste, diminished sense of smell, severe depression, some parts of her body anaesthetized, and semi-conscious convulsions as a way of handling emotional stress. During sleep there was occasional bleeding from the mouth without apparent cause. In the short periods of the second state she was much more lively, cheerful, and indeed healthy. She could not remember second state events when she returned to her first state. Azam remarks that, unlike the Lady of Mac-Nish, there was however no loss of general ideas, ability to write, or whatever.

Félida was made pregnant, during the second state, by a childhood sweetheart; during the second state she quite liked this fact, but she would not acknowledge it during the "normal" first state. When she finally had to do so, she was gravely disturbed. In due course the boy married her and, at age 16, she had their child, which she cared for well.

Félida excited moderate interest, but she got on with her life and Azam attended to other things. Then, after the great wave of interest in multiple personality in 1875, Azam returned. In brief, he found that the cheerful second state now lasted as much as three months, that the first states are now the rare and "abnormal" ones. A cure would have been a final "second" state, with no lapse into the first state. As a teenager the first state was held to be normal and the second one to be morbid. Now in middle age, the reverse is true.

Azam wrote up Félida *as a central case of multiple personality* only in 1876, and provided a sequence of reports after then, chiefly confirming what he had said before.

Naming Multiple Personalities

I began this essay on a nominalist note, suggesting that Foucault taught us to ask how nominal categories had come into being. I suggest that the history of multiple personality illustrates not only that but, more light-heartedly, it also shows us how nominal *people* come into being. I shall use Janet as an example because he is himself extremely aware of how a doctor can foist a set of symptoms on his patient. We now think of his great predecessor Charcot as having trained most of his famous patients to adopt the symptoms that he wanted them to exhibit. Janet is under no such cloud of suspicion. Yet he will tell us, perhaps too honestly, that: "Once baptised, the unconscious [second or third] personality is more determinate and clear-out; it shows more clearly its psychological character." Here is his own example of one such "baptism." He is speaking to Lucie, who engages in automatic writing. Lucie, then, is asked questions to which she replies in writing, without her "normal self" being aware that she is writing.

Janet: "Do you understand me?"
"No." (she writes)
"But to reply you have to understand me."
"Oh yes, absolutely!"
"Then what are you doing?"
"Don't know."
"It is certain that someone is understanding me."
"Yes."
"Who is that?"
"Not Lucie." (*Autre que Lucie*)
"Ah ha! another person. Would you like us to give her a name?"
"No."
"Yes. It would be far easier that way."
"Oh well. If you want: Adrienne." [Footnote: the writer deliberately chose *other* names on other days, which Janet says is "irrelevant"]
"Then Adrienne, do you understand me?"
"Yes."

Now Janet did not invent this practice of naming, which almost certainly was one of the tricks of the hypnotists and charlatans of an earlier date. This is clear in the description of his most famous patient, Léonie, "who had been hypnotized constantly by all sorts of persons from the age of sixteen upwards, and she is now forty-five." Léonie was a peasant who spent a lot of time in the parlors of hypnotists and doctors. In those posher surroundings she acquired the character of a salon wit.

She refuses the name of Léonie and takes that of Léontine (Léonie #2) to which her first magnetizers had accustomed her. "That good woman (Léonie) is not myself," she says, "She is too stupid."

Janet in due course discovered that he could induce yet a third personality, rather dignified, who contrasts herself with the "crazy creature" Léontine. Janet writes,

> This resurrection of a somnambulic personage who had been extinct for twenty years is curious enough; and in speaking to Léonie # 3, I naturally now adopt the name of Léonore which had been given her by her first master.

We shall have to wonder, then, to what extent Janet was the physician attending to his patients, and to what extent, he, too, was their "master."

A Plethora of People

No matter; after 1875 multiple personality was an established phenomenon and the doctors went on to characterize various "types." There were not really all that many cases that popped up into the literature. The number of pieces appearing between 1875–1890 shows how radically the situation had changed.

Certain obvious points are worth recording. In classic multiple personality one state is hysterical. Moreover male hysteria was in general held to be relatively rare (but *not* non-existent, just rare). So all the earlier cases, as we have seen, were women. Sex is made to play an overt role. Félida conceived in State 2, but bore the child in State 1. She had eleven pregnancies (although only two children survived). All the deliveries were in State 1. Were all the conceptions in State 2? Léonine (State 2) denies any knowledge of the husband of Léonie (State 1), but takes Léonie's children as her own. Tales of State 2 rape victims or nymphomaniacs began to appear.

Male multiple personalities do finally get their say and may even have become dominant for a while. One of the first cases is typical: Louis V, first described by Camusat in 1882 and written up by Bourru and Burot in 1888. He is a moral degenerate, a petty thief, who wends his irregular and criminal life between the army, the houses of correction, and the hospitals. He has hysterical anesthesia in consequence of being bitten by a viper at an agricultural House of Correction. His character alternates between moral industriousness and total degeneracy. That inevitably fixes the type: the male multiple personality is a petty criminal, or rather, to use the concept of the time, a *degenerate.* Students of the history of nineteenth century concepts of deviancy will not be surprised, although I did think that the "viper" was too good to be true.

The "degenerate" is perhaps an essential part of the French model of male multiple personality. It is part of the general French theory of deviancy, and was connected with problems about the declining French population — the very race was degenerating by not procreating. But this was hardly an American problem, so on these shores we have a different type, emerging *at exactly the same time.* At the ripe age of 61 the Rev. Ansell Bourne took a second personality upon himself, a Mr. Albert Brown who stole $551 dollars from a bank in 1887. He was hypnotized by William James in 1890, and the two personalities became clearly distinguished.

Sex and crime: what of crime? The good Dr. Azam put that question to the justices of Bordeaux ("in whom I, like all the world, have complete confidence"). It was then a purely hypothetical question *because there had not yet been any male multiple personalities.* But what if Félida were to commit a crime in State 2? Would she then, in her "normal" State 1 be responsible, although unable to remember what she had done? I can see why everyone had confidence in the justices of Bordeaux. They sagely replied that legal responsibility in effect rests on human liberty, and thus requires only a demonstration of voluntary control of one's act. But Azam had said that Felida's "will" was the one thing seemingly unaffected in her two states. Her ability to make choices was unimpaired, even if she afterwards did not remember what she had chosen. So she would be guilty. But there are cases where multiplicity would entail mitigating cirumstances. This would apply if she was made a loan in State 2 and refused to pay it back in State 1 because she had no memory. (More recent justice is less wise, and even in the past decade American murderers have escaped the penalty on grounds of split personality.)

Crime leads to novels. I believe that in the case of the split-mind male criminals, fiction may precede fact. There was a spate of such short stories and novella after 1875. The few written before then usually took hypnotism as the base. The 1886 *Dr. Jekyll and Mr. Hyde* was only a very successful exploitation of the spirit of the times. I have found little evidence that Stevenson had been reading the medical journals. He had little need to. Even that tedious bastion of English philosophy, *Mind*, had started surveying the literature in 1876.

Nominalist Conclusion

Multiple personality is a concept invented by medical people, with the assistance of hypnotists, spiritualists, lawyers, philosophers and novelists. It is not a concept that applied to anyone before 1875 except for a couple of cases whose histories are intrinsically bizarre. But after then it was a possible way for a person to be. There was doubtless some training of patients to behave in this way, but soon the phenomenon became spontaneous enough. It became generally recognized that this was a kind of individual. Deeply unhappy people "chose" — quite the wrong word — to adopt this mode of being disturbed, and were for long reinforced both by medical practitioners and the lay public. The idea was intimately connected with female sex and male crime. The phenomenon was primarily Franco-American. At the start of the twentieth century Morton Prince, in New York, became the king of the multiple personalities and used them as part of his disagreement with Freud. The disappearance of multiple personality as a central psychological concept is partly due to a pair of factors: (i) Multiple personality hardly ever occured in Austrian consulting rooms or asylums, and (ii) Psychoanalysis came to exert a compelling hold on even the broad mass of psychiatrists who rejected it.

Multiple personality, I say, hardly ever occured in Austrian consulting rooms. Certainly it was not diagnosed as such: and so, I say, it did not occur. I am not asserting a "wishful-thinking" nominalism according to which, if people did not name something, then it was not there. This is a dynamic nominalism, about possible ways to be a person, according to which if the classification is not made, then people will not adopt that as a possible way to be.

Multiple personality *as such* did not occur in the Austrian consulting rooms, yet it left deep traces on psychoanalysis. It is bridged in part by an 1890 book by Max Dessoir, *Das Doppel–Ich*. It holds that we are all simultaneous, double personalities. This became a Freudian theme: "Das Es," "Ich," and "Ueberich." Earnest Jones' "id" for "das Es" is less than just: "Das Es" is a neuter noun. As "it," in German it can occur in sentences like "It dreamed to me last night." When these three factions are at war, neurosis results. But there is one-way amnesia between "Ich" and "Es" — an amnesia patterned on Félida X.

The new cure is to remove the repression and inhibition, cure the one way amnesia and integrate the person.

So nowadays self-declared multiple personality lives on chiefly in fiction. I quote from the advertisment for a current best-seller:

> His name is Billy Milligan, and his astounding true story has made nationwide headlines. Now, *The Minds of Billy Milligan* takes up where the news stories left off, to tell how a troubled, physically abused child developed twenty-four separate distinct personalities . . . and how even now, in an Ohio correctional facility, physicians and psychiatrists are struggling to make Billy whole again.

Integration, "making whole again" (again?) is the name of the game.

PART II: PHILOSOPHERS

Why was multiple personality invented, why in 1875, and why in France and America, albeit with different syndromes for males in the two countries? Why did this phenomenon later lose its central place in psychiatric practice? Numerous half-way answers come to mind. Some are suggested by the information provided above. Others will arise from deeper enquiries. Here I turn to a different kind of question. What made this configuration possible at all? In what arrangement of ideas and practices does double personality have a place?

My study is prompted less by an enthusiasm for multiple persons than by a curiosity about the norm of the "whole" human being. There is little room for an obession with multiplicity until integration is a possible measure of mental health. It is not that the integrated person and the split personality come together, the one the evil shadow required by the other. On the contrary, double personality is only one rarified and short lived incident that pits itself against the standard of the whole, unified, person. This latter, however, is in part a philosophical construct. I shall show this by recalling some old philosophy.

Positivism

"This humble person," Pierre Janet said of Félida X, "was the educator of Taine and Ribot. Her history was the great argument of which the positivist psychologists made use at the time of the heroic struggles against the spiritualist dogmatism of Cousin's school." That, then, is plain enough. We need no Foucault to tell us she was part of a power struggle. Since this is no place to recount the history of French positivism after 1870, we are glad that Janet gave us two names to serve as pegs on which to hang a bit of the tale. Taine, Ribot, and Janet himself are sometimes referred to as the founders of the French school of observational psychology.

Hyppolyte Taine (1828–1893) was the notable scientific intellectual, influential critic of art and literature, innovative historian and philosophical psychologist. He was deeply imbued with the positivistic spirit current in the circles in which he moved, and was profoundly attracted by aspects of English culture. He also attempted something of a synthesis of these ideas, with an Hegelian model of spiritual development. He was a determinist, he said, in the sense of Spinoza. He thought he had a determinism consistent with human freedom. Many of his readers did not see things that way. When he was finishing his book *De l'intelligence* he wrote that it was to be his most significant work. We now rank other of his productions higher, but it was undoubtedly one of the founder books of modern psychological theory, and it had an enormous readership. Its psychology is a descendant of Locke, Condillac, and the ideologues. In particular it has no use for a metaphysical or transcendental self: the idea of the self is formed by an awareness of one's capacities and a memory of one's past.

Théodul Ribot (1839–1916) had a more limited repertoire than Taine, whom he much admired. He held the chair at the Collège de France, to which Janet alludes in my quotation — a chair in which Janet followed him, in 1902. His psychological work is highly philosophical in character. He founded the premier philosophical journal, the *Revue philosophique de la France et de l'étranger*. His 1881 *Diseases of Memory* is subtitled *An Essay in Positive Psychology:* the word 'positive', in titles like that, is a declaration of philosophical partisanship. It was soon followed by *Diseases of the Will* (1883) and *Diseases of the Personality* (1884). Each book is full of case histories and theoretical speculation, but each also uses the cases to reject, among other things, metaphysical or transcendental notions of the self and the will. In many ways Ribot's books do little more than amplify the corresponding sections of what Taine had already published in 1870.

Taine's doctrine of the self — *la notion du moi* — is in the associationist Lockeian tradition. Only "if we allow ourselves to slip into metaphysical error do we postulate the *moi* as something complete and independent." In the short term Taine is opposing the historicist and metaphysical previous generation represented by Victor Cousin, but to abbreviate a complex story, recall only two features of the Kantian philosophy that set all this going. There is the transcendental unity of apperception, the "I think" that must accompany all my thoughts in order that my thoughts are thoughts of objects and have objective truth. The "I" is not an *a posteriori* bundle of memories but a preconditon of objective thought in general. Secondly there is the noumenal self which is the basis of human freedom. Although iron necessity rules in the phenomenal realm, there is a noumenal "I" which is free. Let us reflect on why Félida X should have been so valuable in a power struggle against this post-Kantian framework.

Taine and Félida

I have quoted Janet saying that Taine "took possession of" the story of Félida, as Ribot did in turn, so that by now "there is a whole library written about this poor woman." Taine's book was published in 1870. Does not that refute my claim that Félida was fully represented as a multiple personality only after 1875? Not at all. She does not occur in the first edition of 1870. (Azam is mentioned once, in connection with an 1840 paper on hypnosis and surgery). Félida enters the third edition of 1878. Even then she is not exactly outstanding. In the original text Taine had described the Lady of MacNish. He now adds a footnote:

> The case of Félida X is analogous, although less distinctive. The sick woman presents in turn two states, the one sad, the other gay; in the gay state, she remembers all her previous states, gay or sad; in the sad state, she recalls only sad states, and has no memory of gay states.

In the text MacNish's version of the 1816 story is quoted quite fully. That is Taine's *only* example of multiple personality. The lady is compared to somnambulists who cannot recall their sleep-walking, to victims of hypnosis, and to drunks who forget what they did. A pretty strong conclusion is inferred:

> Our images, by connecting themselves, make up a group which in literary and judicial language we call the moral personality. If the two groups are distinctly severed, so that no element of the one calls up any element of the other, as in the case of the lady cited by MacNish, there are two moral personalities in the same individual.

How nice it would be if we could find some more examples like the lady of MacNish! The gods were to provide.

Félida was especially providential. What will a low-grade Kantian say of Mary Reynolds? Maybe that there are two moral personalities, each with its own "I think," maybe each with its own noumenal self. So what? Félida is more embarrassing. Is it the same "I think" that accompanies the gay and the sad states? The "I think" of the gay state must be something that bizarrely includes a different "I think" of the sad state. What of the noumenal free self? As I reported earlier, Azam says that the *will* of Felida is quite unaffected by her disease. Take that at face value: do we then have one noumenal self to do the willing, corresponding to two different "I thinks" that accompany consciousness? Such a chain of reflections is spurious argument but it is wonderful rhetoric to reduce the metaphysical and transcendental self to absurdity. Félida was a godsend to the positivist psychology of the *moi*.

Ribot and Félida

Ribot's 1881 book on memory is full of stories of amnesia, and includes speculative theories of memory. He makes plain that multiple personality, although displaying anmesia, is not particularly instructive about memory. On the contrary.

The study of amnesia in its periodic form is better calculated to throw light upon the nature of the *moi* and the conditions for the existence of a conscious personality than to exhibit the mechanism of memory from a new stand point. . . . I shall be sparing of illustrations: they are sufficiently familiar, the study of the so-called phenomenon of "double consciousness" being a common pastime. The detailed and instructive observations of Dr. Azam in particular have contributed largely to the popular knowledge of periodic amnesia.

Then come the following cases: (1) The Lady of MacNish; (2) a male amnesiac sketched inadequately by Azam, about whom the "psychological details are not always given with precision;" (3) Félida, (4) Dufay's parallel case, reported in 1878, at the same time as the second sighting of Félida; and (5) "The enumeration of certain cases in which [the phases of periodic amnesia] appear in an undeveloped form." One of these is yet another young woman "violated" during one abnormal state who reveals "the fact to her mother in a succeeding crisis." The other is a servant maid who "believed herself every night to be a bishop." On the strength of these examples we may

first reject the idea of a *moi* conceived as a distinct entity of conscious states. Such an hypothesis is useless and contradictory; it is a conception worthy of psychology at its infancy, content to accept superficial observations as the whole of truth and to theorize where it cannot explain. I avow allegiance to contemporary science which sees in conscious personality a compound resultant of very complex states.

He does not conclude that the *moi* is constituted by memory, but that it also needs a certain sense of one's entire organism. "The unity of the moi," he writes, is "that of a very complicated mechanism. It is a concensus of vital properties. . . ."

In the 1881 book Ribot announced that the full study of periodic amnesia requires a work on personality. He published just such a study in 1885. It begins by contrasting the "new school" to which he belongs, and the "old school" which was satisfied with "the hypothesis of an ego, absolutely one, simple, identical." Old school people accuse the new school of "filching their ego." "The language and methods of the two sides are now so different, that all mutual understanding is henceforth impossible."

The new school, the positivist school, has been helped along by new examples. By 1885 Azam is being referred to as a familiar, though thoroughly instructive, old chestnut. The case recited at greatest length is Louis V, the degenerate who at an early age was bitten by a serpent. It had first been described in 1882, and then more extensively in 1883.

Pawns in a Power Struggle

Janet told the true story. Multiple personalities were a valuable tool in a power struggle between the old and the new school of psychology. Battle lines were announced, in a largely theoretical way, by Taine in 1870. At that time he knew of no multiple personality except the 1816 Lady of MacNish. In 1878 he was able to add the sensational cases of Félida. In 1881 Ribot still had a pretty slender list of examples. In 1884 things were looking up.

And thick and fast they came at last,
    And more, and more, and more —
All hopping through the frothy waves
    and scrambling to the shore.

A chair at the Collège de France was created for Ribot in 1888.

University of Toronto

NOTE

* This paper was written in 1981 and submitted during the summer of 1982. It takes no account of studies of multiple personality made after 1981.

ALASDAIR MACINTYRE

POSITIVISM, SOCIOLOGY, AND PRACTICAL REASONING: NOTES ON DURKHEIM'S *SUICIDE*

The time has perhaps come at last when it is possible to evaluate positivism's contribution to the social sciences in a way that does justice to its real achievements while recognizing its philosophical defects. Too many antipositivist writers have been content with cataloguing mistakes when they ought to have learned from Hegel that any sufficiently important error is always a way-station and even perhaps a necessary way-station on the road to some truth. But a condition for recognizing what we have to learn from positivism, in the social sciences as elsewhere, is that we understand that there were many positivisms and not just one or even two. The virtues of positivism emerge only in the detail of particular versions, not in summaries of its most general philosophical theses.

Moreover positivism, like any other philosophical doctrine, can be evaluated only as contributing to specific enquiries. Collingwood's thesis that a condition for understanding a philosophical text is that we identify correctly the questions towards which its assertions are directed as answers needs to be accompanied by the thesis that it is sometimes only by discovering a set of questions to which we need answers that we can give point to a set of assertions which were originally framed for some very different purpose. A doctrine may acquire in new contexts a significance that it originally lacked.

For these reasons the positivism with which I shall be concerned and the context in which I shall evaluate its significance are highly specific. The positivism is that of Émile Durkheim, recognized both by himself and by others as one of Comte's true heirs. And I shall deal only with Durkheim's contribution to the understanding and explanation of suicide, and even then only with certain features of that contribution. I shall not here be at all concerned to place what Durkheim says in its historical context, important thought that is;[1] instead I shall begin from a recent sociological disagreement about the data reported concerning suicide in order to bring out in a contemporary context the enduring importance of Durkheim's positivist contribution — very different in some ways as it is from what Durkheim himself took it to be.

A. Donagan, A. N. Perovich, Jr., and M. V. Wedin (eds.), Human Nature and Natural Knowledge, 87–104.

I

Official suicide rates show remarkable stability over time. Changes in rates are with rare exceptions gradual and the same holds for rates for regional and local communities as well as for the populations of nation-states. Moreover the rank-ordering of the rates of national communities shows the same stability and more strikingly the rank-ordering of suicide rates among immigrant groups in the United States correlates closely with that of suicide rates in the countries from which they come. Evidence such as this led Sainsbury and Barraclough[2] to conclude that, although suicide is "almost certainly underreported," "national and international suicide statistics are sufficiently reliable to be of scientific value. . . . "

In just the period when Sainsbury and Barraclough were defending this conclusion, it was being argued by a number of writers not only that the data presented in official statistics are susceptible to inaccuracy in a variety of ways, but that underlying both the collection and the presentation of the data is the false assumption "that suicide is a unidimensional and unvarying form of behavior," whereas "the meanings associated with suicidal behavior in fact vary both within a single society and . . . between different cultures."[3] Thus at the very least, on the view taken by Douglas, Atkinson, and others, we do *not* have good reason to affirm the conclusion argued for by Sainsbury and Barraclough.

How ought this issue to be decided and what is at stake in deciding it? I take it that the interesting question is not *whether* the clearly impressive evidence cited by Sainsbury and Barraclough is sufficient to establish their conclusion but rather *what* the conclusion is which it is sufficient to establish. What they demonstrate is that official suicide rates certainly provide a far from inadequate measure of *something*; what they do not tell us is what this something is. Let us call it 'suicide*'. What the considerations adduced by Douglas and Atkinson show is that whatever the class of events picked out by 'suicide*' those events cannot be ones characterized in terms of what Douglas and Atkinson call 'meanings', that is, in terms of the specific intentions and beliefs of suicidal individuals, for these vary too largely, both from individual to individual and from culture to culture, to provide us with criteria for membership of that class. The causal explanation of varying rates of suicide* in different populations at different times will have to be in terms of factors which are themselves relatively invariant in respect of the highly specific cultures of the different populations and so also of the beliefs and intentions which vary from culture to culture.

Two points are evident at once. The first is that the concept of suicide* is in fact a concept which Durkheim himself employed and bequeathed to many later sociologists. He has been criticized for not recognizing that his reliance on official statistical data committed him in practice to the definition(s) of suicide used by those who collected and presented the official figures, thus rendering his own initial reflections on definition irrelevant and redundant. But since one central task that Durkheim set himself was to define 'suicide' without making any reference to the particular reasons, intentions, and beliefs of particular suicidal individuals and since the argument so far has made it clear that if 'suicide' is to be defined in such a way as to name that kind of event the incidence of which is measured by official statistical methods, it must be defined in just that way, this criticism was in part at least misdirected. Durkheim's concept of suicide is — very often at least — 'suicide*.'

Secondly Durkheim's project of founding a positive science of sociology required the employment of concepts of just this kind not only for the reasons that he himself adduced, but also for the reasons which Quine was to give.[4] The elimination from scientific discourse of all intentional idioms is necessary if the causes and effects studied by the sociologist are to be properties of social orders and collectivities and not the reasoning and intention of individual agents; *and* such elimination is also necessary if the generalizations linking these causes and effects are to have a law-like character, that is, if they are to entail well-defined sets of counterfactual conditionals. Thus Durkheim's positivism, although formulated in late nineteenth century terms, is perhaps not only less conceptually confused but also less archaic than it has been sometimes supposed to be.

To this it may be retorted that I have already conceded that whatever the phenomenon named by 'suicide*' is it certainly is not suicide as suicide is understood within any particular social group in any particular culture. The intention embodied in suicidal acts must always presuppose some more or less determinate conception of what it is to die and what it is to bring about someone's death. As such conceptions vary, so will the phenomenon of suicide. 'Suicide' and cognate expressions in natural languages pick out a number of types of act, not just one, and none of them are suicide*, for their identification as what they are depends on a correct understanding of their belief informed character.

It is not difficult to construct a positivist answer to this objection. It is that scientific progress often requires the replacement of some ordinary language concepts by concepts specifically designed for the purpose of constructing the kind of law-like generalizations which provide explanations and predictions; so for example, the ordinary concepts of 'heat' and 'cold' had to be replaced by more precise conceptions of temperature. But this answer does indeed bring matters to a decisive test. For it makes it clear that the replacement of pre-scientific conceptions of suicide by some concept such as 'suicide*' is justified if and only if it issues in the formulation of highly confirmed, genuinely law-like generalizations entailing clearly defined sets of counterfactual conditionals. One of the major contributions of Douglas's book is its presentation of the history of the attempts to formulate and to provide adequate confirmation for such generalizations from Durkheim's onwards. What Douglas's narrative shows is that all such attempts have failed: the generalizations formulated simply do not have sufficient predictive power to provide them with adequate confirmation and the theories of which they are a part are unable to deal with counterexamples or anomalies in any convincing way.

The sociological reviewers of Douglas's book tended to emphasize his doubts about the validity of official statistical findings. But on the view that I am taking here even if the answer to those doubts made by Sainsbury and Barraclough is accepted, Douglas has still provided a sound basis for rejecting much that is central to the Durkheimian and postDurkheimian treatments of suicide. The only reply left to the systematic defender of positivism is to suggest that all that Douglas's narrative of failure entitles us to assert is that the Durkheimian project has failed *so far*. The attempt to explain different rates of suicide in different social groups by appeal to the differing degree of *anomie* in such groups, or to the incidence of status change or status integration or to ecological factors, may all have failed up till now; but at some point in the future a new theory or a new formulation of some old theory may yet accomplish what their predecessors failed to achieve.

It is very much to the point that precisely the same reply, *mutatis mutandis*, would be available to contemporary defenders of both witchcraft and astrology. They too are able to produce significant correlations between a variety of items, but are unable to formulate adequately confirmed law-like generalizations. And there came a point in the history of each at which this failure made it reasonable to reject their claims and unreasonable for their defenders to argue that all that we were entitled to assert was that they had failed *so far*. At some point therefore, whether past or future, positivist sociological enquiry in suicide studies may turn out to have been or to be at one with witchcraft and astrology in this respect.

To say this is not to deny that the accumulation of significant correlations of rates of 'suicide*' with other phenomena has been of immense value; it is to suggest that the correlations discovered by sociologists in this area are compatible with, and may require for their explanation no more than, a social universe whose regularities are not an expression of law-governedness, but of a kind of order captured by generalizations prefaced by "characteristically and for the most part . . . " whose subject-matter is accessible only through the prescientific concepts that Durkheim and his heirs sought to replace. What nonintentional sociological concepts such as that of suicide* may indeed identify are stable, measurable large-scale social effects produced by varying belief-informed intentional behavior in different social settings; so that the *same* nonintentional effect is produced and sustained by a changing range of intentional causes. If social life is structured causally in this way, this is in itself a striking phenomenon, requiring futher explanation. But such explanation, requiring as it will the use of intentional idioms even to identify what is to be explained, will not be positivist in form or substance. The project of discovering genuine law-like generalizations in this area will still be what it is now, a failure.

What is the importance of that failure? Durkheim's vision of the social order was one of law-governed structures producing deterministically such effects as changing suicide rates independently of the beliefs and reasoning of the individuals who inhabit those structures. So notoriously he explained the fact that higher suicide rates are found in Protestant groups than in Catholic not by reference to the different moral beliefs about the wrongness of suicide in the two groups, but by the lesser degree of *anomie* in the social group that happened to be Catholic. Individuals of course differ in their degree of psychological vulnerability to the impact of social causes and it is not those causes which determine *which* individuals commit suicide. But that a greater or lesser proportion of the group will commit suicide is determined independently of individual beliefs and reasoning. Hence what is at issue is a certain kind of sociological determinism. And it is just because of the failure of the Durkheimian project and of its successors that we have good reason to reject the claims of such a determinism.

That could not have been known *a priori*; it had to be learned from the failure of the Durkheimian research program. But it follows of course that in one very important way that research program was not barren, although its outcome has not been what Durkheim hoped. For its failure is, if my argument is correct, the *only* way of learning why a particular account of the relationship of the individual to society lacks justification. This is not something that could have been learned by purely philosophical reflection, if we understand by such reflection a form of thinking that does not appeal to the results of particular empirical enquiries. In this area — and, I believe, also in others — it has been the working out of the positivist program for the social sciences in detail which has provided the central ground for rejecting it. But this is not our only debt to Durkheim; for if we now put on one side the positivist framework which Durkheim used to define his overall enterprise, then much of what he has to say about suicide will take on a quite different kind of significance. For this to emerge it is necessary to find a quite different point of entry into the discussion of suicide.

II

It is a striking feature of *Suicide* that Durkheim simultaneously insists upon the social character of suicidal acts and yet neglects the social character of the *thinking* of suicidal agents. But all reasoning and more particularly all practical reasoning has a social character. That Durkheim omitted to notice this doubtless partly derived from the tradition of conceiving thinking as an activity private and internal to individuals, whether inside the Cartesian mind or inside the materialist head. But it was also due to the way in which he thought about social structures. In *Suicide* at least — the same is not true of all his other works — his initial methodological stance ensured that he took too little cognizance of the fact that social structures are themselves embodied sets of more or less systematic, more or less coherent beliefs and that to do what a social role requires involves treating certain actions by others and certain preceding states of affairs as providing good reasons and sometimes conclusively good reasons for doing this rather than that. The activities conducted within trade unions, insurance companies, automobile factories, sports clubs, and churches all presuppose certain beliefs about and standards for practical reasoning, beliefs and standards that particular individuals who do what their roles in such organizations require may dissent from as private persons. But even apart from social role-playing an individual can only act intelligibly in the eyes of others if in large part at least his or her actions can be interpreted as flowing from good reasons for that kind of person in that kind of situation to act in that kind of way. An inability to impute such reasons to an agent entails an inability to understand what that agent is doing of a kind that makes it impossible to know how to respond. And that is to say, intelligible action is in general possible only where there are shared, socially established standards as to what counts as a good reason. The existence of such standards is of course perfectly compatible with the existence of disagreement about what counts as a good reason. But the expression of such disagreement will itself only be intelligible in the light of shared, established standards as to what counts as a good reason for disagreeing. And of course many social institutions and perhaps all social institutions of any complexity embody possibilities of disagreement. The beliefs embodied in a social institution are not necessarily or even usually a coherent set. Social structures, as both Hegel and Marx understood, may embody contradictions.

Yet what makes thinking social is not just the fact that social structures are partially constituted by shared beliefs and standards of reasoning. That would perhaps at best show that *some* thinking must be social. It is also a matter of the character of thinking itself. Saul A. Kripke has recently advanced a cogent intepretation of Wittgenstein's arguments in the *Investigations* against the possibility of a "private language"[5] in which he treats those arguments as the application to a special case of a more general Wittgensteinian thesis about rule-following. On Kripke's account Wittgenstein is considering how a particular kind of sceptical challenge could be met. Suppose that I have said that 98 + 27 = 125. When asked how I know that this is the correct answer, I say that I am applying the rule of addition that I learned to apply in previous cases. The sceptic then enquires how I know that it is *this* rule, yielding this answer '125' as the only correct answer, that I learned in the previous cases from which I am extrapolating in using the rule of addition. The importance of this sceptical challenge is that, unless it can be met, we shall be unable to specify the difference between merely thinking that one is applying some rule correctly and actually being justified in so thinking. Wittgenstein's thesis is, as Kripke expounds it, that this challenge cannot be met and this difference adequately specified "if we confine ourselves to looking at one person alone, his psychological states and his external behavior . . . The situation is very different if we widen our gaze from consideration of the rule follower alone and allow ourselves to consider him as interacting with a wider community."[6] For it is only where "others will have justification conditions for attributing correct or incorrect rule following to the subject"[7] that is it possible for there to be such an activity as checking whether someone has followed a rule correctly. This highly abbreviated summary of Kripke's lengthy and lucid exposition of Wittgenstein's lengthy and yet highly compressed discussion in *Investigations* §§198–243 brings out at least minimally, I hope, the point of Wittgenstein's rhetorical question in §199 "Is what we call 'obeying a rule' something that it would be possible for only *one* man to do . . . ?"

It is important that Wittgenstein's point has application to far more than rule-following. Someone who uses a new metaphor, advances an original hypothesis, or suggests an innovative course of action is not following *rules* in the sense in which someone who does an addition sum is. But what makes what they do intelligible as what it is is precisely the same appeal to and dependence upon usually implicit and tacit social agreement upon the relevant kind of standard. Once again it is important to stress that not only does this not rule out the possibility of widespread disagreement within one and the same culture and social order, but it also does not rule out — it is in fact what makes possible — particular individuals using the resources of reasoning which they have learned in the context of socially agreed standards to put those same standards in question. Thus *being a good reason* for either a belief or an action is a social status accorded to some thoughts and not to others.

One corollary is of central importance for my present argument. The fact that what counts as a good reason is determined by public standards entails that if I cannot make my reasoning intelligible as reasoning to others I shall be equally unable to make it so intelligible to myself. If it is impossible that others should know what I am doing, then it will also be impossible that I should know what I am doing. And this is of course a possible state of affairs. Unintelligible actions and transactions do take place and when they do, it is one symptom of nonrational causality in the social world. So, for example, severe depressive mental illness, a consequence perhaps of some biochemical cause, may produce actions — including suicidal actions — which are certainly actions, but which lack all intelligibility. And there is no doubt that unintelligible actions can have social as well as biological causes. But such causality has to be established case by case on the basis of empirical evidence. Any general contrast between explanation in terms of social structure and explanation in terms of the agent's reasons, the type of contrast so crucial to Durkheim's initial methodological and epistemological stance in *Suicide*, cannot be sustained.

One of the features of social life which Durkheim's stance is apt to obscure is the way in which some kinds of social structure may embody a solution to moral and metaphysical problems and the way in which different social structures may embody rival solutions to one and the same problem. Consider for example one general problem area for a variety of human practices, that which concerns the relationship of behaving well to doing well, of excellence to success. In war, in games, in areas of scientific enquiry or artistic achievement when excellence has for one reason or other become competitive (consider, for example, the race to discover the structure of the DNA molecule or competition for patronage among painters or architects), it becomes important and sometimes difficult both to acknowledge the claims of excellence for its own sake and to recognize the importance of winning. How is honorable defeat (where the honorable man or woman is one who does what excellence requires even when it cannot or can no longer lead to success or victory) to be ranked in relation to victory? This problem understood in a variety of ways lies at the heart of a number of cultures we may note in passing that it is a problem that cannot even arise within either a Kantian or a utilitarian framework and that therefore such cultures presuppose the falsity of both these moral philosophies. In traditional Japanese culture[8] there was a stock of saga narratives in which ritual suicide played a key part in defining the relationship between excellence, honor, and victory. So, the twelfth century hero Minamoto no Yoshitsune[9] who had devoted his whole life to vindicating the honor of his clan and to the pursuit of excellence in "the way of the bow and the sword," but had incurred the implacable enmity of his power-hungry half-brother, finally met his brother's army of thirty thousand with only nine friends. While the nine fought to the end, Yoshitsune calmly read the Buddhist scriptures and then committed a scrupulously enacted ritual suicide, a suicide which made his military honor undefeatable by mere military power.

The explanation of this type of suicide can be put either in terms of the individual's reasons for action *or* in terms of role, status, function, and conflict. But it will turn out that these are no more than two idioms for expressing one and the same explanation. Doing what role, status, and situation require and doing what the individual has the best of reasons for doing are one and the same in this highly integrated type of culture. It would be quite wrong to think of an individual such as Yoshitsune as determined or even constrained by social forces or cultural forms. For to speak in that way presupposes a type of distinction between what is social and what is individual and a corresponding possibility of an external causal relationship between them which has no application to the kind of social order and individual character portrayed in the saga.

Yet here it is worth remembering that what the saga presents is an ideal type. Even in actual traditional Japanese society this degree of integration of the individual and the social could not have been found except at one end of the spectrum of the social order; and the function of such ideal types is very largely to enable us to map the degree of deviation from them. Consider how very different the relationship of individual to society is in a certain kind of modern Japanese suicide.[10] An adolescent student whose family have placed their hopes in his success in entering some private or public bureaucracy through doing well in highly competitive academic examinations fails those examinations. He cannot move forward into the place in society to which he aspired; the disgrace is such that he cannot return to his former place within his family and his local community; there is no social space left within which he can exist and he can have no hope of finding such a space. He commits suicide. About this kind of suicide two preliminary observations need to be made.

The first is that whereas the kind of suicide typified by the story of Yoshitsune represented a triumph over imminent defeat, this kind of modern suicide represents total defeat, an enactment of hopelessness. While the older type of suicide provided a way of not despairing, this more modern type is a form of despair.[11] The older type of suicide could complete a life that was in Aristotle's sense *eudaimōn*: the act of suicide expressed the achievement of a supreme good. But the newer type expresses an inability to move towards such a completeness. It expresses the frustration of the individual's practical reasoning by external social circumstance.

Secondly, the beliefs embodied in the traditional form of suicide were parts of a coherent, rationally articulate, and rationally defensible whole. They took their place in an overarching theory of human goods and of *the* good for man. But the beliefs of the population from which those individuals who commit the modern kind of suicide with which I am concerned are much more of a mishmash. Attenuated notions of honor and disgrace, a Japanese version of the puritan work-ethic, ideas of the family in process of change, bureaucratized standards and modes of achievement all somehow coexist and become manifold sources of inconsistent attitudes. Why does this matter?

There is an important distinction between two different ways in which an action may be done for a reason. There is on the one hand the type of case where the reason for which some particular action is done is part of some determinate structure of practical reasoning. A reason can be identified as part of such a structure if and only if two conditions are satisfied. The first concerns the structure. When I speak of structure here I mean the kind of form that Aristotle ascribed to the practical syllogism and much of what Aristotle said about practical syllogisms does indeed hold of all genuine practical reasoning. Its major premise is that something is a good for some class of human being of which the agent who is rehearsing the syllogism believes himself to be a member; its minor premises specify what means or opportunity are afforded to the agent here and now for achieving or realizing that good; and its conclusion is an action. A statement of the conjunction of the premises of a piece of practical reasoning provides one type of primary background against which an action can be seen as intelligible. And it is precisely because intention-embodying actions can be understood as the conclusions of pieces of practical reasoning that they can be seen as consistent or inconsistent with the beliefs of the agent and by extension with the beliefs of others and with the beliefs presupposed in particular social orders.

An action issues from a piece of practical reasoning as one part of a performance from the preceding parts. But it does not of course follow that every premise has been recited, every inferential bridge crossed on that particular occasion. Much may have been left implicit or taken for granted, much may have been rehearsed on some previous occasion to which the present reasoning is merely a sequel or addendum. Hence when an agent apparently reasons "*p*, . . . (performs action *Q*)" it may be necessary to treat this as an enthymeme in order to interpret the agent's reasoning correctly. But if it is not so necessary, the '*p*' may still be functioning as a reason, but it is not functioning as part of some structure of practical reasoning.

Secondly, if the reason for which some particular action is done *is* part of such a structure, then two sets of counterfactual conditionals must hold of the agent. First it must be true of the agent that if he or she had learned that one of the premises of his or her practical reasoning was false (or correspondingly that one of his or her inferences was invalid) then he or she would not have done that particular action — unless of course he or she had also learned the truth of some other statement which, conjoined with his or her other beliefs, entailed the same practical conclusion. And secondly it must also be true of the agent that if he or she had learned the truth of some other set of relevant statements, had learned for example of the possibility of achieving some hitherto unrecognized greater good, which made it more imperative to perform some other action, the doing of which was contingently incompatible with doing what his or her reasoning hitherto required, then he or she would have acted differently.

If either of these two conditions is not satisifed, then an agent who performs an action and cites some reason to him or herself and others as his or her reason for doing what he or she does, cannot truly be said to have been guided by his or her reason *qua* good reason, *qua*, that is, part of a piece of valid practical reasoning. In this connection the consistency of the agent's beliefs is important. For when the agent's beliefs are in relevant part inconsistent, the logical consequences of denying a premise or of learning some additional truth about human goods will not be clearly specifiable and we shall have no warrant for asserting any clearly defined sets of counterfactual conditionals and thus for ascribing genuinely rational causality.

When by contrast an action is reason-guided but the reason is not part of a structure of practical reasoning and we have no warrant for asserting such sets of counterfactual conditionals, then the consistency of the agent's beliefs is irrelevant to the question of whether or not the agent's reason was or was not the cause of his action. For if and insofar as the reason functioned as a cause, it did so in virtue of something other than the logical relationships specifiable only in terms of the structures of practical reasoning. And that is to say that when we ask why that particular reason was causally productive of that particular action, the answer will not be in terms of the more extended reasoning and beliefs of the agent, but will have to be framed instead in terms of underlying psychological and social causes. Notice that in this latter type of case learning what reason caused the agent to do what he or she did need not be in any way a movement towards making his or her action intelligible. Indeed to learn that the agent performed this particular action for that particular reason may be to make the action less rather than more intelligible. And we may come to conclude that the complex of action-and-reason cannot itself be made intelligible as part of the actions and transactions of a group of agents, but must rather be understood as a kind of happening, brutely unintelligible as an action, which needs to be understood and explained in terms of impersonal psychological and social causation.

What type of case will be susceptible to explanation in terms of practical reasoning and of social structures which embody structures of practical reasoning and what type of case will be susceptible to explanation in terms of impersonal psychological and social causation? Genuine practical reasoning is not restricted to, but is characteristically and paradigmatically at home in the context of practices (the word 'practice' is used as defined in *After Virtue*.[12]). For it is in the context of practices, such as those of the arts and the sciences or such practices as gardening or farming or games such as chess and football, that we are educated not only so that certain kinds of reason become the causes of certain kinds of action by ourselves and so that we learn how to cause certain kinds of action on the part of others by affording them reasons, but also so that we learn to interpret extended and complex courses of action by others as reason guided in such a way as to commit ourselves to the truth of conditionals of the form "he or she would not have done that if . . . ," where the "if" clause is counterfactual. And in so doing we learn how to distinguish more generally between warranted and unwarranted counterfactual conditionals of the relevant kind.

This is made possible by the way that within practices a shared recognition of goods and a shared acknowledgment of standards provides socially established forms of practical reasoning, a way that is paralleled nowhere else in social life. And hence characteristically and paradigmatically to understand an action as caused by a piece of structured practical reasoning will be to understand its place within some form of socially established practice. Conversely when some type of action in a given social order is reason-guided, but does not occur within the context of a practice, its explanation will characteristically and for the most part be in terms of what I have called impersonal psychological and social causation.

It is, I hope, now clear that the two different types of Japanese suicide which I characterized earlier, the traditional and the particular kind of modern student suicide, are to be distinguished in just this way. That typified by the story of Yoshitsune represents a kind of action identifiable only within the context of a complex set of socially established practices informed by just that kind of coherent set of beliefs and standards which make genuine practical reasoning possible. That typified by the suicides of a certain kind of modern Japanese examination failure occurs within a context of incoherent beliefs where although reasons may be proximate causes of action explanation must look to the psychological and social factors productive of this kind of complex of reason and action. The reasons of agents are in this kind of case part of what is to be explained rather than part of the explanation.

At the beginning of this section I drew attention to the contrast between Durkheim's treatment of suicide as a *social* phenomenon and his failure to treat the reasons given by suicidal agents for their actions as similarly social. I then argued, following Wittgenstein and Kripke, that the having and giving of reasons for beliefs and actions is an essentially social phenomenon. What now emerges is that the having and giving of reasons is not always one and the same. There are at least two quite distinct types of phenomenon to be identified and they find their place in quite different types of social order. But this in turn suggests that there may be something more to be said for Durkheim's account of suicide than I have so far allowed. For the type of suicide with which Durkheim was especially concerned, the modern European or North American suicides that he called *anomic*, is clearly not at all of the same kind as traditional Japanese suicide and indeed such suicides tend to possess the characteristics which I ascribed to one specific type of modern Japanese suicide — occurring outside the context of established social practices, against a background of incoherent beliefs and values, and not the outcome of well-defined practical reasoning — to an even higher degree. My argument has moved towards the conclusion that this type of suicide is not to be *explained* in terms of the agent's reason-giving. And just this was after all what Durkheim argued on very different grounds. The question therefore arises of whether part at least of Durkheim's argument in *Suicide* cannot be restated in a defensible form.

III

"The reasons ascribed for suicide, therefore, or those to which the suicide himself ascribes his act, are usually only apparent causes . . . they may be said to indicate the individual's weak points, where the outside current bearing the impulse to self-destruction most easily finds introduction. But they are no part of this current itself and consequently cannot help us to understand it."[13] Durkheim reaches this conclusion about modern European suicides by a very different route from that which I followed, but it is sufficiently similar to suggest a striking thesis. Although Durkheim's general conception of the relationship of individuals to the social order which they inhabit is seriously flawed, it may nonetheless be the case that he succeeded in characterizing the causal order of certain areas of life in specifically modern societies. On Durkheim's view social orders exert constraints upon and regulate the behavior of individuals. The degree and character of this constraint may vary; what is constant is the presence of a set of external causal relationships between characteristics and states of social orders and characteristics and states of individual persons. As the degree of this contraint weakens in modern societies so the individual is loosed from his or her regulating social bonds. Where previously the social order dictated the forms of and set bounds to their behavior, they must now find the resources for decision-making and action within themselves. In so far as they are freed from such external constraint, they suffer from *anomie*, normalessness.

To say this is to say that such individuals inherit a type of social order which to some degree in certain areas at least lacks socially established shared standards by which the practical reasoning of individual agents may at once be guided and made intelligible to others. And in such a type of social order I have already suggested that two distinctive characteristics of action will be found. Individuals in areas of life where they lack the means to make their actions fully intelligible to others will also lack the means to make them fully intelligible to themselves. And their giving of reasons for their actions, whether to themselves or to others, will not provide us with the causes which explain their actions, but rather with further features of their actions which stand in need of explanation.

This suggests that in some areas at least of distinctively modern cultures, such as our own, which perhaps have gone even further than contemporary Japan in abandoning traditional patterns of action, we might expect to find suicidal acts exhibiting precisely these two characteristics. Those who commit them will do so being far from clear as to what precisely they are doing; and those who commit them will cite as reasons for their actions reasons which we shall find it impossible to use in *explaining* their suicidal acts. Empirical findings suggest that modern suicides do in fact often enough exhibit just those characteristics. The reasons that are provided in suicide notes tend to puzzle rather than to illuminate; for what is cited as a reason is characteristically some kind of event or state — the onset of an illness, the death of a spouse, the loss of a job — which does not seem to provide a good reason for killing oneself to the vast majority of those experiencing such events or states. It may be of course that such reasons function as minor premises in some pieces of practical reasoning whose major premise has not been made available; but no one has yet supplied us with grounds for believing that in the vast majority of reason-giving suicides such a major premise can be supplied.[14]

Moreover those sociologists who have, since the publication of Douglas's book, directed their attention systematically towards the processes by which certain deaths come to be classified as suicides[15] have tended to emphasize how the process of classification tends to impose on suicidal actions a determinateness which is not a feature of those actions themselves. And this reinforces the view that Durkheim was right in connecting specifically modern suicide with *anomie*, although in error in the way that he made this connection. Lack of socially established forms of practical reasoning, that is, of evaluative and normative reasoning, accompanied by the mélange of assorted fragments which furnished so much of modern moral belief,[16] will indeed be apt to render many individuals at key points in their lives *anomic*, and as such unable to make fully intelligible to themselves or to others in light of established standards of evaluation and interpretation either precisely what they are doing or precisely why they are doing it. And the explanation of what they are doing will have to lie beyond themselves in impersonal features of the social order. In this type of society an external causal relationship will hold between characteristics of the social order and types of event in individual lives.

That there are such causal relationships can be established only by means of appeal to invariant law-like regularities and thus this causality is very different indeed from that exemplified in practical reasoning. Humeans and their positivist succesors have of course equated this type of causality with causality as such. And sociological positivists, such as Durkheim, naturally enough supposed that if social phenomena were to be susceptible of scientific investigation then social phenomena must exemplify this type of Humean causation. That in so doing they made a philosophical mistake will still be denied by pertinacious Humeans, but perhaps by few others. It has been the thesis of this paper, however, that Durkheim's error in this respect in *Suicide* was partly at least the outcome not of a philosophical mistake, but of a genuine perception of some salient characteristics of specifically modern societies. The type of theoretical explanation which Durkheim in *Suicide* at least supposed to be the key to an understanding of all social orders is in fact peculiarly and specifically appropriate to one very special type of social order, that with which Durkheim himself was concerned. Positivism in sociology turns out to be, unsurprisingly, one of the philosophical shadows cast by the social order of modernity.

Vanderbilt University

NOTES

[1] See Lukes (1972).
[2] Sainsbury and Barraclough (1968), p. 1252.
[3] Atkinson (1971) and above all Douglas (1967).
[4] Quine (1960), chapter 6.
[5] Kripke (1981).
[6] *Ibid.*, pp. 283–284.
[7] *Ibid.*, p. 284.
[8] See Morris (1976).
[9] See *Ibid.*, chapter 5.
[10] See De Vos (1962) and Iga (1967).

[11] See the remarks of Dr. Foyoda, a Japanese psychoanalyst, reported in Riesman and Riesman (1967).
[12] See MacIntyre (1981), p. 175.
[13] Durkheim (1951), p. 151.
[14] For the type of evidence that is available about such suicide notes see Jacobs (1967).
[15] See especially Atkinson (1978).
[16] See MacIntyre (1981), chapers 2 and 17.

BIBLIOGRAPHY

Atkinson, J.: (1971), 'Societal Reactions to Suicide: the Role of Coroners' Definitions', in Cohen, 1971, pp. 165–191.

Atkinson, J.: (1978), *Discovering Suicide: Studies in the Social Organization of Sudden Death,* University of Pittsburgh Press, Pittsburgh and London.

Block, I. (ed.): (1981), *Perspectives on the Philosophy of Wittgenstein,* Basil Blackwell, Oxford.

Cohen, S. (ed.): (1971), *Images of Deviance,* Penguin Books, Harmondsworth, England.

De Vos, G.: (1962), 'Deviancy and Social Change: A Psychocultural Evaluation of Trends in Japanese Delinquency and Suicide', in R. Smith and R. Beardsley, 1962, pp. 153–171.

Douglas, J.: (1967), *The Social Meanings of Suicide,* Princeton University Press, Princeton.

Durkheim, É.: (1951), *Suicide: A Study in Sociology,* trans. J. Spaulding and G. Simpson, ed. G. Simpson, The Free Press, Glencoe, Illinois.

Iga, M.: (1967), 'Japanese Adolescent Suicide and Social Structure', in Shneidman, 1967, pp. 224–250.

Jacobs, J.: (1967), 'A Phenomenological Study of Suicide Notes', *Social Problems* **15**, pp. 60–72.

Kripke, S.: (1981), 'Wittgenstein on Rules and Private Language: An Elementary Exposition', in Block, 1981, pp. 238–312.

Lukes, S.: (1972), *Émile Durkheim,* Harper and Row Publishers, Inc., New York.

MacIntyre, A.: (1981), *After Virtue: A Study in Moral Theory,* University of Notre Dame Press, Notre Dame.

Morris, I.: (1976), *The Nobility of Failure: Tragic Heroes in the History of Japan,* New American Library, New York.

Quine, W.: (1960), *Word and Object,* MIT Press, Cambridge, Massachusetts.

Riesman, D. and Riesman, E.: (1967), *Conversations in Japan,* Basic Books, New York.

Sainsbury, P. and Barraclough, B.: (1968), 'Differences between Suicide Rates', *Nature* **220**, No. 5173, p. 1252.

Shneidman, S. (ed.): (1967), *Essays in Self-Destruction,* Science House, New York.

Smith, R. and Beardsley, R. (ed.): (1962), *Japanese Culture: Its Development and Characteristics,* Aldine Publishing Co., Chicago.

Wittgenstein, L.: (1958), *Philosophical Investigations,* trans. G. E. M. Anscombe, 3rd ed., Macmillan, New York.

PART II

INTERPRETING THE TRADITION

JANET BROUGHTON

ADEQUATE CAUSES AND NATURAL CHANGE IN DESCARTES' PHILOSOPHY

I

Descartes uses versions of the principle that a cause must contain what comes about in its effect in his proofs that God exists and that the external world exists. I want in this paper to show that Descartes does not regard this adequacy principle simply as a formal device useful in constructing a special kind of argument, but rather regards it as a principle true of natural change in general. I shall support this claim by showing that it is that principle which generates central problems for Descartes's treatment both of causal connections between body and mind and of causal connections among bodies. My method, then, is indirect. I will be looking at these problems, and at the ways Descartes handles them, as symptoms; and I will be arguing that they arise from an underlying and thoroughgoing commitment to the conception of causality articulated in the adequacy principle.

Let me make a few preliminary remarks about this adequacy principle. The version of it which first appears in the *Third Meditation* proof of God's existence says that a cause must contain at least the *degree of reality* that comes about in its effect,[1] but Descartes slides in the *Third Meditation* into the unrestricted claim that a cause must contain *everything* that comes about in its effect;[2] and it is this unrestricted principle he later invokes in the *Sixth Meditation* proof that the external world exists.[3] Elsewhere, Descartes does not seem concerned to distinguish the two principles, and the one explanation he gives of why causes must be adequate to their effects, in the *Second Replies*, is an explanation of the unrestricted principle.[4] In what follows, I will be concerned only with the unrestricted adequacy principle.

The full statement of that principle says that a cause must contain what comes about in its effect in the same or a higher form (formally or eminently). The only clear examples Descartes gives of eminent containment are the possible causation of my sense-idea by God or an angel[5] and the creation by God of *res extensa*.[6] Apparently Descartes

A. Donagan, A. N. Perovich, Jr., and M. V. Wedin (eds.), Human Nature and Natural Knowledge, 107–127.

thinks that being an infinite thinking thing is "higher" than being an idea of an extended thing or being an extended thing. But, as I will show later, we need not puzzle over eminent containment in order to talk about the adequacy of causes in cases of psychophysical or physical change. The formal containment clause is the one which is supposed to apply to such cases. In the hope that coining a term will make matters clearer, I shall call the unrestricted adequacy principle with just the formal containment clause the *formal containment principle.* This principle says that a cause must be characterized by the same property that comes to characterize its effect.

Descartes believes that the causes of ideas must, like all causes, be adequate to their effects; and he thinks this implies that the cause of an idea must have, formally or eminently, the same sort of "formal" reality that the idea has of "objective" reality. I will presently take up the question of just what this means. For now, I want only to stress that the formal containment principle can apply to the causes of ideas; it rules out "eminent" containment, not "objective" reality.

Let me briefly sketch what will follow. In sections two through four, I will take up the case of sensation. In section two, I will show that the problem with mind-body interaction Descartes acknowledged in his correspondence with Elizabeth is most likely a problem generated by the formal containment principle. In section three, I will trace out several moves Descartes makes which go some way toward solving that problem, and in section four I will claim that Descartes also tries out a far more radical solution. In section five I will turn to physical change. I will exhibit a pair of contradictory claims Descartes makes about motion transfer and try to show that one of them most probably issues from the formal containment principle. Section six is a brief conclusion to the paper.

II

Descartes claimed that the human mind and body interact — that motions in the brain cause sensations, memories, emotions, and imaginings, and that the mind's volitions cause motions in the brain and, ultimately, our voluntary actions. His critics thought Descartes failed to explain how this interaction is possible; indeed, they thought the failure inevitable. Those Cartesians who developed various doctrines of occasionalism did so in part because they thought mind-body interaction to be impossible.

I think Descartes himself detected a problem with his account of mind-body interaction and tried to solve it in several ways. In what follows I will focus on the causal connection between brain states and sensations, and I will first show what Descartes thought was problematic about that connection and then trace the ways he tried to solve the problem. Although his final position fell short of full-blown occasionalism, it was in its own way extraordinary.

Let me begin by sketching the picture Descartes has of human sensation.[7] External bodies, whose properties are extension, figure, and motion, come in contact with the human body. They may make direct contact with the skin, or emit light rays which enter the eye, or agitate the air next to the ear, or emit particles which strike against the tongue or the inside of the nose. In any case, they impinge upon nerves extending from the sense organs to the brain. These nerves, Descartes says, are tiny tubes containing "animal spirits;" and when a nerve is struck, its animal spirits are set in motion. This motion travels to the brain, causing the animal spirits in the inner part of the brain to move in certain patterns, these patterned motions impinge upon the pineal gland, and the resulting motions of that gland cause sensation. Among our sensations are our various ideas of pleasure, pain, hunger, thirst, extension, figure, motions, hardness, heat, light, color, scents, and sounds.[8]

If two brain movements are exactly alike, they will cause exactly similar sensations; and if two sensations are exactly alike, they are the effects of exactly similar brain movements. "Each of the movements which are in the portion of the brain by which the mind is immediately affected brings about one particular sensation only."[9]

A Humean might not have found any special difficulties accompanying the claim that brain movements cause sensations. Events of one sort regularly follow events of another sort, and the conjunction of sensations and brain movements can be as constant as any.

But in 1643 Princess Elizabeth asked Descartes how mind, being unextended, and body, being extended, can interact;[10] and Descartes answered, " . . . the question you ask is the one which may most properly be put to me in view of my published writings."[11] Descartes thinks there *is* something problematic about the claim that brain movements cause sensations; and he must be thinking about some way in which a causal claim amounts to more than a claim about constant conjunction.

What is this troublesome aspect of causation? Elizabeth's question gives us a clue; it is the *distinction* between mind and body that bothers her. She is bothered not by their numerical difference but rather by their radical difference in *kind.* So we may suppose that the problem Descartes is acknowledging is this: Mind and body are entirely different kinds of things, but entirely different kinds of things cannot cause changes in one another.

This supposition is confirmed by a passage in the *Principles.* Descartes is criticizing the view that real qualities (e.g., heaviness) exist, and he says,

> We can very well conceive how the movement of one body can be caused by that of another, and diversified by the size, figure, and situation of its parts, but we can in nowise understand how these same things (viz. size, figure and motion) can produce something entirely different in nature from themselves ...[12]

Here Descartes is explicitly stating that causal interaction between very different kinds of things is inconceivable. That, plus Cartesian dualism, does indeed create a problem for the claim that brain movements cause sensations.

Descartes's attempts to solve this problem in his letters to Elizabeth seem unsuccessful. He first says that mind and body interact in the same way heaviness and body do.[13] When Elizabeth properly points out that Descartes himself thinks the heaviness-body case is an impossible one,[14] he tries to mitigate the distinction between mind and body:

> I beg [Your Highness] to feel free to attribute matter and extension to the soul because that is simply to conceive it as united to the body. And once she has formed a proper conception of this and experienced it in herself, it will be easy for her to consider that the matter attributed to the thought is not thought itself, and that the extension of the matter is of different nature from the extension of the thought....[15]

Elizabeth is unhappy with this proposed solution; she quite rightly thinks Descartes cannot in any way allow matter to be attributed to mind or thought to body.[16]

Why doesn't Descartes solve the problem by simply allowing that very different kinds of things can interact? He cannot do this, I think, because to do so would be to give up a principle he holds dear, the principle that a cause must contain, formally or eminently, whatever comes about in its effect.[17] This principle, the unrestricted adequacy principle, is one which, Descartes says, applies to *all* causal situations.[18] Here is Descartes's commentary on it in the *Second Replies*:

That there is nothing in an effect which did not pre-exist in some similar or higher manner in its cause, is a first notion than which none clearer can be had; and this vulgar notion — nothing comes into existence from nothing — does not differ from it; for if it is conceded that there is something in an effect which did not exist in its cause, then it has also been conceded that this something has come into existence from nothing; nor is it clear why nothing cannot be the cause of a thing, unless from the fact that in such a cause there does not exist the same thing as in the effect.[19]

Fortunately, we need not consider in what follows the problematic case where a cause contains eminently "the same thing as in the effect;"[20] as I shall show, mind-body causal interaction involves only formal containment. Where mind and body are concerned, then, the relevant adequacy principle is what I am calling the formal containment principle. It is the principle that a cause must contain formally everything that comes about in its effect.

We can now see why Descartes was concerned about the great difference between mind and body. When a change in a brain state causes a change in a mental state (i.e., causes a sensation), a change in an extended thing is causing a change in a thinking thing. The formal containment principle implies, at least *prima facie*, that the brain must contain exactly what comes about in the mind. It would follow first that the brain must have a mental state. But the physical and the mental are so radically different that a physical thing cannot have a mental state.[21] And second, for Descartes it would seem to follow that a physical thing must contain pains, colors, odors, etc. But the only qualities physical things have are extension, motion, duration, and the like.[22] The problem Descartes acknowledged when he wrote to Elizabeth is, I think, really the one generated by these apparent conflicts between the claim that mind and body interact causally and the claim that the formal containment principle applies to their interaction.

III

The first move available to Descartes for solving this problem depends on his distinction between an idea as "an act of my understanding" and as "the thing which is represented by this act."[23] Descartes does not have to claim that sensations as acts of mind are caused by brain movements. The mind itself is the cause of its acts.[24] It is only sensations as the objects of thought, as the content of sense-ideas, that are the effects of brain movements.[25] When I have a sensation, the sheer consciousness which occurs is the effect of my mind; but insofar as the sensation is that of which I am conscious, it is the effect of brain movements. So Descartes need not take the formal containment principle as implying that the brain must contain a mental state. He could argue that the 'mentalness' of a sensation — its being thought of — is contained in the mind itself.

But there still remains a problem. The containment principle seems also to imply that the *objects* of sensations, including pains, colors, etc., must be contained in the brain. This problem is highlighted by Descartes's proof in the Sixth Meditation that physical objects exist. In it, Descartes applies the containment principle to the objects of my "ideas of sensible things."[26] He appeals not to eminent containment but to *formal* containment, and he applies the principle to the contents of *all* my sense-ideas. During the course of the argument, he claims that the cause of the "objective" reality of all my sense-ideas contains what is in them formally.

I said earlier that when a cause formally contains what comes about in its effect, it contains exactly what comes about in its effect. This, Descartes makes clear, is just as true of effects which are objects of ideas as it is of other kinds of effects. He explains how to apply 'formal containment' to cases involving the objective reality of ideas as follows:

> To exist *formally* is the term applied where the same thing exists in the object of an idea in such a manner that the way in which it exists in the object is exactly like what we know of it when aware of it. . . .[27]

Can the brain contain formally what is in the objects of sense-ideas? Descartes could say that it contains some of those objects. When I sense extension, figure or motion, it is certainly true that my brain contains extension, figure, and motion. But as I have said, Descartes cannot allow that the brain contains formally any *other* qualities presented in the objects of sense-ideas.

To avoid ascribing colors, pains, etc. to the brain, Descartes can argue that brain movements *and the nature of the mind* cause the objects of our sensations. He says in the *Principles*,

. . . our mind is of such a nature that motions which are in the body are alone sufficient to cause it to have all sorts of thoughts, which do not give us any image of any of the motions which give rise to them.[28]

He adds,

. . . our mind is of such a nature that the diverse motions of the body suffice to produce in it all the diverse sensations that it has.[29]

So whatever is in the objects of our sensations but is not contained formally in brain movements can now be ascribed to the nature of the mind. In other words, Descartes is resolving the conflict between the formal containment principle and the claim that brain movements cause the objects of sensations by acknowledging that brain movements are not their *total* cause. A parallel resolution appears in Replies II. In Objections II, the formal containment principle was denied on the grounds that sun, rain, and earth are the causes of animals although the former do not contain life, while the latter do.[30] Descartes resolved the conflict between the principle and the apparent anomaly by saying that if animals possess something which sun, rain, and earth do not, then sun, rain, and earth are not the sole causes of animals.[31]

The formal containment principle will not conflict with the causal claim about sensations if everything but the extension, figure, and motion presented in the objects of sensation can somehow be ascribed to the nature of the mind as a cause. A natural way to explain how the mind causes sense—ideas of color, pain, etc. depends on the distinction between ideas which are clear and distinct and ideas which are not.

Insofar as I clearly and distinctly apprehend the objects of my sense-ideas, what is in them *is* contained formally in brain movements. This is the line Descartes takes in the argument for the existence of corporeal substance in the *Principles*.[32] He reasons there as follows: I clearly and distinctly apprehend that the objects of my sense-ideas are extended, figured, and capable of motion; thus extension, figure, and motion must be contained formally or eminently in the cause of the objects of my sense-ideas; therefore the cause of those objects contains extension, figure, and motion formally, i.e., that cause is body.

Next, Descartes must assign a cause to objects of sense-ideas insofar as they do *not* contain extension, figure, and motion. Their cause, of course, cannot be brain movements alone. As the quotations from the *Principles* suggest, Descartes tries instead to say that brain movements *and* the nature of the mind share causal responsibility for the objects of sense-ideas like those of pain, color, etc. When we ask just how they share causal responsibility, we can again exploit Descartes's distinction between clear and distinct ideas and obscure ones.

Descartes often seems to apportion causal responsibility for, say, my sense-idea of red in this way: the redness which exists in corporeal objects is the "disposition" of those objects to move my nerves in a certain way.[33] Let us call this physical state of affairs, "scientific red." Scientific red can be completely described in terms of extension, figure, and motion. If my idea of red were clear and distinct, it would be an idea of scientific red.[34] The obscuring of this clear and distinct idea is the cause of my idea of non-scientific red. I myself am the cause of the obscurity; the corporeal object, on the other hand, is the cause of my idea insofar as I have not obscured it. Brain movements contain scientific red formally, and my mind contains obscurity formally.

The trouble with this explanation is that Descartes cannot really hold that my sense-idea of red is an obscured idea of scientific red. First, an idea of scientific red, obscured or not, is not a sense-idea at all. As Descartes implies in the *Principles*, we perceive non-scientific red by means of our senses but acquire the idea of scientific red by means of a proof.[35] We sense the former but conclude the latter. Second, a sense-idea need not be obscured: "we have a clear or distinct knowledge of pain, colour, and other things of the sort when we consider them simply as sensations or thought."[36] If scientific red cannot be the object of a sense-idea, and if a sense-idea need not be obscured, then Descartes has failed to explain how what is in the object of a sense-idea like that of red exists formally in brain movements and the nature of the mind.

Let me now summarize what I have said so far. Descartes claims that

(1) Brain movements cause sensations,

but he acknowledges in a letter to Elizabeth that claims, like (1), for mind-body interaction conflict *prima facie* with the principle that

(D) A cause cannot be of a sort very different from its effects.

This principle is itself simply a consequence of the unrestricted adequacy principle:

(UAP) A cause must contain, formally or eminently, everything that comes about in its effect.

The alternative which applies to sensation, which I have been calling the formal containment principle, is

(FCP) A cause contains formally everything that comes about in its effect,

and (FCP) clearly conflicts with (1). Descartes can resolve some of the conflict by changing (1) into

(2) The mind causes sense-ideas *qua* mental acts,

and

(3) Brain movements cause the objects of sense-ideas.

But even (3) conflicts with (FCP), except for sense-ideas of extension, figure, and motion. So Descartes tries changing (3) into

(4) Brain movements cause the objects of sense-ideas of extension, figure, and motion,

and

(5) Brain movements and the nature of the mind cause the objects of all other sense-ideas.

But this attempt also fails.

IV

In order to make the causation of sensation consistent with the formal containment principle, Descartes assigns an increasingly large causal role to the mind. But, we have seen, as long as the mind shares its role with brain movements, the formal containment principle creates problems. I want to show in what follows that that principle drove Descartes to claim that the mind *alone* causes sensations.

Descartes hints at a position like that in the following passage from the *Passions*:

> [the soul's] nature is such that it receives in itself as many diverse impressions, that is to say, that *it possesses as many diverse impressions*, as there are diverse movements in this gland [i.e., the pineal gland].[37]

A portion of a letter to Mersenne is more explicit:

> I hold that all those ideas which involve no affirmation or negation are innate: for the organs of our sense do not bring us anything which is like the ideas which arise in us at their stimulus.[38]

We cannot be sure what Descartes means here by "innate" or "stimulus," so let us turn to the passage in *Notes Against a Programme* which presents a similar view more fully. Let me quote the passage in its entirety:

> . . . any man who rightly observes the limitations of the senses, and what precisely it is that can penetrate through this medium to our faculty of thinking must needs admit that no ideas of things, in the shape in which we envisage them by thought, are presented to us by the senses. So much so that in our ideas there is nothing which was not innate in the mind, or faculty of thinking, except only these circumstances which point to experience — the fact, for instance, that we judge that this or that idea, which we now have present to our thought, is to be referred to a certain extraneous thing, not that these extraneous things transmitted the ideas themselves to our minds through the organs of sense, but because they transmitted something which gave the mind occasion to form these ideas, by means of an innate faculty, at this time rather than at another. For nothing reaches our mind from external things through the organs of sense beyond certain corporeal movements . . . but even these movements, and the figures which arise from them, are not conceived by us in the shape they assume in the organs of sense, as I have explained at great length in my *Dioptrics*. Hence it follows that the ideas of the movements and figures are themselves innate in us. So much the more must the ideas of pain, colour, and sound and the like be innate, that our mind may, on occasion of certain corporeal movements, envisage these ideas, for they have no likeness to the corporeal movements. Could anything be imagined more preposterous than that all common notions which are inherent in our mind should arise from these movements, and should be incapable of existing without them? I should like our friend to instruct me as to what corporeal movement it is which can form in our mind any common notion, e.g., the notion that "*things which are equal to the same thing are equal to one another*," or any other he pleases; for all these movements are particular, but notions are universal having no affinity with movements and no relation to them.[39]

In this passage, Descartes is arguing that sense-ideas are innate, in the sense that they owe their origin to the mind alone. It is not clear whether Descartes means the same by 'cause' as he does by 'origin'. On the other hand, he is prone to using the two terms interchangeably.[40] On the other hand, he might hold that something must migrate from its origin but not from its cause, although it is difficult to understand how something could migrate from an extended substance to an unextended substance.[41] A third reading, compatible with the first, is that 'origin' designates a cause of something other than its immediate cause. To be on the safe side, I will assume in what follows that 'origin' and 'cause' are not synonymous. This cautious assumption will make my analysis of the passage longer but, I hope, stronger.

The general argument which Descartes uses in this passage runs as follows: If the idea of *X* owes it origin to the senses, then the senses must present *X* to the mind. But the senses do not present *X*; they present something else. Thus the idea of *X* does not owe its origin to the senses. Descartes applies this argument to these three cases: ideas of figure and motion; ideas of pain, color, sound, etc.; and ideas of common notions (e.g., of things being equal to each other if equal to the same thing). In each case, the argument rests on showing that the senses present something different from the content of the given idea.

As for ideas of figure and motion, Descartes says that what the senses present to the mind is similar to, but not the same as, the figure and motion in our ideas. For example, light rays from an external object strike the optic nerves at the back of the eye in patterns having certain shapes. In the *Optics*, to which he refers in the *Notes* passage, Descartes says that the figures at the back of the eye "usually contain only ovals and diamond shapes" in the cases where our corresponding sense-ideas are ideas of "circles and squares."[42] I think Descartes's optical theory should allow that sometimes the figure at the back of the eye is square when the corresponding sense-idea is an idea of something square. But Descartes omits the qualifying "usually" in the *Notes* and speaks as if the senses always present figures and motions different from the ones we have ideas of.

The senses do not even present something similar to our ideas of common notions and of pain, color, sound, etc., because the motions and patterns the senses present are quite unlike the contents of those ideas. They, too, are ideas which do not owe their origin to the senses.

I want to claim that in this *Notes* passages, Descartes not only has denied that sensations *originate* from brain movements but also has denied that they are *caused* by brain movements. We have seen in the general argument in the *Notes* that if sensation originates from the senses, this requirement must be met: The senses must present to the mind exactly what is in the objects of sensation. Descartes shows that this requirement is not met by showing that brain movements do not contain exactly what is in the objects of sensation. Thus if *X* is presented to the mind by the senses, *X* must be contained in brain movements. The reverse is not always true. In sleep, for example, brain movements contain what has not been presented by the senses. But where sensation is concerned, brain movements cannot contain what the senses have not presented, and the senses cannot present what brain movements do not contain.[43]

But we have also seen that the formal containment principle applied to brain movements as the cause of sensation yields this requirement: Brain movements must contain exactly what is in the objects of sensation. Thus Descartes has shown in the *Notes* passage that a necessary condition for causation, as well as origination, has not been met. By showing that sensations do not originate from brain movements, Descartes has also shown that they are not *caused* by brain movements.

I have cautiously assumed that 'cause' and 'origin' are not synonymous for Descartes, and so one might grant that Descartes *ought* to say sensations are not caused by brain movements, yet wonder whether he *knows* this is what he ought to say. But I think there is considerable evidence that Descartes himself means in this passage to deny that brain states cause sensations.

First, all throughout the *Notes* passage in which he denies that sensations originate from brain movements, Descartes carefully refrains from saying that brain movements cause sensations. He reverts to the term 'cause' shortly afterwards, but he does so just when he becomes sloppy about the claim that sensations do not originate from brain movements: "sight, of itself and by its proper function, presents nothing beyond *pictures*, and hearing nothing beyond *voices* or *sounds*."[44]

There is a second piece of evidence that Descartes is aware that he denied that brain movements cause sensations. He has shown that for ideas of figures and motions, of pains, colors, etc., and of common notions, an idea does not owe its origin to brain movements. The case of ideas of common notions, however, is different from the first two cases. A common notion has "*no* affinity with movements and *no* relation to them."[45] Descartes must specify some affinity or relation which holds between brain movements and sense-ideas and which does not hold between brain movements and ideas of common notions. By specifying this relation, though, he must deny what he has just proved: that none of these ideas owes its origin to brain movements.

Now if Descartes were not aware that the *Notes* argument denies that brain movements cause sensations, then surely he would take the obvious course and specify the relation of causation. He could say that sense-ideas of motion and color are caused by, but do not originate from, brain movements; whereas common notions are neither caused by nor originate from brain movements.

But Descartes fails to provide this obvious solution, and I take this failure to be evidence that he recognized this: The formal containment principle plus the *Notes* argument entail a denial that brain movements cause sensations.

The solution Descartes does provide is extraordinary. First, he says that brain movements are the occasions of sense-ideas. Then in order to explain how a particular kind of brain movement can regularly occasion a particular kind of sense-idea, he introduces an innate faculty: "extraneous things . . . transmitted something which gave the mind occasion to form these ideas, by means of an innate faculty, at this time rather than another."[46] This innate "meta-faculty," then, regulates the mind's faculty for having sense-ideas so that just the right sense-ideas appear at just the right times.

Now the mind alone causes sensations, and brain movements do not. We have followed the course of conflict between Descartes's claim that brain movements cause sensations and the formal containment principle. We have seen the causal claim modified to ease the conflict, and we have finally seen that claim vanish. In what I would call a most honest moment, Descartes recognizes that either the causal claim, however modified, or the adequacy principle, must go. His decision to reject the former but not the latter is perhaps a tribute to the tenacity with which he has followed that "principle than which none clearer can be entertained," no matter where it might lead him.

V

Let me now turn to a puzzling inconsistency in what Descartes says about causal interaction among bodies, an inconsistency which arises, I think, from Descartes's adherence to the unrestricted adequacy principle. Bodies interact only when they collide with one another. The basic physical law which governs this interaction at impact, which "covers all the particular causes of corporeal change — so far as they are themselves corporeal,"[47] runs as follows:

> When a moving body collides with another, then if its own power of going on in a straight line is less than the resistance of the other body, it is reflected in another direction and retains the same amount of motion, with only a change in its direction; but if its power of going on is greater than the resistance, it carries the other body along with it, and loses a quantity of motion equal to what it imparts to the other body.[48]

The inconsistency appears in passages where Descartes talks about the situation in which one body "loses a quantity of motion equal to what it imparts to the other body." Sometimes he endorses and sometimes he denies what I will call the migration theory of motion.

According to the migration theory, when a body "loses a quantity of motion equal to what it imparts to the other body," its motion leaves it and migrates into the second body. There is something (a quantity of motion) which the first body has; and that very self-same something is lost by the first body and gained by the second. So during collision, the one body loses some motion, the other gains the same motion, and that motion loses one "subject" and gains another by migrating from the one to the other. It is this theory which Locke comments on as follows:

> ... in the communication of motion by impulse ... we can have no other conception, but of the passing of motion out of one body into another; which, I think, is as obscure and inconceivable as how our minds move or stop our bodies by thought ...[49]

We can find this migration theory in a number of Descartes's writings, ranging from the early *Le Monde* through at least the *Principles.* The following quotations are typical:

> ... it is impossible that ... motions should change in any other way than with regard to the subject in which they are present ...[50]

> ... when one body moves another, it loses as much of its own motion as it gives to the other ...[51]

> ... [God] does not preserve [motion] as a constant inherent property of given pieces of matter, but as something passing from one piece to another as they collide.[52]

It is not always obvious, however, that Descartes endorses the migration theory. There are many passages which are metaphysically more guarded than the three above – the statement of the third law quoted earlier is one such passage. But more importantly, there are passages in which Descartes *denies* that the migration theory is true and that he ever espouses it. The most famous of these passages occurs in a letter Descartes wrote late in his life to More. Descartes says,

> You observe correctly that a *motion, being a mode of a body, cannot pass from one body to another*. But that is not what I wrote . . .[53]

Of course, Descartes *had* written just that; indeed, earlier in the very same letter to More he seems practically to endorse the migration theory.[54] I want first to discuss why Descartes ever held the migration theory, and then I will explain what very good reason he had for thinking himself that it is "obscure and inconceivable."

Let me begin by merely mentioning a rather uninteresting way of explaining why Descartes held the migration theory. He might simply have been taken in by a manner of speaking. It is, after all, easy to slide from the fairly neutral locution, "The motion which *A* loses is equal in amount to the motion which *B* gains" to the loaded locution, "The motion which *A* loses is (identical with) the motion which *B* gains."[55] I think, however, that we should pass over this superficial reason for Descartes's endorsement of the migration theory and look instead for a deeper one.

I want to suggest that we should understand the migration theory as Descartes's way of answering the question, "Where does the motion in an impelled body *come from*?" And I believe Descartes's general answer to the question is that whatever comes about in an effect must be contained, formally or eminently, in its cause; that is, his general answer is given by what I have called the unrestricted adequacy principle. So I think the migration theory is the outcome of applying the unrestricted adequancy principle to causal interaction between bodies.

For reasons independent of the adequacy principle itself, Descartes thinks that the quantity of motion gained by one body during collision is the *same* as the quantity of motion lost by the body which hit it. Thus no matter what eminent containment might amount to,[56] we need not take it into account here. Once again, then, it is what I have been calling the formal containment principle which concerns us, the principle that a cause must contain formally everything that comes about in its effect.

Prima facie, it seems that the formal containment principle would be satisifed if the one body simply acquired a quantity of motion equal to that which the other body lost. The principle does not seem to require that the motion gained be the self-same motion lost; the principle does not seem to require that motion maintain identity throughout change.

But let us notice first that the question whether an adequacy principle requires this really emerges only in this collision case. When God is the cause and his effect something creaturely, the cause is perfectly infinite and so surely contains what comes about in its effect *eminently*; furthermore, because the cause is infinite, it is not in the slightest diminished by its creative activity. So the question of whether what was brought about pre-existed self-identically in its cause does not arise there. Nor does that question ever really emerge explicitly about mind-body interaction, although perhaps it should.[57] There the pressing problem is that the properties of the cause are not *like* enough to the properties of the effect. The question whether properties maintain *identity* throughout change thus seems moot.

Gassendi objected to the unrestricted adequacy principle as follows:

> — that common saying — there is nothing in the effect which is not in the cause — seems to be understood of the material, rather than of efficient causality. For the efficient cause is something external and frequently of a diverse nature from the effect.[58]

And indeed it seems that an adequacy principle gets whatever plausibility it has from its affinity with the idea that some material stays the same during change. But since Descartes will only allow efficient causes into his ontology, it should not be surprising to find that he packs the notion of something staying the same during change into the notion of efficient causation.

From a certain point of view, the common saying that there is nothing in the effect which is not in the cause, looks commonsensical. Leaving aside the intricacies of scholastic metaphysics, it sounds perfectly sensible to say that if I make a clock out of pile of clock parts, the pile must have contained all the parts in the finished clock. Or if the action of the ocean on the ocean floor results in a sand bar, all the sand in the bar must have pre-existed on the ocean floor. And the reason this sounds so sensible is that neither gears nor grains of sand can pop into existence. By this I mean at least that there is some description of the finished product as composed of parts, such that each part, so described, maintains identity throughout the producing of the product.

According to Descartes, each of these situations is correctly described in terms of the rearrangement of corpuscles, the microscopically small parts of which any larger physical thing is composed. And corpuscles certainly maintain their identity throughout change. They do not pop into or out of existence. But to describe correctly and scientifically each of those changes, one must also, of course, mention the rearranging of the corpuscles. Thus one must also inevitably describe their quantities of motion, collisions, and changes of motion.

But the unrestricted adequacy principle applies to properties as well as to things; properties cannot pop into existence any more than things can. It should not be surprising that if *properties* are not to pop into existence, they as well as things, must maintain identity throughout change. Hence the migration theory of motion: During collision the *self-same* motion migrates from one body into another.

There are a number of puzzles and problems connected with the claim that a property maintains identity throughout change. I think *Descartes's* major difficulty with this view is that it makes a mode too like a substance. Motion, on this view, maintains its identity *independently* from the particular bodies it belongs to. To be sure, Descartes need not be claiming that there could be motion if there were no bodies at all; but the less extravagant claim seems to create problems enough. For example, Descartes says of modes,

> . . . if we wish to consider them apart from the substances in which they are, that will have the effect of our taking them as self-subsisting things and thus confounding the ideas of mode and substance.[59]

It seems that Descartes wants both (a) to consider motions "apart from the substances in which they are" and yet (b) to call them modes. He needs (a) to make motion conform to the formal containment principle, and he needs (b) for a very straightforward reason. A motion cannot, for Descartes, be identifiable independent of its particular subject, for he *defines* motion as the transference of one part of matter or one body from the vicinity of those bodies that are immediate contact with it, and which we regard as in repose, into the vicinity of others.[60]

He emphasizes the modal dependence of motion so defined in the same passage:

Further, I understand that it is a mode of the mobile thing and not a substance, just as figure is a mode of the figured thing, and repose of that which is at rest.[61]

But, as I have tried to show, the modal dependence of motions on bodies conflicts with the independence required of them by the formal containment principle.

It is this conflict, I think, that accounts for Descartes's inconsistency about endorsing the migration theory of motion. When he applies the formal containment principle to the cause-effect relation occuring during collision, he ends up endorsing the migration theory. But when he recognizes that motion must be a mere mode, he disavows the migration theory:

You observe correctly that a motion, being a *mode* of a body, cannot pass from one body to another. But that is not what I wrote . . .[62]

Had Descartes not applied the adequacy principle as far-reachingly as he did, he might well never have endorsed the problematic migration theory at all.

VI

I have been arguing that for Descartes the adequacy principle goes very deep, that it is not just a handy premise in existence-proofs, but also shapes his fundamental views about natural change. As I suggested in section five, we can see this principle as Descartes's way of answering the question, "Where does what comes to be come *from*?" And, as I suggested, Descartes answers as he does in part because he is rejecting the Aristotelian analysis of change, with its complex structure of form, end, matter, and mover.

Descartes rejects the traditional answer, but not the the question. It should perhaps, then, be unsurprising that his own novel answer generates its own new set of problems and questions.

University of California, Berkeley

NOTES

[1] Descartes (1972), vol. 1, p. 162; Descartes (1897–1910), vol. 7, p. 40.
[2] Descartes (1972), vol. 1, pp. 162 ff.; Descartes (1897–1910), vol. 7, pp. 41 ff.
[3] Descartes (1972), vol. 1, p. 191; Descartes (1897–1910), vol. 7, p. 79.
[4] Descartes (1972), vol. 2, pp. 34–35; Descartes (1897–1910), vol. 7, p. 135.
[5] Descartes (1972), vol. 1, p. 191; Descartes (1897–1910), vol. 7, p. 79.
[6] Descartes (1970), p. 251.
[7] See Descartes (1972), vol. 1, pp. 289–293; Descartes (1897–1910), vol. 8, pp. 315–319.
[8] Descartes (1972), vol. 1, p. 187; Descartes (1897–1910), vol. 7, pp. 74–75.
[9] Descartes (1972), vol. 1, p. 197; Descartes (1897–1910), vol. 7, p. 87. See also Descartes (1972), vol. 1, p. 289; Descartes (1897–1910), vol. 8, p. 316.
[10] In her first letter, Elizabeth refers only to the action of mind upon body (Descartes (1897–1910), vol. 3, pp. 660–661), but later she makes it clear that her question applies to the action of body upon mind as well (*ibid.*, p. 685). Descartes clearly was assuming that Elizabeth meant to be asking the more general question in her first letter.
[11] Descartes (1970), p. 137; Descartes (1897–1910), vol. 3, p. 664.
[12] Descartes (1972), vol. 1, p. 295; Descartes (1897–1910), vol. 8, p. 322.
[13] Descartes (1970), p. 139; Descartes (1897–1910),vol. 3, pp. 667–668.
[14] Descartes (1897–1910), vol. 3, p. 684.
[15] Descartes (1970), p. 143; Descartes (1897–1910), vol. 3, p. 694.
[16] Descartes (1897–1910), vol. 4, p. 2.
[17] E.g, Descartes (1972), vol. 2, p. 34; Descartes (1897–1910), vol. 7, p. 135.
[18] Descartes (1972), vol. 2, pp. 34–35; Descartes (1897–1910), vol. 7, p. 135; cf. Descartes (1972), vol. 1, p. 162; Descartes (1897–1910), vol. 7, pp. 40–41.
[19] Descartes (1972), vol. 2, p. 35; Descartes (1897–1910), vol. 7, p. 135; translation mine.
[20] See the examples mentioned in section one.
[21] E.g., Descartes (1972), vol. 1, p. 190; Descartes (1897–1910), vol. 7, p. 78.
[22] E.g., Descartes (1972), vol. 1, pp. 193–194; Descartes (1897–1910), vol. 7, pp. 82–83.
[23] Descartes (1972), vol. 1, p. 138; Descartes (1897–1910), vol. 7, p. 8.
[24] Descartes (1972), vol. 1, p. 162; Descartes (1897–1910), vol. 7, p. 41; and Descartes (1972), vol. 2, p. 11; Descartes (1897–1910), vol. 7, p. 103; and Descartes (1970), p. 235; Descartes (1897–1910), vol. 5, p. 221
[25] Descartes (1972), vol. 1, p. 191; Descartes (1987–1910), vol. 7, p. 79; and Descartes (1972), vol. 2, p. 11; Descartes (1897–1910), vol. 7, p. 103–104.
[26] Descartes (1972), vol. 1, p. 191; Descartes (1897–1910), vol. 7, p. 79.
[27] Descartes (1972), vol. 2, p. 53; Descartes (1897–1910), vol. 7, p. 161.
[28] Descartes (1972), vol. 1, p. 294; Descartes (1897–1910), vol. 8, p. 320.
[29] Descartes (1972), vol. 1, p. 295; Descartes (1897–1910), vol. 8, p. 322.
[30] Descartes (1972), vol. 2, p. 25; Descartes (1897–1910), vol. 7, p. 123.
[31] Descartes (1972), vol. 2, p. 34; Descartes (1897–1910), vol. 7, p. 134.
[32] Descartes (1972), vol. 1, pp. 254–255; Descartes (1897–1910), vol. 8, pp. 40–41.

[33] Descartes (1972), vol. 1, p. 296; Descartes (1897–1910), vol. 8, pp. 322–323.

[34] This, I think, is part of what Descartes means when he writes,

> . . . we perceive that sensations such as pain are not pure thoughts of a mind distinct from a body, but *confused* perceptions of a mind really united to one. For if an angel were in a human body, he would not have sensations as we do, but would simply perceive the motions which are caused by external objects.

(Descartes (1970), pp. 127–128; Descartes (1897–1910), vol. 3, p. 493; emphasis added.)

[35] Descartes (1972), vol. 1, p. 296; Descartes (1897–1910), vol. 8, p. 323.

[36] Descartes (1972), vol. 1, p. 248; Descartes (1897–1910), vol. 8, p. 33.

[37] Descartes (1972), vol. 1, p. 347; Descartes (1897–1910), vol. 11, p. 355; emphasis added.

[38] Descartes (1970), p. 108; Descartes (1897–1910), vol. 3, p. 418.

[39] Descartes (1972), vol. 1, pp. 442–443; Descartes (1897–1910), vol. 8, pp. 358–359.

[40] E.g., Descartes (1972), vol. 1, p. 160; Descartes (1897–1910), vol. 7, p. 38.

[41] Indeed, on Descartes's view it is difficult to understand how anything but a substance can migrate. Descartes does sometimes deny that a mode can migrate from one substance to another. See section five below for a discussion of the migration of motion.

[42] Descartes (1965), p. 107; Descartes (1897–1910), vol. 6, pp. 140–141.

[43] The term 'present' cannot mean 'present for inspection', because Descartes holds that the mind does not inspect the brain when it senses (Descartes (1972), vol. 1, p. 342; Descartes (1897–1910), vol. 11, p. 346). Nor could "the senses present *X* to the mind" be interpreted as a sufficient condition for sensing. Consciousness is a contribution of the mind, not of what the senses present.

[44] Descartes (1972), vol. 1, p. 444; Descartes (1897–1910), vol. 8, p. 360; emphasis added.

[45] Descartes (1972), vol. 1, p. 443; Descartes (1897–1910), vol. 8, p. 359; emphasis added.

[46] Descartes (1972), vol. 1, p. 443; Descartes (1897–1910), vol. 8, p. 359.

[47] Descartes (1897–1910), vol. 8, p. 65.

[48] *Ibid.*

[49] Locke (1975), p. 311 (II, xxiii, 28).

[50] Descartes (1897–1910), vol. 11, p. 11.

[51] Descartes (1970), p. 64 (slightly altered); Descartes (1897–1910), vol. 2, p. 543.

[52] Descartes (1897–1910), vol. 8, p. 66.

[53] Descartes (1970), p. 258; Descartes (1897–1910), vol. 5, p. 404.

[54] More had tried to show that "the same substance can occupy different amounts of space at different times" (Descartes (1970), pp. 256–257; Descartes (1897–1910), vol. 5, p. 402) by arguing that "numerically the same motion can occupy now a larger body, now a smaller one, on [Descartes's] own principles" (Descartes (1970), p. 257; Descartes (1897–1910), vol. 5, p. 402). One would expect Descartes to criticize this argument by denying that the same motion could migrate from one body to another. But he seems content to let that part of the argument stand and instead denies that More has demonstrated anything about *substances,* since his argument is only about modes (quantities of motion). I shall return to the substance-mode distinction later.

55 This mistake is just as easy to make in French and Latin as it is in English.
56 See the examples mentioned in section one.
57 Cf. Descartes (1972), vol. 2, p. 231; Descartes (1897–1910), vol. 7, p. 387.
58 Descartes (1972), vol. 2, pp. 159–160; Descartes (1897–1910), vol. 7, pp. 288–289.
59 Descartes (1972), vol. 1, p. 246; Descartes (1897–1910), vol. 8, p. 31.
60 Descartes (1972), vol. 1, p. 266; Descartes (1897–1910), vol. 8, p. 53.
61 Descartes (1972), vol. 1, p. 266; Descartes (1897–1910), vol. 8, p. 54.
62 Descartes (1970), p. 258; Descartes (1897–1910), vol. 5, p. 404; emphasis added.

BIBLIOGRAPHY

Descartes, R.: (1897–1910), *Oeuvres*, ed. C. Adam and P. Tannery, 11 vols., Cerf, Paris.

Descartes, R.: (1965), *Discourse on Method, Optics, Geometry, and Meteorology*, ed. and trans. P. Olscamp, Bobbs-Merrill, New York.

Descartes, R.: (1970), *Philosophical Letters*, ed. and trans. A. Kenny, Clarendon Press, Oxford.

Descartes, R.: (1972), *Philosophical Works*, ed. and trans. E. Haldane and G. Ross, Cambridge University Press, Cambridge, England.

Locke, J.: (1975), *An Essay Concerning Human Understanding*, ed. P. Nidditch, Clarendon Press, Oxford.

DOROTHEA FREDE

HEIDEGGER AND THE SCANDAL OF PHILOSOPHY

The topic of this paper is a criticism ventured by Heidegger against Kant's attempt to prove the objective reality of the objects of experience. This proof is given by Kant in the *Critique of Pure Reason* as a refutation of psychological idealism.[1] It has been further elucidated by Kant in the preface to the second edition, where he calls it "a scandal of philosophy and of human reason in general that the existence of things outside us must be accepted merely on faith, and that if anyone thinks good to doubt their existence, we are unable to counter his doubt by any satisfactory proof."[2] And Kant expresses the hope to have made up for this scandal by a *strict* proof.[3] It runs as follows:[4]

Thesis
The mere, but empirically determined, consciousness of my own existence proves the existence of objects in space outside me.
Proof
I am conscious of my own existence as determined in time. All determination of time presupposes something *permanent* in perception. This permanent cannot, however, be something in me, since it is only through this permanent that my existence in time can itself be determined. [But this permanent cannot be an intuition in me. For all grounds of determination of my existence which are to be met with are representations; and as representations themselves require a permanent distinct from them, in relation to which their change, and so my existence in time wherein they change, may be determined.][5] Thus perception of this permanent is possible only through a *thing* outside me and not through the representation of a thing outside me; and consequently the determination of my existence in time is possible only through the existence of actual things which I perceive outside me. Now consciousness in time is necessarily bound up with consciousness of the possibility of this time-determination; and it is therefore necessarily bound up with the existence of things outside me, as the condition of time-determination. In other words, the consciousness of my existence is at the same time an immediate consciousness of the existence of other things outside me.

Since unanimity among scholars on the interpretation of this proof, its stringency and function within the *Critique*, has not been achieved (in fact, no two interpretations I have looked at are quite in agreement[6]), I will not even try to give an interpretation of my own but merely want to draw attention to some of the problems involved in the

A. Donagan, A. N. Perovich, Jr., and M. V. Wedin (eds.), Human Nature and Natural Knowledge, 129–151.

proof so that we can see more clearly what Heidegger refers to in his criticism. Various questions come to one's mind when one looks more closely at Kant's proof.

What kind of "proof" is it? Is it a deduction, and if so, is it an empirical, metaphysical, or transcendental one? Or, to use Kant's own terminology, is he dealing with a *quaestio iuris* or a *quaestio facti*?[7] Since he expresses himself in a quite general way, i.e. that experience of one's own self necessarily presupposes the experience of the external world, the former seems to be indicated. And, indeed, the often arbitrary or even chaotic order of our internal psychological states and changes seems to be in need of a stable framework, a coordinate system, such that an objective experience of one's own self within time can be reached or else one would forever be lost in the maze of one's own representations. A proper order presupposes something permanent, something "to count on" to allow for the unity and objectivity of experience quite generally.

What would the sceptic say to Kant's "strict proof"? He would probably claim that Kant is begging the question and merely appealing to the *facticity* of our experience, an experience which is based on the coexistence of different kinds of appearances and wherein the relatively stable can be used to sort out the changing; but the *relatively* stable or periodically recurrent would be quite sufficient even if we lived in a more obviously Heraclitean world than our present one. The plausibility of Kant's argument seems to rest on the dualism of internal and external experience, a division which the sceptic may refuse to accept. Kant is mistaking, so the sceptic may maintain, his own basic tenets when he makes use of the concept of the permanent in his analysis of sense-experience. To do so must, of course, be tempting on account of the central role which the "permanent" plays in the transcendental deduction of the category of substance, the schematism of substance and the first analogy of experience. The sceptic who does not accept the division of empirical reality/transcendental ideality, however, may maintain that Kant can in fact only point to the receptivity of our sense-experience, while any further insistence on more than a relative permanence contained in the external appearances would be a sin against the Kantian spirit for it would be backed up only by the assumption of something like the "thing in itself." Whether Kant's "strict proof" can be defended against such criticism of the sceptic empiricist cannot be discussed here. I just wanted to mention some of the sensitive points and now want to turn to Heidegger's criticism of Kant's dealing with the "scandal of philosophy".

Heidegger discusses Kant's proof in connection with the problem of the concept of reality in §43 of *Being and Time* under the title "Dasein, Worldhood and Reality."[8] His objections to Kant's treatment of the problem of the reality of the external world, as we shall see, are not addressing themselves so much to the actual proof itself but rather to the *demand* for such a proof.[9] These objections are worth studying because not only do they seem justified, in my opinion, if one accepts Heidegger's analysis of the notion of being, but — if properly understood — they also help us to achieve a better understanding of Heidegger's own enterprise of *Being and Time*. Heidegger never denied his indebtedness to traditional philosophy, and his treatment and criticism of his predecessors mirror the development of his own thought. This fact is often overlooked by modern interpreters who conceive of Heidegger only as an existentialist philosopher rather than, as he himself preferred, as the originator of a general ontology based on a new understanding of the meaning of *Being*.

What is the question of being? Heidegger is the only philosopher of whom one can say that one and the same question occupied him very early on till the end of his life.[10] As he frequently claims, what went wrong with traditional western philosophy was due to a misunderstanding of the meaning and importance of the question of being and, finally, forgetfulness of them altogether. As he claims right at the beginning of *Being and Time*, Plato and Aristotle still understood the problem of the meaning of being; this understanding gradually got lost in the history of ontology until "being" was regarded as the most universal, indefinable, emptiest, and self-evident of concepts, therefore neither needing nor allowing any further elucidation.[11] This neglect is the reason why Heidegger imposed the mission upon himself to "destroy the history of ontology"[12] and give back to the concept its old importance. And this mission explains why especially the younger Heidegger remained under the influence of traditional philosophy, its problems and attempted solutions, just as much as he criticized them.[13]

When we nowadays discuss the question of a thing's being (if we do that at all) we normally just mean whether it "can be found," "is real," "occurs somewhere," etc., and are usually not able to give any more concrete meaning to the term. "Existence," so we have learned from

Kant, "is not a predicate" and cannot be defined; its function is usually taken to be as wide as that of the existential operator. When we look at the ontology of Plato and Aristotle, as to the authorities to whom Heidegger appeals, we can see that Heidegger's complaints were not just the complaints of someone who was "das Land der Griechen mit der Seele suchend." Plato and Aristotle did indeed understand the notion of being in a sense different from the one described by my vague terms above. For them, "to be" was always "to be something"; there was no genuinely absolute use of "being". A thing's *being* was for them "τὸ τί ἦν εἶναι," "that which it was to be," or in other words, what it consists in, the thing's nature and essence. This use of being is not unfamiliar to us; for example, we may say of someone that he "feels questioned in his very being" and not mean that he feared for his life, but challenged in what he felt essential about himself.[14] In this sense, then, being may have as many different meanings as there are different kinds of natures.

Taken in this sense, the concept of being presents us with a metaphysical problem, for it could, then, have (a) indefinitely many meanings — if one held that every individual is essentially different from all others — or (b) as many as there are different kinds of entities. Plato and Aristotle each tried solutions in their own way; for Plato (at least in his middle period) there must have been as many meanings as there were forms and the form of the Good provided the overall unifying metaphysical principle. Aristotle tried to solve the problem by assuming as many meanings as there are categories: for whatever is, has being in the sense of being a substance, or a quality, quantity, relation, etc., of some sort. The members of different categories do then have a fundamentally different kind of being; the concept of "being itself" when transcending the categories would, as Heidegger claims, be entirely devoid of content. It seems, however, that Aristotle did not let matters rest with such a homonymy or analogy of the meaning of being in the categories but eventually introduced a device which does establish some kind of unity of being. This device has been called "focal meaning"[15]; it assigns the central meaning of *being* to *substance* and a secondary one to all the *qualities* in the other categories in the sense that whatever else exists, exists by inherence in a substance. Just as there are many different meanings of "healthy" (such as a healthy color, healthy food, healthy climate) which focus on a common central concept, namely "health", so there are many meanings of "being" which focus on the central concept of "substance". And on this focus, if on anything, the hope for metaphysics as a master-science is based.[16]

I do not know whether Heidegger ever took notice of this change and refinement in Aristotle's doctrine of being. But it is clear that it was the alleged *homonymy of being* which attracted his attention at a very early age when someone presented him with a copy of Brentano's dissertation, *On the Several Senses of Being in Aristotle.*[17] The problem interested young Heidegger at once and kept him under its spell for his lifetime.[18] As Heidegger saw it, the history of philosophy was to a large extent determined by this question, and different philosophers tried different answers. Descartes's division of all being into *res extensa* and *res cogitans* is a witness to this, as is Kant's attempt to give a complete account of the categories of human reason which constitute our understanding of the different kinds of entities as presupposed and "given" in experience.[19]

And Heidegger's own task, as he had set it out for himself in *Being and Time,* was designed to work out a new analysis of the manifold meaning of being and its unification and foundation in the notion of *time* as the horizon of all understanding of being. As it stands, Heidegger never fulfilled his task to his own satisfaction and therefore never published *Being and Time,* vol. I, division iii, nor the whole of vol. II.[20] What we do have was originally only meant as the first step in Heidegger's fundamental ontology, namely the analysis of that entity which has in its very nature a concern for and understanding of the question of being — man ("Dasein") himself. I will now try to give a very brief résumé of Heidegger's analysis of man's existence insofar as this is necessary for our understanding of his discussion of the concept of "reality" and his denial that anything like the proof of the existence of the external world is necessary or possible.

Man, according to Heidegger, is never an isolated entity, an independent substance, that eventually can take up contact with the things outside itself by the aid of its perceptive and cognitive faculties — as depicted in traditional philosophy. Instead, there is always already some kind of understanding of being, and what Heidegger does, according to his own view, is merely to unfold what this understanding of being already contains and to provide a fundamental analysis of the structure which underlies this normally vague and inarticulate understanding. This understanding always already displays, as Heidegger explains at great length, an intrinsic relationship to both the self and the world. The relatedness to the self ("in-each-case-mineness") and to the world ("being-in-the-world") together constitute human existence, and what Heidegger does in *Being and Time*, as we have it, is just to work out what this being-in-the-world consists in, how we encounter the things in the world, and how we understand ourselves in this concernful dealing.[21] In order to avoid all prejudices about the constituents of human nature which we may have inherited from traditional philosophy, Heidegger starts in his analysis with a mode of being that we are familiar with before any kind of theorizing and that should be immune to cultural and historical difference, namely our *average everdayness,* which is what characterizes us normally and for the most part. This attempt to give a phenomenological description of man in the most low-keyed way possible also explains part of the much maligned Heideggerian terminology. For since he takes it that even our pre-ontological concepts are heavily (yet inconspicuously) laden with meaning allegedly "natural" but in fact derived from tradition, he often coins his own.[22] The importance Heidegger attributes to the approach via the state of everydayness and to a clarification of our implicit understanding is due to the fact that for Heidegger the meaning of being is constituted by this very understanding and is to be elicited from it. The structure of being cannot be elicited in any other way, and Heidegger hopes to be able to unearth gradually what is behind the everyday understanding. For though we have no other access to "being," this does not mean that we are already conscious of the structure of being; in fact, a lot of archaelogical work has to be done before Heidegger can bring to light the *a priori* conditions of human understanding which underlie our everyday existence.[23]

The upshot of this analysis can only be indicated here. 1. Man at first does not approach the things around him as mere indifferent "objects" given by sense-perception but always in a practical context which determines their function. They are "equipment" for something and their readiness-to-hand is their way of being.[24] Our whole environment, in fact, can be compared to a huge workshop wherein everything has its use, its purpose, its place. This "practical meaningfulness", as I prefer to translate "Bewandtnis" (rather than by "involvement"), in its totality constitutes the intelligibility of our everyday world, lets us find our way around in it, and assigns things their place and being. 2. Heidegger does not deny that things can also be encountered without any such practical reference or function, as merely "occurring" objects ("occurrence" seems to be a much less confusing translation of "Vorhandenheit" than "presentness-at-hand").[25] He claims, however, that leaving aside the practical aspect under which the things are usually encountered by us offers only a secondary way of approaching the things in our world. It is in this way that theoretical investigation studies its objects and their attributes; and because philosophers, like scientists, leave aside the practical meaningfulness of the everyday world they have taken pure indifferent occurrence ("reine Ding-Vorhandenheit") as the natural ontological determination of things.[26] 3. "Readiness-at-hand" and "occurrences" are not the only two ways we understand entities in the world. The being of our fellow-men is based on a yet different way of understanding, which Heidegger calls "being-with," claiming that we have in our very being a definite relationship to, and dependence on, other entities like ourselves.

I cannot go into a further discussion of Heidegger's analysis of our everyday existence, how our being-in-the-world which contains readiness-at-hand, mere occurrence, and being-with constitute a meaningful whole, and how all this is ultimately explicable in terms of *time.* It is sufficient for our understanding of Heidegger's discussion of Kant's proof to see what Heidegger's categories of being are and how he derived them, i.e., that they are the elements of our understanding of being. Although this introduction may be too brief it should be clear where and why the question of the reality of the things in themselves might be brought in. One just cannot help wondering if it matters to Heidegger whether the things we encounter in our concernful dealing with the world can be proved to exist or not. Would the concept of our being-in-the-world be the same if a sceptic could maintain that the entities dealt with might be only the product of our imagination, hallucination, dream, or, *horribile dictu,* that we might be brains in vats? Furthermore, one wonders, does it make any difference to Heidegger's ontology whether we can or cannot accept as certain that the things really do have the properties which we attribute to them. In other words, do we encounter them as they are *in themselves*?

Heidegger seems to have anticipated that his readers might see a difficulty here and therefore included a discussion of the problem of *reality* in *Being and Time.*[27] He tries to point out that the whole question of whether the existence of the external world can be demonstrated is, if properly understood, a pseudo-problem, which could only arise on the basis of a misconceived ontology and an unquestioned concept of "reality." Against Kant he ventures to say that the real scandal consists not in the *lack* of a proof of the existence of things outside us, but rather in the fact that such proofs are still *demanded.* And he repeatedly asserts that if our ontological presuppositions were properly understood it would be clear that such a proof is neither necessary nor possible.[28] Yet, although Heidegger discusses the whole matter at some length, it is not immediately clear *why* he thinks that a properly worked out ontology makes any proof of the existence of the external world superfluous.

A first scrutiny of Heidegger's argument here and elsewhere in *Being and Time* seems to suggest that Heidegger only wanted to discard this concept in a very special sense, a sense which is not necessarily presupposed by all philosophers who demand a proof for the reality of the external world. Heidegger, in fact, thinks that the very expression "re-ality" ("*res*") shows what kind of ontology is behind such a demand: "Real" things are totally independent objects which merely occur alongside one another and with us.[29] As Heidegger sees it, this concept of reality which is based on the notion of "substantiality of independent self-subsistent objects"[30] led to all the difficulties which have troubled philosophers for centuries. All these difficulties, such as the question how the subject can have an object, how there can be any connection between the psychical and the physical, the material and intelligible, could only arise because the whole phenomenon of our being-in-the-world and the ontological status of things in the world was passed over. And they were passed over precisely because reality was conceived as Dingvorhandenheit or the indifferent occurrence of independent things.

Now, even if one accepts Heidegger's criticism and his analysis of our being-in-the-world and the ontological status of the things therein and promises never again to use the term "reality" in the criticized sense, the question would still remain whether the entities encountered in our dealings with the world are in fact "there," i.e. whether they exist in a strong sense, and whether we can have any certainty about them and their properties. Furthermore, there is the question of the nature of the "things in themselves" and whether we can have access to them and describe them in the way Heidegger claims for phenomenological ontology.[31] For even if we agree with Heidegger that we always already have a world in which we live and in whose terms we understand ourselves and the other entities, it would seem to make all the difference if we can suppose that the entities encountered are *there* (and not only on faith) to be encountered and are not just the products of our disposition to make up such a world and fill it with what is "Zuhanden," "Vorhanden," and "Mitsein." And it is hard to imagine that Heidegger does not see the difference.

Now, one plausible explanation is that for Heidegger the very fact that we have a world wherein we encounter the various things in the way described guarantees that the things we deal with are there to be encountered. In this case Heidegger himself would maintain a "realist" position, a position of "critical" realism, however, which is not based on the assumption of indifferently occurring substances.[32] On this interpretation Heidegger would hold the view that if there were no things to be encountered there could be no everydayness as he described it. For we could not move around in various places, deal with things ready-to-hand for this or that purpose, find those things either fit or unfit for our purposes, etc. A being-in-a-dream-world, Heidegger might claim, would be totally different from our being-in-the-world in everydayness. Thus, if we accepted his phenomenological description as fitting our intuitions of everdayness, we would also have to accept for certain that there *are* all those inner-worldly entities.

A defender of a "realist" interpretation of Heidegger can count on considerable evidence for support. First Heidegger's own terminology seems to suggest it, e.g., when he calls it the task of phenomenology merely to "'free" things, "set them free in their being," "let them be seen in their being," "disclose being," etc.[33] Besides his terminology there are other indications. In the section where he discusses Kant's proof and the notion of reality he concedes that even if there no longer were any human beings who could encounter other entities in the way described, there still would *be* those entities. "But the fact that reality is ontologically grounded in the Being of Dasein does not signify that only when Dasein exists and as long as Dasein exists can the real be as that which in itself it is."[34] And in the same context he admits that even after any understanding of Being has ceased to exist, "it can indeed be said that then entities will still continue to exist." Thus Heidegger's realism seems to be reduced to a disagreement about what "to be real" means, and one might want to assume that

Heidegger would accept the realist's position as soon as he gives up the old notion of "substantiality" and "objective existence" as the basis of his ontology.

This view can be fortified further by the fact that Heidegger himself states that there is a *doxographical* agreement between his ontology and the assumption of the realists, and "in so far as this existential assertion does not deny that entities within the world are occurrent (vorhanden) it agrees doxographically with the thesis of realism in its results."[35]

If Heidegger held such a position of "reformed" realism, it would seem understandable that he rejects Kant's proof, claiming that "if Dasein is understood correctly, it defies such proofs, because in its Being it already *is* what subsequent proofs deem necessary to demonstrate for it."[36] This and many similar passages suggest that Heidegger holds that since we have a world, the existence of the entities within the world can be deduced (in the Kantian sense) *a priori* from the character of our existence which thus defies or denies the very demand for such a proof.

It is easy to see, however, that if Heidegger did hold such a realistic position with respect to the existence of the external world, his justification for the rejection of Kant's proof would not do.

For it would not be clear why he claims that such a proof would be neither necessary nor *possible.* At best he could maintain that it would not be *necessary,* because the existence of the inner-worldly things can be inferred already from our being-in-the-world. But this would not make the proof *impossible,* for Heidegger's arguments, as reconstructed by me, would be something like a proof.

Furthermore, though I cannot fully discuss the problem of imaginary worlds, it seems possible that all the phenomena which Heidegger takes to be characteristic of our being-in-the-world (the handling of things ready-at-hand successfully or unsuccessfully; the being with others; the theorizing about merely occurring things; etc.) could somehow be reproduced in a dream-world as mere creations of our imagination.

What seems to be even more important, however, is the fact that according to Heidegger all the structures of our being-in-the-world are founded on our nature, our understanding. Consequently, one would assume that for Heidegger other entities with differently structured minds would have a different world, differently furnished — if they have a world at all. It is clear that this must present a difficulty to a *realist* Heidegger. The acceptance of *any* kind of world would, perhaps, still allow the inference that there exist some things outside, whether outside of us or outside of the other intelligent entities. But this does not answer the questions (a) whether things really exist and (b) whether they can be known "in themselves" as Heidegger claims.[37] For how could there be an in-itself if differently structured intelligent beings could encounter totally different entities or encounter them *as* totally different?

Since Heidegger emphasizes the fact that he is describing the having-a-world of a particular kind of entity, i.e., man, he cannot have simply overlooked this point, even though he never discusses the possibility of different sorts of minds. It seems therefore unlikely that he held that the fact that we have a world guarantees that we have access to things unqualified. At best it could be said that they somehow "allow" us to encounter them in such a way.[38] When Heidegger straight-forwardly rejects any proof he must therefore have more in mind than the fact that we move around with relative certainty in our everyday-world. To do this, "faith" in the sense criticized by Kant would be sufficient.

There is, indeed, much in the text which runs contrary to an interpretation which ascribes a "reformed" realism to Heidegger. For example, he mentions it as one of the points in which idealism is superior to realism that the idealist at least locates being and reality within consciousness. And it seems that Heidegger disagrees with the idealist only because the latter has omitted any attempt to give a proper ontological analysis of "consciousness" and its special relation towards the concept of being.[39] I have to admit, in fact, that I had to treat the text in quite an eclectic way in order to construct a realist interpretion of Heidegger. For at the same time that he asserts a certain *in*dependence for the entities outside us he also stresses a *de*pendence of their being on our understanding. "Of course only as long as Dasein *is* (that is, only as long as understanding of Being is ontically possible) 'is there' Being. When Dasein does not exist, 'independence' 'is' not either, nor 'is' the 'in itself'."[40] And Heidegger explains further that "Being (not entities) is dependent on the understanding of Being; that is to say, reality (not the real) is dependent upon care. . . ."[41] There are many similar passages which confirm that Heidegger wanted to maintain a dependence of the being of all things on human understanding.[42]

Admittedly, for those not familiar with Heidegger's philosophy the distinction between entities and their Being and his talk of the latter's dependence and the former's independence must sound quite puzzling. I hope to be able to work out the rationale behind that distinction.

When I mentioned the importance of the notion of Being for Heidegger's philosophy I claimed that he takes it in a meaning similar to the one presupposed by Plato and Aristotle, that is, that to be always is to be something, that a thing's being is determined by what-it-is.[43] That is to say, for them "being" did not confer such a vague sense of "existence" that all sorts of different entities would be regarded as "existing" in the same sense. As Aristotle stresses time and again: "τὸ ὂν λέγεται πολλαχῶς. . . ."[44]

For Heidegger, and this is an important point, it must further be understood that a thing's being is what is intelligible in its "what it is." This is not to say that for Plato and Aristotle intelligibility is excluded: in pre-Kantian, or better pre-sceptical times there was just no question of a possible discrepancy between being and intelligibility and therefore also no question of whether a thing's being *depended* on its intelligibility or not.

Let me illustrate what this means for Heidegger: When he defines man's everyday-being as "being-in-the-world," what he means is that this is what is contained in our *self*-understanding (and not just a matter of fact), that is, that we understand ourselves as being in our very essence related to the world. And this is the principle which allows Heidegger to determine what the being of all the entities encountered within the world is: because this is how we understand them. We understand something either as being for some use, or as existing in its own right — such as the entities occurring in nature — or as another such as ourselves (if we are able to see it in this light).

Heidegger himself often does not express himself in so simple a way, but there are several passages which make it quite clear that this is what he has in mind: "When entities within-the-world are discovered along with the Being of man [Dasein], that is, when they have come to be understood, we say they have meaning. But that which is understood, taken strictly, is not meaning but the entity, or rather, its Being."[45] And this, as mentioned earlier, is Heidegger's mission: to reawaken our understanding of this meaning of being. "In the question which we are to work out, what is asked about is Being — that which determines entities as entities, that on the basis of which entities are already understood."[46]

Once this point about Heidegger's philosophy is understood, many other points become clearer and also more acceptable; e.g., his talk about man's "disclosedness," the notion that we have a "lumen naturale" which we carry with us such that we can bring light to wherever we are and to whatever we are doing — and to ourselves in the situation, too.[47] And it also becomes clear why Heidegger sees himself as the heir of Parmenides, the first philosopher who emphasized the connection between thought and being.[48]

What is important for the present discussion, however, is that this notion of being allows Heidegger to maintain that there *is* no being unless there is some entity who has an understanding of being; in other words, every thing's being depends on the structure of the entity which has this understanding of its being. This explains why Heidegger always emphasizes the fundamental difference between the things and their being. If a thing's being consists in its *meaning* then it only has a being when there is someone for whom this *is* its meaning. And it becomes clear why the entity's being but not the entity itself depends on understanding: One and the same thing may for one person *be* a rusty tool, for another a piece of modern art. We all know the frustration over ontological debates of this kind where we cannot agree on the object's *being* (leaving its physical properties aside).[49] In this context it also becomes intelligible why Heidegger can claim, what at first sight seems quite unintelligible, that we encounter things in everyday-life as they are *in themselves*.[50] The ready-at-hand just *has* this in-itself that one can hammer with it, cut, read, etc. It can be used in a definitive way and is always already understood in terms of its "practical meaningfulness." This usability is, on the other hand, not something which we can attribute to the equipment at will. There are often painful limits to our license: A hammer is too heavy, a nail too short, a screw-driver missing.

Heidegger's position can, of course, be called subjectivistic and he in fact acknowledges this at various stages of his analysis[51] by asserting,

> Of course, only as long as Dasein is, is there Being. When Dasein does not exist, "independence" is not either, nor is the "in itself." In such a case even entities within the world can be neither discovered nor lie hidden. In such a case it cannot be said that entities are, nor can it be said that they are not ...[52]

But the term "subjective" can only be applied if the qualification is made explicit that it is the "what the things are understood *as*" which depends on the subject, viz. on the *a priori* conditions of his mind while sheer arbitrariness is excluded. Whether Heidegger's position with respect to *what* he states as the basic structure of our being-in-the-world is plausible, whether his fundamental analysis is acceptable is quite another matter and cannot be dealt with here.

From all this it should be clear how Heidegger's rejection of Kant's proof should be understood: things have an "in themselves" only if there is some understanding within which they are what they are. The question what they are apart from this meaning turns out, then, to be senseless. Thus, there is no "in itself" if one transcends the entity who has this understanding, that is, there is no such thing as "indifferent occurrence," "substantiality," "autonomy," etc., if there is nobody who conceives of things in these terms. And the question whether things really do have the physical qualities they seem to have to us would have to be answered in the same way: there are physical qualities only if there is someone who has a concept of "physical qualities." Heidegger illustrates this by referring to Newton's laws. "To say that before Newton his laws were neither true nor false cannot signify that before him there were no such entities as have been uncovered and pointed out by those laws. Through Newton those laws became true; and with them, entities became accessible in themselves to Dasein."[53]

We can now also understand what Heidegger meant when he said that "doxographically" his interpretation of the things-in-the-world agrees with the thesis of realism that the external world exists as "occurring" (vorhanden). There is, in fact, only a verbal agreement. For Heidegger "Vorhandenheit" (occurrence) does not mean existence independent of any understanding (which is what the realist is trying to establish) but "indifferent occurence" *in* our understanding. For it is the ontological status *we* ascribe to some kinds of entities, e.g., in the case of things in nature it may mean that they can be found, studied, admired, etc., that they last for some time, some longer, others shorter, that they behave in a certain way, occur and flourish under certain conditions. We can say that they occur "indifferently" because they seem to posses all these attributes whether or not someone takes any notice of them. With things ready-at-hand it is different. They do lose their ontological status as soon as there is nobody who *could* make use of their practical meaningfulness or at least thinks that they have their functions.

This distinction is behind Heidegger's otherwise rather paradoxical remark that even after mankind has ceased to exist, though there will be no more being, there still will be the entities; "But now, as long as there is an understanding of Being and therefore an understanding of 'occurrence' it can indeed be said that *in this case* entities will still continue to be."[54] The point is really quite simple. Since the things' occurrence, as *we* understand occurrence, does not depend on our noticing or using them, it is clear that those things which apparently last for a long time will still "occur" after all of us have closed our eyes for good. To express it in a different way: *ontically* such things do not depend on us and can be manipulated only as far as our powers reach; *ontologically* their being or essence does depend on us because that is what "being" means!

Given the basis of Heidegger's concept of being it seems, then, justified to reject the demand for a proof of the existence of the external objects. Such a proof is not necessary because the thing's being is already "given" or constituted by our understanding; it is not possible because there is no being which *means* anything aside from the one which we already understand. The only thing one can reasonably try to do is to see whether *ontologically* our understanding is well worked out, whether we have the right categories and criteria for determining them. Then there might also be the ontical (factual) problem whether we are actually dreaming, hallucinating, etc. The modern scientist may indeed find it sometimes embarrassing to have to classify his objects, not just ontically, but ontologically too. What kind of things are quarks, for instance? Are they hypothetical entities but the subject of empirically verifiable or at least falsifiable theories? But these problems are not what is the issue when a proof of the existence of the external world is demanded.[55]

It would seem, however, that this rejection of the Kantian proof would only be accepted by philosophers who *share* Heidegger's conception of being. Others might want to maintain that it does make sense to use the term "exists" in a much wider sense, e.g., to distinguish everything there is in the external world from the products of our mind. And it might seem that in this weak sense "there is" is innocuous enough to make even Heidegger accept it alongside his strong concept of being. For he might have to admit that there is a difference between what can be verified externally and what depends only on our imagination (such as owls and rabbits on the one side, unicorns and chimaerae on the other). And this would bring back the old question as to what this distinction amounts to and how we can be sure that the external world is *there*.

I think that Heidegger would readily admit the difference between the imaginary and the external but would point out that this very distinction shows that it depends on our *understanding* of "being" in each case. And in order to give a proof of the existence of the external world we would have to know already what "being" means in the case of external objects, such as rabbits and owls, and how their being differs from that of the imaginary objects, such as unicorns and hydras. We would thus be in Heideggerian waters again, for we would clearly assume what we are going to demonstrate.[56]

Let me illustrate this further, though the point of my example is only the question of *subsuming* a given entity under given kinds of being. For someone who does not know what "mammoths" are, the assertion "there are mammoths" does not have any precise meaning. As long as he does not know whether mammoths are a religious sect, a kind of garment, a type of car, or a species of animals now extinct, he will not know what he is asked to do if someone demands of him to prove that there are mammoths. Only after consulting an encyclopedia and finding out what they are, where they have been found, where they lived, where one can find their remains, does the question make any sense. Before that he will feel as embarrassed as Pooh Bear was when asked to search for Small, because with Rabbit's friends and relations one never knew what kinds of animals they were and whether one ought to be looking at the top of a tree or in the petal of a buttercup.

The difficulty of someone asked to give a proof of a vague kind of existence without any precise kind of knowledge seems to be quite similar: The only kind of proof one could reasonably demand is either that a particular kind of entity in *fact*, ontically, has this or that being, or that our ontological classifications make sense. The latter is what the phenomenologist is trying to do.

I will now, finally, try summarily to state what inconsistency Heidegger finds Kant to have committed when he demands the proof for the existence of the external world. When Kant asks whether the external objects of experience are "real" he does not doubt that they are given to us in experience. They are, he asserts, immediately given to us.

Kant, of course, realized the "subjectivity" of the objective. As Körner expresses it, "there can be no *it* unless there is an *I* which could be aware of it and thereby of itself.[57] What Kant did not realize is that "objectivity" and "reality" must remain suspect as long as they are not further differentiated and that this is not achieved in his analysis of what is perceived by the senses and thought by reason. In fact, Kant merely appeals to an already established conception of the empirically *given* which centers around the notion of substance and pure "Dingvorhandenheit." Since this undifferentiated notion is already presupposed and cannot be demonstrated, some Kantian scholars have suspected that he may have had more in mind than the already given, thereby transgressing the borders he himself tried to establish in the *Critique*, namely that human reason in its dogmatic use has to stay within the limits of experience. But the dogmatic use itself stands in need of qualification and must remain vacuous until one introduces the appropriate differentiations.

Heidegger does not deny that we can be wrong in our understanding of particular things (he is quite in agreement with Kant about this point),[58] nor that it is difficult sometimes to draw the line between the "real" and the imaginary. But these are ontic questions and settled by ontic means such as consistency, coherence, etc. Then there are ontological problems which are often tied together with factual ones. The whole situation is complicated by the fact that there is no presupposition-free understanding; we cannot escape Meno's contention that all inquiry is necessary circular because we cannot get to know something we do not already (at least implicitly) understand. As Heidegger sees it, all we can do is to get into the "hermeneutic circle" in the right way, by clarifying what has been understood indistinctly beforehand.[59]

Applied to Kant this means that if the concept of the "permanent" is an empirical one, it is already understood and cannot be used to prove that what seems to be given is in fact "there". If being has many meanings then "reality" cannot be treated as an unambiguous concept. What the many meanings are, and whether we find Heidegger's own analysis plausible, we cannot discuss here. The notion of "Vorhandenheit" seems especially to be in need of more refined distinctions, and the scientist may have to revise the concept in the course of his research all the time. *The* external world, taken as a whole, seems to be an elusive lady when we do not confine ourselves to the everyday-world. The philosopher is therefore well advised to revisit Heraclitus's kitchen once in a while.

Heidegger's criticism of Kant should not just be regarded in a negative sense. His main reproach is that Kant was simply not Kantian enough but still misguided by traditional philosophy, the notion of the "inner" versus the "outer," the "psychological" versus the "physical," "things" and their "properties."[60] That this is Heidegger's main concern comes more to the front in his two monographs on Kant which could not be discussed here.[61] A more penetrating analysis would show how Kantian (although in a critical sense) Heidegger himself was and understood himself to be in his earlier philosophy.

Postscript: It should be noted that Kant himself saw the need of a clarification of the notion of being, and that he did so in his discussion of the ontological argument for the existence of God, the very place where he contends that existence is not a predicate. And he later affirms that the notions of reality, substance, and causality, when attributed to the highest transcendental object of our understanding, can have no meaning. It is therefore understandable that Heidegger remains ominously silent about questions of God in his earlier "critical" philosophy while in his later philosophy some mystic divinity seems to be attributed to *Being* itself. Both the younger and the older Heidegger realized with Plato that there can be no definitive talk of what *is* beyond being in a meaningful sense, about what is "ἐπέκεινον τῆς οὐσίας".

Rutgers University

NOTES

1 Kant (1929), B 274–279.
2 *Ibid.*, B XXXIX–XLI.
3 *Ibid.*, B XXXIX.
4 *Ibid.*, B 275.
5 In brackets is the altered version suggested by Kant in the preface to the second edition, p. XXXIX.
6 I just want to enumerate the most relevant literature: Vaihinger (1981), Kemp Smith (1923), Paton (1936), Ewing (1938), Körner (1955), Kaulbach (1958–1959), Lehmann (1958–1959), Strawson (1966), Prauss (1971) and (1974), and Wilkerson (1976).
7 Kant (1929), A 84, B 116.
8 Heidegger (1962a), pp. 244–256.

9 *Ibid.*, p. 248.

10 See "My Way into Phenomenology" in Heidegger (1972), p. 81, for his early fascination with the problem. See also Heidegger's letter to W. Richardson, printed in the preface to Richardson (1963), pp. viii–xxiii.

11 Heidegger (1962a), pp. 21 ff.

12 See *ibid.*, pp. 41 ff.

13 In recent years more attention has been paid to the development of Heidegger's earlier years since his earlier work is now becoming accessible in the piecemeal edition of his Nachlass under the editorship of F. W. von Hermann.

14 As Heidegger insists, the most crucial wrong step in traditional philosophy was the separation of being and essence. Already the title of *Being and Time* indicates how Heidegger (in all due modesty) saw his task in history: to put together again on the horizon of time what Thomas Aquinas has separated in *On Being and Essence* (see (1962a), §1). (Sartre again tried to give a different turn to the history of philosophy by shifting the emphasis to *Being and Nothingness.*)

15 See Owen (1960).

16 See Aristotle, *Metaphysics*, 1003a33 ff. Cf. Ross (1953), p. 254, and Kirwan (1971), p. 1.

17 Brentano (1960).

18 See Heidegger (1972). See also the preface to the editions of (1962a) after 1953.

19 The question of the similarities and dissimilarities of the categories of being in Aristotle, Descartes, and Kant is a very difficult one and cannot be dealt with here.

20 We may forever remain in the dark about the reasons for Heidegger's "holding back" volume I, 3, and the entire volume II, since he gave explicit instructions not to include it in his Nachlass even though the manuscripts existed.

21 See Heidegger (1962a), chapters 2–4. These chapters are the most vivid and accessible of *Being and Time*. See Prauss (1977).

22 Very often it is just a question of getting used to the terminology, a task which is much easier in German than in English because there are peculiarities of connotations, etymologies (even if they are false) which are not noticeable in their English counterparts, such as "Dasein" for man, "Zeug" for equipment, "Bewandtnis" for involvement, and the untranslatable "Das Man."

23 See Heidegger (1962a), §§5–7, especially §7c, "The Preconception of Phenomenology."

24 The analysis in terms of "equipmentality" is one of the reasons why Heidegger is often compared to the American pragmatists. See Rorty (1976–1977); Wilshire (1977); and Zimmerman (1978).

25 The translation of "Zu*hand*enheit" and "Vor*hand*enheit" is a problem because the two words which look so much alike have had a different history in German. While "Zuhandenheit" kept the notion of "handiness," "Vorhandenbeit" lost it and does not imply nearness. One can speak of things' Vorhandenheit, e.g., of some star in some distant galaxy, or of a fish somewhere at the bottom of the sea, or some old relic in an attic. This is not conveyed by the translation "present-to-hand." I therefore prefer the translation of "indifferent occurrence" even though the etymology gets lost. I also prefer to translate "Bewandtnis" by "practical meaningfulness" because "involvement" is ambiguous. When we know a thing's "Bewandtnis" we know what "is the matter with it," e.g., in the case of Aladdin's magic lamp.

[26] Because "indifferent occurrence" designated the being of entities when one looks aside from their practical meaningfulness, Heidegger calls it a "founded mode" ((1962a, §13) of the understanding of the world. Ever since philosophers have moved away from Heraclitus's kitchen (Aristotle, *De partibus animalium,* 645a17), they have been oblivious to the primary, i.e., practical, meaning of things. Instead, they adopted the ontology of "substantiality" and indifferent occurring thinghood. This ontology is the result of the philosophers' fallenness: since they are preoccupied with "theoria," sheer indifferent "looking" at things, they have overlooked the primary givenness of the ready-to-hand. The scientists inherited this notion, and the little man in the street was much too much in awe of both philosophers and scientists to regard practical meaningfulness and everydayness as anything but trivial. It takes a lot of subtlety (although the philosophers could have learned something about it from Socrates) to realize that the hammer's in-itself consists in its hammering.
[27] Heidegger (1962a), §43, pp. 244–256.
[28] *Ibid.*, pp. 249 and 207.
[29] See *ibid.*, pp. 201 ff and 89 ff in criticism of Cartesian ontology.
[30] *Ibid.*, p. 22.
[31] *Ibid.*, pp. 49 ff.
[32] A realist, it seems, could easily accommodate the distinction of readiness-to-hand, indifferent occurrence, and being-with. In fact, almost any kind of metaphysical system could adopt this division.
[33] Heidegger (1962a), pp. 87 ff; 32 et passim.
[34] *Ibid.*, p. 255; cf. pp. 227–228.
[35] *Ibid.*, p. 207.
[36] *Ibid.*, p. 249; cf. p. 246.
[37] *Ibid.*, pp. 245–246.
[38] One is strongly reminded of Kant's contention that our senses in their receptivity are somehow affected by whatever the unknown and unknowable thing-in-itself may be – and what the difficulties of this position are.
[39] Heidegger (1962a), p. 246.
[40] *Ibid.*, pp. 255–256.
[41] *Ibid.*, p. 255.
[42] See *ibid.*, p. 228: "Entities *are*, quite independently of the experience by which they are disclosed, the acquaintance in which they are discovered, and the grasping in which their nature is ascertained. But Being *is* only in the understanding of those entities to whose Being something like an understanding of Being belongs."
[43] For a discussion of the predicative, copulative, and absolute use of 'to be', see Kahn (1973).
[44] See Aristotle, *Metaphysics*, 1003a33 and 1028a10.
[45] Heidegger (1962a), pp. 192–193; cf. pp. 198–199.
[46] *Ibid.*, pp. 25–26.
[47] *Ibid.*, p. 171.
[48] See *ibid.*, pp. 341–348 and 256 et passim.
[49] The three modes of "readiness-to-hand," "indifferent occurrence," and "being-with" are probably not the only ones. Heidegger mentions the problem of understanding of "nature" (see (1962a) §43 c). The question of the proper interpretation of works of art is explicitly discussed in "Der Ursprung des Kunstwerkes" in (1950), pp. 7–69.

Heidegger also mentioned the difficulty of classifying animals in a private discussion reported in Neske (1977).

[50] Heidegger (1962a), pp. 105, 106, 120–121, and 255.

[51] *Ibid.*, pp. 140–141 and 417 et passim.

[52] *Ibid.*, p. 255. Earlier Heidegger had given a very suggestive discussion of how objects can defy us, how they can become conspicuous, obtrusive, and obnoxious because they lose their handiness and become mere things. The question of how much one can change the ontological status of things becomes urgent in the question of what counts as art and pieces of art and how much liberty the artist can take. If Christo should finally manage to wrap up the "Reichstag" in Berlin is the whole thing a piece of art? What does this imply? One can see that the artist has an ontological problem just as much as the more or less sympathetic spectator does on his side. Legal problems arise here, too, and the Heideggerian knows why.

[53] *Ibid.*, p. 265. Heidegger's insistence that "truth" should be understood in its primordial sense as "unhiddenness," while the "truth" as presupposed in a correspondence theory is only a derivative mode becomes intelligible on this background. See (1962a) §44 and (1947).

[54] Heidegger (1962a), p. 255.

[55] It is clear, then, why the question of the certainty of the existence of the external world is intractable. It is intractable as long as we do not treat the *ontical* and the *ontological* question separately. For what is ontically explicable may be ontologically confused and vice versa. As Heidegger sees it, all major crises of sciences are ontological ones and are fundamentally different from problems we may have concerning ontic questions. In this connection we can understand why Heidegger claims that there cannot be a genuine sceptic ((1962a), pp. 271–272), for if he understands himself he has to negate, and eradicate, his own "givenness".

[56] In the case of the imaginary we may need more than one kind of ontology, e.g., in the case of mythology we may need one kind, in fiction another, with symbols yet another, etc.

[57] See Körner (1955), p. 62 (his italics).

[58] See Kant (1929), B XLI: " . . . we must in each single case appeal to the rules according to which experience . . . is distinguished from imagination." But these rules presuppose already the appropriate ontological understanding.

[59] See Heidegger (1962a), §2. The notion of the hermeneutical circle has become fruitful in theology and in the interpretation of literature, especially after Gadamer's (1975) became widely known.

[60] Heidegger (1962a), pp. 248–249.

[61] See Heidegger's (1962b) and the little treatise (1963). A lot of literature on Kant has appeared during the last years which I could not discuss here. I just want to refer to the contributions to *L'héritage de Kant* (1982), especially section IV ("Kant et Heidegger"), and the new bibliography by H. Sass (1982).

BIBLIOGRAPHY

Brentano, F.: (1960), *Von der mannigfachen Bedeutung des Seienden nach Aristoteles*, Wissenschaftliche Buchgesellschaft, Darmstadt.

Düring, I., and Owen, G. (eds.): (1960), *Aristotle and Plato in the Mid-Fourth Century*, Elanders, Boktryckeri, Aktiebolhe, Göteborg.
Ewing, A.: (1938), *A Short Commentary on Kant's Critique of Pure Reason*, University of Chicago Press, Chicago.
Gadamer, H.: (1975), *Truth and Method*, ed. G. Barden and J. Cumming, The Seabury Press, New York.
Heidegger, M.: (1947), *Platons Lehre von der Wahrheit*, Francke, Bern.
Heidegger, M.: (1950), *Holzwege*, Klostermann, Frankfurt am Main.
Heidegger, M.: (1962a), *Being and Time*, trans. I. MacQuarry and E. Robinson, Harper and Row, New York.
Heidegger, M.: (1962b), *Kant and the Problem of Metaphysics*, trans. James Churchill, Indiana University Press, Bloomington.
Heidegger, M.: (1963), *Kants These über das Sein*, Klostermann, Frankfurt.
Heidegger, M.: (1967), *What Is a Thing?*, trans. W. Barton, Jr. and V. Deutsch, Henry Regnery Co., Chicago.
Heidegger, M.: (1972), *On Time and Being*, trans. J. Stambaugh, Harper and Row, New York.
L'héritage de Kant: mélanges philosophiques offerts au P. Marcel Régnier (1982), Beauchesne, Paris.
Kahn, C.: (1973), *The Verb "To Be" in Ancient Greek*, D. Reidel, Dordrecht, Holland.
Kant, I.: (1929), *Critique of Pure Reason*, trans. N. Kemp Smith, Macmillan Co. Ltd., London.
Kaulbach, F.: (1958–1959), 'Kants Beweis des Daseins der Gegenstände im Raum ausser mir', *Kant-Studien* **50**, pp. 323–347.
Kemp Smith, N.: (1923), *A Commentary to Kant's Critique of Pure Reason*, Macmillan and Co., Ltd, London.
Kirwan, C.: (1971), *Aristotle's Metaphysics, Books* Γ, Δ, E, Clarendon Press, Oxford.
Körner, S.: (1955), *Kant*, Penguin Books, Ltd., Harmondsworth, England.
Lehmann, G.: (1958–1959), 'Kants Widerlegung des Idealismus', *Kant-Studien* **50**, pp. 348–362.
Neske, G.(ed.): (1977), *Erinnerungen an Martin Heidegger*, Neske, Pfullingen.
Owen, G.: (1960), 'Logic and Metaphysics in Some Earlier Works of Aristotle', in Düring and Owen, 1960, pp. 163–190.
Paton, H.: (1936), *Kant's Metaphysic of Experience*, 2 vols., George Allen and Unwin, Ltd., London.
Prauss, G.: (1971), *Erscheinung bei Kant*,Walter de Gruyter, Berlin.
Prauss, G.: (1974), *Kant und das Problem der Dinge an Sich*, Bouvier, Bonn.
Prauss, G.: (1977), *Erkennen und Handeln in Heideggers Sein und Zeit*, Alber, München.
Richardson, W.: (1963), *Heidegger: Through Phenomenology to Thought*, The Hague, Martinus Nijhoff.
Rorty, R.: (1976–1977), 'Overcoming the Tradition: Heidegger and Dewey', *Review of Metaphysics* **30**, pp. 280–305.
Ross, D.: (1953), *Aristotle's Metaphysics*, Clarendon Press, Oxford.
Sass, H.: (1982), *Heidegger Bibliography*, Bowling Green State University Press, Bowling Green, Ohio.
Strawson, P.: (1966), *The Bounds of Sense*, Methuen and Co., Ltd., London.

Vaihinger, H.: (1981), *Kommentar zu Kants Kritik der reinen Vernunft*, 2 vols., W. Spemann, Stuttgart.

Wilkerson, T.: (1976), *Kant's Critique of Pure Reason: A Commentary for Students*, Clarendon Press, Oxford.

Wilshire, B.: (1977), 'James and Heidegger on Truth and Reality', *Man and Reality* **10**, pp. 79–94.

Zimmerman, M.: (1978), 'Dewey, Heidegger, and the Quest for Certainty', *Southwestern Journal of Philosophy*, pp. 87–95.

RUTH BARCAN MARCUS

SPINOZA AND THE ONTOLOGICAL PROOF*

I should like in this paper to characterize the "ontological proof" for the existence of God. In so doing I will discuss Wolfson's[1] analysis of the proof and argue that his characterization is not, as has been claimed, wholly mistaken, but is insufficient. I will also argue that the more adequate characterization here proposed gives us a good account of Anselm's and Descartes's proofs but that on that account, there is no "ontological proof" in Spinoza or at least not in the text where it has usually been thought to be located. In conclusion, I will touch upon the confusion of metaphysical notions and epistemological notions such as that between conceivability and possibility, which has conflated quite different sorts of arguments under the heading "ontological proof." It is clear from these preliminary remarks that the term "proof" is being used in the wide sense of argument and not in the narrow sense of valid or successful argument. Nor will I be concerned with the validity of those arguments which qualify as ontological.

I

In *The Philosophy of Spinoza*, Wolfson, after a lengthy discussion of texts, concludes:[2]

> W1 . . . none of the ontological proofs in their various forms as given by its three main exponents, Anselm, Descartes and Spinoza prove directly that God exists. What they prove is that the existence of God *is known to us by a certain kind of immediate knowledge.* Their various proofs can be reduced to the following syllogism: If we have an idea of God as the greatest, the most perfect, or a self caused being, then God is immediately perceived by us to exist. But we have an idea of God as the greatest or as the most perfect or as a self caused being. Therefore God is immediately perceived by us to exist.

It should be noted that in the quoted passage and throughout, Wolfson uses 'is perceived by us' and 'is known to us' interchangeably. What

A. Donagan, A. N. Perovich, Jr., and M. V. Wedin (eds.), Human Nature and Natural Knowledge, 153–166.

Wolfson says in the quoted passage is that the ontological proof is not a direct proof of God's existence. What he can be understood to be saying is that it is an indirect proof, but indirect proofs cannot be excluded as genuine. *Reductio ad absurdum* is often called "indirect proof." However, I am not suggesting that it is this logical mode of indirection which is the basis of Wolfson's claim. It is rather, for want of some standard locution, what I will call an epistemological mode of indirection. In the quoted passage, Wolfson tells us that what is proved *directly* is that the existence of God *is known* to us by a certain kind of immediate knowledge. Suppose Wolfson's claim is correct; that what is proved directly is, in part, the proposition

It is known to us that God exists.

Now a plausible analysis of statements like

It is known that *P*

or

x knows that *P*

is that a necessary condition for their being true is that *P* be true. It is incoherent for *P* to be known and *P* to be false. There are some attempted analyses which challenge the claim that the truth of 'It is known that *P*' entails the truth of *P*. Nevertheless the more generally accepted analysis is persuasive, and it does take

P is known and *P* is false

as inconsistent.

If therefore we can prove *directly* that

It is known that God exists

as Wolfson is claiming in the quoted passage, then it requires one further step to arrive at

God exists.

The validity of that further step is not peculiar to the theological content.

Now we do not usually demonstrate propositions by first demonstrating that they are known and then showing that as a consequence of their being known, they are true. But it is interesting to examine why, for it illuminates Wolfson's claims. What would a demonstration of, for example,

> It is known that the sum of the angles of a triangle consists of two right angles

consist in? Well, one obvious way is for some epistemological subject to produce a proof. On the common analysis of 'It is known that *P*' it is not merely a necessary condition that *P* be true but the knower, the epistemological subject, must be able to justify his claim. In mathematics that usually and perhaps always consists in providing a proof. I say usually because some mathematical realists sometimes claim mathematical knowledge intuitively and that such intuitions may count as justifications. The claim is abetted by some of Gödel's results for one can interpret those results as having the consequence that there are mathematical truths for which there never will be a proper proof, which are undecidable, but for which intuitive knowledge might be claimed. But, setting those considerations aside for the moment, let us suppose that for mathematical propositions, demonstrating that *P* is known consists in showing directly that someone has in fact produced a proof of *P*. But for that, we would, in effect, have to display the proof. And displaying a proof is simultaneously a proof of *P* as well as of

> It is known that *P*.

If our interest is in proving *P*, then the further step of asserting that I know it is otiose.

Now what counts as a mathematical proof can be made explicit by setting out canons or rules. There are, of course, in contemporary foundations of mathematics, disagreements about appropriate canons, constructivist vs. non-constructivist canons, for example. Still, whatever one's position on rules of proof, it is agreed that they must be wholly explicit so that a proof can be checked against them. There can therefore be a mechanical check against the canons and there is a sense in which we can say of a mathematical demonstration that it is possible for some *P* that it not merely be provable, but be proved, yet not known. A machine can be programmed to prove simple and even moderately complex theorems, some of which may be genuinely in doubt. If no epistemological subject examines those proofs and printouts are discarded *P* remains proved but not known. I deny (which I will not defend here) that any machine capable of generating such proofs can also be claimed to know that *P*.

Mathematical demonstrations, like most demonstrations, are not routed through an epistemological step of showing first that they are known. The cosmological arguments, the arguments from first cause for the existence of God, are such for which validity is claimed independent of any step about what is conceived or known. What distinguishes the ontological argument is the indispensibility of the epistemological subject in arriving finally at the conclusion that God exists. In this respect, the ontological argument is like the *cogito.* The conceiving, understanding, knowing subject is indispensible to the demonstration. In the ontological argument, what is concluded directly is that something is knowing to us, i.e., the existence of God. As noted above, it is a trivial step from

It is known that God exists

to

God exists.

What makes the ontological proof indirect is that it is routed through epistemological claims about what we conceive, understand, or known.

Although Wolfson is not uniformly clear about what was or was not intended in the ontological argument, it is quite consistent with his view that the ultimate intention of at least some of the philosophers cited was to prove that God exists and to present an argument in which the statement of God's existence is the conclusion in the strictest sense of the term.

Indeed, Wolfson says as much in a passage immediately preceding the one already quoted (W1). In that passage "God exists" is finally in the conclusion. Wolfson says[3]

> W2 Truly speaking, if the ontological proof were to be put into a syllogistic formula in such a way as to bring out its *entire* force, it would have to be as follows
>
> Everything which is immediately perceived by us to exist exists.
>
> God is immediately perceived by us to exist.
> Therefore, God exists.

The second premise, "God is immediately perceived by us to exist" is the conclusion of Wolfson's first quoted syllogism (W1). In W2 we have, in accordance with our gloss, the move from its being known or perceived that God exists to the conclusion that he does. The two syllogisms taken together form a sorites in which ''God exists'' *is* finally a conclusion in the strictest sense of the term. But it is the first quoted syllogism (W1), the movement from having an idea or concept of God to knowing that the *ideatum* to which it is adequate exists, which marks the peculiarity of the ontological argument.

Having defended Wolfson so far, I should like to go on to discuss a deficiency in his characterization, for I am in agreement that the account is not wholly adequate. It is rather, insufficient. Unless Wolfson tells us more explicitly what the special features of the "certain kind of immediate knowledge" are, too many arguments will count as "ontological." Now the syllogism in W1 which Wolfson says displays the structure of the ontological argument is a hypothetical syllogism of which the conditional premise is

C If we have an idea of God as the greatest, the most perfect, or a self caused being, then God is immediately perceived by us to exist.

The antecedent of *C* is generally taken as non-controversial. It just says that we have the concept, i.e.,

> We have an idea of God as the greatest, or as the most perfect, or as a self caused being

from which it follows by *modus ponens* that

> God is immediately perceived by us (or known by us) to exist.

The deficiency in Wolfson's characterization is of course in failing to say more about the conditional premise which must be true if the conclusion is to follow. He does in W1 say that the immediate perception, the immediate knowledge is of "a certain kind" but he seems not to distinguish that kind from claims to immediate knowledge such as that of the mystic or other psychological claims of immediacy. Wolfson's failure to differentiate kinds of immediate knowledge is seen from his remarks on Anselm's[4] response to Gaunilo's objections.

Anselm's response appeals to Gaunilo's faith and conscience and Wolfson says:[5]

> Is it not possible that in appealing to faith and conscience Anselm is really invoking the argument from revelation as attested by the tradition by which the existence of God is established as a fact of immediate personal experience? Such an argument is common Jewish philosophy, and it may be considered as partly psychological, insofar as the proof from revelation derives its validity from the fact that it is an immediate experience, and partly historical and social, insofar as the truth of the fact of revelation is attested by an unbroken chain of tradition universally accredited within a certain group.

It is correct to point out that the texts of Anselm and Descartes do not support Wolfson's suggestion that the kind of immediate knowledge referred to in the schema of the ontological argument (W1) is akin to revelation. But that does not wholly vitiate Wolfson's account. Rather it shows it to be insufficient.

The insufficiency of Wolfson's characterization is then in his failing to say more about the conditional premise. He allows that the premise might be established by revelation or by undifferentiated claims of psychological immediacy. In any case, an argument of the form of Wolfson's syllogism in which the conditional premise is unsupported by further rhetoric of a logical or rational sort, should not be included among the ontological arguments. The assertion of the mystic who insists without further explanation that to have the vision of God is to know that he exists *simpliciter*, is not in the spirit of those arguments of Anselm and Descartes which have been singled out as ontological. In stopping where he does, Wolfson does seem to classify among the ontological proofs those in which further rational support or justification need not be given for the conditional premise. But what is crucial in Descartes's and Anselm's arguments and which marks them as ontological is that they *do* attempt to give rational support to the conditional premise. They try to persuade us that a rational intelligence cannot deny that premise without incoherence. A rational intelligence cannot have the concept of the greatest or most perfect being and yet fail to know that God exists for such a conjunction, they claim, leads, if not to outright contradiction, then perhaps to what

Moore has called a pragmatic contradiction. A fool, of course, devoid of some rational powers, may fail to appreciate the incoherence and therefore, in the absence of faith and revelation, may say sincerely in his heart that there is no God.

If Wolfson is suggesting, and he seems to be, that the conditional premise which goes from the idea of God to immediate knowledge of God's existence is the same kind of certain knowledge which the mystic claims, then, as has been pointed out, there would be no accounting for some of the texts which purport to give an ontological argument. For those texts, *do* attempt to give support for the conditional premise of a rational kind. When Anselm appeals to Gaunilo's faith and conscience it is surely an appeal to the faith and conscience of a man guided by reason. Gaunilo is no fool.

My disagreement with Wolfson's critics is therefore on where to locate Wolfson's failure. It is the rational support for the conditional premise which Wolfson has failed to see as the special feature of an ontological argument. That rational support is manifestly clear in Descartes.[6] Consider the passages from the Fifth Meditation.

> D1 It is certain that I no less find the idea of God, that is to say, the idea of a supremely perfect Being, in me, than that of any figure or number whatever it is, and I do not know any less clearly and distinctly that an actual and eternal existence pertains to this nature than I know all that which I am able to demonstrate of some figure or number truly pertains to the nature of this figure or number . . .

and

> D2 . . . when I think of it with more attention I clearly see that existence can no more be separated from the essence of God than can its having its three angles equal to two right angles be separated from the essence of the rectilinear triangle, or the idea of a mountain from the idea of a valley; and so there is not any less repugnance to our conceiving a God that is a Being supremely perfect to whom existence is lacking (that is to say, a certain perfection is lacking) than to conceive of a mountain which has no valley.

In the first quotation (D1), Descartes simply says that he finds the idea of God in himself and he knows from that idea clearly and distinctly that an actual and eternal existence pertains to this nature *and that he knows that no less clearly and distinctly than he knows that some property he is able to demonstrate about a figure or a number pertains to the nature of that figure or number.* Now Descartes is claiming an immediacy for that connection which those who criticize Wolfson for placing a premium on immediacy have failed to appreciate. There is in the quoted passage not even a *reductio* argument. What Descartes is saying is that, without further demonstration, and from the idea of God, he knows that God exists with *at least* the clarity and distinctness that he knows a demonstrated mathematical truth. He is claiming that the knowledge, albeit immediate, is rational.

Again, in the second quotation from Descartes, although there is the hint of a *reductio*, the immediacy of the move from a rational subject's idea to knowledge of existence is highlighted. We cannot, says Descartes, have the idea of a mountain without at the same time having the idea of a valley. Similarly, he says, it is repugnant, which I read as "repugnant to reason," to conceive of a God to whom existence is lacking. A reasonable man cannot have the concept of God and yet fail to know immediately that God exists and that is, after all, Wolfson's conditional premise.

What I have claimed in the foregoing is that it is correct to fault Wolfson for being untrue to the texts, but it is not because Wolfson claims that the conclusion of the ontological argument is that God is known to us by a certain kind of *immediate knowledge.* It is rather Wolfson's failure to say something more about the conditional premise, his failure to distinguish the "immediacy" of the knowledge claimed in the ontological argument from that claimed for revelation or for other kinds of psychological immediacy. The "certain kind of immediate knowledge" by which a knowledge of God, is claimed in the ontological arguments should have been singled out as *rational* intuition. Having the idea, grasping the essence of God is supposed to lead us to know, with clarity and distinctness, that he exists. Given that the conditional premise asserts immediacy between antecedent and consequent we cannot press the sorites back further in an obvious way. There is no further premise. That is, after all, what immediacy entails. But we can bolster support where rational intuition is somewhat blunted. We can get such a man to see it for himself akin to Socrates's efforts in the *Meno.* We can point to an analogy with mathematical insights or with concepts which entail one another such as the concept of mountain and valley. There are *reductio* supports; we can try to show that denying the conditional leads to incoherence.

According to Anselm,[7] "No one in fact, *understanding* what God is, can conceive that God does not exist." Since anyone who claimed he could not conceive *not-P* would also claim to know *P*, Anselm's assertion fits Wolfson's schema of the conditional premise in W1. More precisely, Anselm's claim entails Wolfson's conditional premise: for the knowledge claimed by Anselm is not merely that the non-existence of God is known to be false; propositions can be known to be false yet their truth still be conceivable. What he is asserting is that an epistemological subject, upon understanding what God is, will know that God exists with such certainty that conceiving of his non-existence would be incoherent. Anyone who fails to make the connection is accused by Anselm of lacking "a rational mind" or of being "stupid and a fool." Here we see the departure from Wolfson. Stupidity, foolishness, or irrationality have never been claimed as impediments to mystical insight or revelation.

II

The question remains whether there is in Spinoza's writings an ontological proof as here characterized. Let us even suppose that Spinoza has succeeded in showing that God and substance are identical. Is there an argument which goes from the idea of God or substance and the claim that from that idea it is known that God or substance exists, to ultimately the existence of substance? Is there supporting rhetoric for the conditional premise such as mathematical analogies, reductions, and the like?

Now it is possible that from Spinoza's complete texts one might be able to construct such an argument. But there is one segment of the *Ethics*[8] which has been singled out by Wolfson and others as constituting an ontological argument. Broadly speaking it is all of Part 1 terminating in Proposition XI which says that God or substance exists necessarily. My claim is that in Part 1 of the *Ethics* or indeed elsewhere in the *Ethics*, there is no ontological proof to be found as here characterized.

The *Ethics* is patterned after Euclid's geometry. At the outset that geometry presupposes the existence of certain objects such as points and lines. The definitions and axioms define some of their essential properties and some of the relations between those geometrical objects. Analogously, Spinoza in the *Ethics* begins with the presumption of the existence of things, substance, modes, God, and relations between them, including the relation of identity. The *Ethics* is largely an unpacking and elaboration of those relationships. Spinoza's definitions with which he begins are like Euclid's "real" definitions. It is not merely "ideas" which are being defined which must *then* be shown to correspond to their ideata. Euclid's definitions were understood as telling us what objects geometry is about, and they define or mark out some of their essential properties. Spinoza's eight definitions are about things, finite and infinite, substance, modes, God, free things, compelled things, and so on. It is *supposed* simultaneously that we have ideas of those things which are true to them. That those ideas are adequate to their ideata is never in question. Some of those presumptions are spelled out in the axioms. Axiom I already tells us that everything is either in itself or in another. Given the definitions, existence of examples of each kind are presupposed. For example, Definition VI supposes that there are things which exist from the necessity of their own nature. They are free things. Compelled things are those which are determined to existence by another. Note that the demonstration of Proposition VII, the assertion that it pertains to the nature of substance to exist, is not a *proof* of the existence of substance. That is already supposed. It purports to show that the existence of substance, unlike other existences, follows from its own nature. Indeed, Spinoza wonders whether a demonstration of Proposition VII is even required. He says in the second Scholium of Proposition VIII,

> If men would attend to the *nature* of substance they could not entertain a single doubt of the truth of Proposition VII. Indeed, this proposition would be considered by all to be axiomatic and reckoned among common notions.

As for Proposition XI, which is usually singled out as the conclusion of the "ontological proof," i.e.,

> God or substance consisting of infinite attributes . . . , necessarily exists,

it is a direct consequence, and instance, of Proposition VII. One cannot locate in Part 1 of the *Ethics* any demonstration that from the *idea* or *concept* of God it follows that he is known to exist or that he exists. What is asserted in Proposition XI is that he exists in a certain way — necessarily. His existence, *simpliciter* is never in question.

The demonstrations that substance and God exist in a certain way, i.e., necessarily, may misleadingly suggest the form of an ontological argument to those who do not trace it from the definitions. The Scholium of Proposition VIII does say that "if men would attend to the nature of substance they could not entertain a single doubt that the existence of substance follows from its nature." Proposition XI does ask us to conceive, if we can, that God does not exist. But those urgings are used not to show that He exists or that He is known to exist but to show that we would be forced to deny that His essence involves His existence, which is one of the essential or defining properties of God or substance. To argue that the existence of substance or God follows from its nature is quite a different matter from arguing that the existence of substance or God follows from our concept of it. That our idea of God has an ideatum (as all ideas do for Spinoza) and is furthermore adequate to its ideatum is, at least in the *Ethics*, taken as true at the outset. In summary, although Wolfson was correct but incomplete in his characterization of an ontological argument, he was wrong in claiming that such an argument could be found in the *Ethics*. The place for an ontological argument in the *Ethics* would have been in support of some of the axioms and definitions.

III

I should like in conclusion to say something about the uses of modal logic in our understanding of proofs which have been classified as ontological. Some and perhaps much of the recent revival of interest in such proofs stems from the revived interest in modal logic and the belief that ontological arguments can be translated into modal arguments. I do not propose to go into all of those various efforts although the extent to which those efforts succeed, fail, or are misapplied is itself illuminating. I should like just to mention in passing two such attempts.

In a paper 'Spinoza's Ontological Argument' Charles Jarett[9] argues that if we interpret the first part of the *Ethics* as claiming that

(1) It is possible that God exists
(2) Necessarily (If God exists then he exists necessarily)

it follows in one of the strong modal systems (with some constraints on what is known as the Barcan formula) that

(3) Necessarily God exists.

I do not propose to discuss whether such as argument can be culled from the *Ethics.* As it and many such attempts stand, it is not an ontological argument as we have characterized it which is being presented. It does not have the requisite epistemological features. There is no conditional which takes us from what is *conceived* to what is known or to what exists. There is no argument to warrant the identification of 'conceivable' with 'possible'. To convert the ontological argument as here characterized to a modal argument like the one presented above supposes that certain epistemological notions can be translated into logical or metaphysical ones; that for example to say of something that it is conceivable is to say that it is possible. And to say of something that it is inconceivable that it not be the case is to say of something that it is necessary. Plantinga,[10] for example, in his protracted modal transformation of Anselm's argument, raises no questions about the legitimacy of such a transformation. He asserts, without argument, from Anselm's texts, "And when [Anselm] says that a certain state of affairs is conceivable he means to say (or so at any rate I shall take him) that it is a logically possible state of affairs — possible in our broadly logical sense. So for example . . . 'God's existence in reality is conceivable' is more clearly put as 'It is possible that God exists in reality.' "

On the most common and plausible use of such epistemological concepts as "conceivable," there is no justification in the identification. Conceivability requires an epistemological subject; someone to do the conceiving. A proposition may be taken as conceivable to an epistemological subject if it is consistent with the subject's beliefs or, more weakly, if he believes it to be consistent with his beliefs. On such an account there are powerful counterexamples to the identification of epistemological concepts with logical, or what Kripke has lately called metaphysical, modalities. A mathematical claim for example, if true, is necessarily true, and if false, is necessarily false. It is not controversial to suppose that when a mathematician conjectures that *P* where *P* is a mathematical proposition he finds it conceivable that *P*. Yet *P* may be false and hence impossible. Indeed such was the case with at least one of Hilbert's long list of conjectures. Or consider the more mundane case where someone finds it conceivable that Hesperus is not the same as Phosphorus yet, as an empirical matter, "they" are found to be identical. But if *x* and *y* are identical, "they" are necessarily identical and hence it is impossible that they not be the same although it was conceivable to some epistemological subject in ignorance of that discovery that Hesperus was different from Phosphorus. Furthermore, and apart from such counterexamples, modal arguments like Jarett's and Plantinga's, if valid, would remain valid in the absence of *any* epistemological subject or an idea (or concept) of God altogether. Modal arguments for the existence of God, arguments which purport to prove God's existence from essence, are interesting in their own right but are not to be conflated with those paradigm examples of the ontological argument such as Anselm's and Descartes'. Perhaps "epistemological proofs for the existence of God" would have been a more appropriate designation for Anselm's proof. What remains to be explored is the set of presuppositions connecting those metaphysical and epistemological concepts which lead to the confusion and conflation of the two kinds of arguments.

Yale University

NOTES

* This paper overlaps in part with a paper 'Bar-On on Spinoza's Ontological Proof' presented in 1976 at a conference on Spinoza of the International Institute of Philosophy. The *Proceedings* of that conference will be published.

1 Wolfson (1958), vol. 1, pp. 158 ff. Italics added.

2 *Ibid.*, p. 176. Italics added.

3 *Ibid.*, Italics added.

4 Anselm (1965), p. 169.

5 Wolfson (1958), p. 171.

6 Descartes (1955), vol. 1, pp. 180–181.

7 Anselm (1965), chapters III and IV.

8 Spinoza (1930), (*Ethics*, trans. W. White).

9 Jarett (1976).

10 Plantinga (1974), p. 199.

BIBLIOGRAPHY

Anselm: (1965), *St. Anselm's Proslogion with a Reply on Behalf of the Fool by Gaunilo and the Author's Reply to Gaunilo*, ed. M. Charlesworth, Clarendon Press, Oxford.

Descartes, R.: (1955), *Philosophical Works*, trans. E. Haldane and G. Ross, 2 vols., Dover Publications, Inc., New York.

Jarett, C.: (1976), 'Spinoza's Ontological Argument', *Canadian Journal of Philosophy* **5**, 685–691.
Plantinga, A.: (1974), *The Nature of Necessity*, Clarendon Press, Oxford.
Spinoza, B.: (1930), *Selections*, ed. J. Wild, Charles Scribner's Sons, New York.
Wolfson, H.: (1958), *The Philosophy of Spinoza*, 2 vols., Meridian Books, New York.

MICHAEL V. WEDIN

TRACKING ARISTOTLE'S NOÛS*

By common account *De Anima's* treatment of *noûs* is pulled in opposite directions by Aristotle's finitistic and his transcendentalistic tendencies. His finitistic side wants an account of *noûs* to be an account of strictly individual noetic activity without mention of factors or entities extrinsic to embodied persons. Despite stirring about the possibility that *noûs* is independent from body, the finitistic side dominates the work. Indeed, with appropriate weighting the independence of *noûs* can, for the most part, be assimilated to the dominate strain. In Γ.5, however, Aristotle's transcendentalistic side allegedly surfaces in the figure of productive mind [νοῦς ποιητικός]. Here commentators have taken up, roughly, three lines of interpretation. For some, Γ.5 remains exclusively interested in the individual mind and simply, in fact of course not so simply, countenances the immortality of one of its parts.[1] For others, curiously undaunted by Aristotle's silence on the point, productive mind, while not properly a part of the individual mind, is required to complete the account of individual noetic activity.[2] And some, finally, retreat to the position that in Γ.5, Aristotle gestures toward the transcendent intelligences of *Metaphysics* Λ without intending any connection whatever with individual noetic activity.[3]

None of these three routes is successful in restoring an easy wholeness to the *De Anima* account. Nor does any one of them appear, beyond a doubt, to capture what Aristotle wants to say. Arguing this to full effect would obviously require detail that is better left for another occasion. Here I shall proceed directly with interpretation.[4] The target is mainly *De Anima* Γ.4 & 5 and the interpretation's thrust is that together these chapters provide the essentials of a thoroughly finitistic account of individual noetic activity. More particularly, I shall suggest that because productive mind figures in a general finitistic account of mind, it must itself yield to a non-transcendentalistic analysis. I begin with some remarks on Γ.4 and continue at somewhat greater length in Part II with Γ.5.[5]

A. Donagan, A. N. Perovich, Jr., and M. V. Wedin (eds.), Human Nature and Natural Knowledge, 167–197.

I

Aristotle is the first philosopher whose views are sufficiently sophisticated to constitute a genuine theory. As such he is the first serious functionalist in the philosophy of mind. Primarily, an account of the soul is to concern itself with what various sorts of creatures can do. Those capable of more intricate, linguistically informed behavior are said to be endowed with *noûs.* A functionalist account of *noûs* will not, of course, consist solely in a list of kinds of noetic behavior but rather will attempt to explain how an entity must be structured to be capable of producing such behavior. Whether we must introduce into the explanation a separate entity as the doer of noetic acts is an interesting but, at the outset, independent question.

In general Aristotle downgrades the ontological question in favor of the functionalist program less because that question is especially difficult than because answers, if any, to the ontological question will follow from the demands of the functionalist theory.[6] Thus, at the start of Γ.4 (429a10–12) he proposes to examine the distinguishing features of *noûs* and to say something about how thinking comes about, *whether or not* mind is separate extensionally. So it is clear that Aristotle thinks the functionalist program can be pursued without settling the ontological question. Especially suggestive for my purposes is the division of labor indicated in the line. For the question of what features are distinctive of *noûs* is kept separate from the question of how thinking occurs and thus the possibility arises that Γ.4 is meant to address the first and Γ.5 the second question. It is this suggestion that I want to exploit.

Among the features Γ.4 marks out as distinctive of mind are the following.

1. Thinking is something like being affected [429a14–15]

and

2. One can think of what one wishes [429a7].

1 follows from the analogy, adduced early on in Γ.4, between perceiving [αἰσθάνεσθαι] and thinking [νοεῖν].[7] The analogy, often granted strict governance over the Γ.4 account, is in fact crucially imperfect. For, as 417b19–21 makes plain, what produces the activity [τὰ ποιητικὰ τῆς ἐνεργείας] of perceiving is invariably something

external while in the case of thinking it is somehow something in the soul. Since this in turn is held at 417b21–23 to imply 2, which has no counterpart at all for perception, the analogy itself should not be taken too literally. On one point, however, Aristotle takes the analogy quite seriously, namely, on the causal role of the object of thought. This is clear from 429a14's requirement that thinking be (something like) being affected by the object of thought [ὑπὸ τοῦ νοητοῦ] together with 429a17–18's remark that the relation between mind [τὸ νοῦς] and objects of thought [τὰ νοητά] is similar to that between that which can perceive [τὸ αἰσθητικός] and objects of perception [τὰ αἰσθητά]. (see also *Metaphysics* 1072a30 where mind is said to be moved by the object of thought [νοῦς δὲ ὑπὸ τοῦ νοητοῦ κινεῖται].) Thus, just as, causally, it is the object of desire or perception that produces actual desiring or perceiving, so also it is the object of thought that produces actual thinking. Let us then enter as a third feature of mind

3. Actual thinking is produced by the object of thought [429a13–14].

Three further features central to the Γ.4 account are

4. Mind is nothing actual until it thinks [429a24],
5. Mind is identical with its object [430a3–5],

and

6. Mind thinks itself [429b9].

4 is said to follow from (at least) three other theses. Ordered in terms of consequence, they are that mind's range is unrestricted, that it is unmixed [ἀμιγής] with anything and that its sole nature is that it is potentiality. This last thesis most likely means simply that mind's sole nature is the potentiality to be identical with any possible object of thought. After all its nature could hardly consist in being identical with any given set of objects. Nor can its nature consist in some noetic activity that is invariant over variation in content or object of thought, for, as Aristotle insists at 430a5–6, we do not always think. So the account of mind's nature is functionally in line with his account of other faculties. Both are to be defined as potentialities$_2$ (developed potentialities) or, in his alternative idiom, as actualizations$_1$ of certain potentialities$_1$ (undeveloped potentialities). The difference is just that

faculties other than mind are actualizations$_1$ of distinct physical structures, namely, sensory organs. It is this that explains the limitation on the range of objects accessible to the non-noetic faculties.

Note that it need not, indeed cannot, be implied that the mind is devoid of all modifications, particularly those connected with learning, memory, and the like. How else could the theoretical man come to contemplate this rather than that truth? The point rather is that while receptive mind is modified, such modifications are no part of its nature. Unlike the physical structures that partially define other faculties' natures these modifications in no way set boundary conditions on what the mind can do.

Finally, the fact that the mind is not defined as the actualization$_1$ of any set of physical structures explains why it is indeed nothing actual until it thinks. For all other faculties there is something actual even when the faculty is not operative, namely, the particular physical structures over which they are defined. Notice that the mind may well depend on a complex of physical structures, say in virtue of its dependence on images, without being the actualization$_1$ of any such structures. So 4 need not be made into the paradoxical claim that mental content is somehow created out of nothing.

Let me now turn to 5 and 6. 6, the thesis that the mind thinks itself, does not enjoy a settled interpretation among Aristotle's commentators. I shall try to avoid the thick of the debate by restricting comment on 6 to the discussion at hand. Notice first that in finding 6 at 429b9 I reject the translators' popular substitution of "thinks by itself" for "thinks itself." The former enables one to take Aristotle as claiming only that when one is a fully capable noetic agent, one can think by oneself. It cannot, of course, be denied that this makes sense in context:

> When mind has become each thing [ἕκαστα] as one who actually$_2$ knows [ὁ ἐπιστήμων ὁ κατ' ἐνέργειαν] is said to be and this happens when he can actually$_2$ exercise his potentiality$_2$ by himself — it is still in some sense a potentiality$_2$ but not in the same way as before it learned or discovered (i.e., not as a potentiality$_1$). Then it is capable$_2$ of (actually$_2$) thinking itself [αὑτὸς δὲ αὑτὸν τότε δύναται νοεῖν] [429b5–9].

There are, however, cogent reasons to resist the substitution. Not the least of these is thorough manuscript agreement for reading "δὲ αὑτόν" rather than the needed "δι' αὑτοῦ" in the final line of the passage. The emendation also makes the line curiously otiose, for

Aristotle has just said that a subject can exercise his mind by himself. Finally, the fact that 5 is introduced in the first line indicates that the topic of discussion is the relation between mind and its objects. So 6 is hardly out of place here.[8] Aristotle appears to regard the scope of 429b5–9 as entirely general, certainly so for 5. It follows then that 6 also is asserted as a general thesis. So implicitly the passage is committed to something like the following.

7. Whenever one actually thinks (i.e., when the mind is identical with the object of thought), then the mind thinks itself.

Although I shall later have more to say that bears on 7, let me now comment further on 6. Notice that in 429b5–9 Aristotle asserts but does not explain 6. Indeed at the chapter's end 6 is regarded as something of a puzzle. 1 plus mind's simplicity makes it problematic how the mind can think at all. To this is immediately attached the question how mind can think itself. In light of our preceding paragraph this is an appropriate query: If for the mind to think an object is for it to somehow think itself, then perplexity regarding the first will spill over to the second. The two problems require joint illumination.

Aristotle in fact offers a solution of sorts in Γ.4. The problem was that, ordinarily, in cases of being affected two actual things are involved, a first thing that is affected and a second thing that causes the first to be affected.[9] In the case of mind, however, there is only a single actual thing, for in any episode of thinking the mind is actually identical with the object that produces the thinking. Since whatever else it *may* be affected by, namely any other potential object of thought, is no actual thing, the problem of how mind can think and yet be simple is supposedly met. The problem of mind thinking itself submits to similar resolution: Since, in the case of things without matter, mind is identical with its object, it follows that in such cases mind thinks itself.

Notice, however, that this solution solves the pair of puzzles strictly in terms of the language in which they are set. What it does not do is provide any picture of how the mind can do what the solution requires of it. In effect we still lack a theory about the functional organization of the mind. This, I suggest, is part of what Γ.5 is meant to provide. So let us look at that perplexing chapter.

II

What sort of picture of mind underlies an account containing the features we have just discussed? In particular can we sketch a picture that gives productive mind a natural and finitistic role? So far as I can tell this is possible only by taking seriously Aristotle's distinction in Γ.5 as a distinction between receptive and *productive* mind. The standard translation of "νοῦς ποιητικός" by "active mind" is, I suggest, off the mark doctrinally as well as linguistically (here "creative mind" fares much better). So I shall assume that there is a quite definite sense in which one aspect or part of the mind is productive or creative and that, as the explicanda, 1–7 figure as constraints on what is reasonable to say about this aspect.

Very roughly, the underlying view I have in mind (call it *M*) is this:

> At *t a* thinks [noetically][10] of something *P* if, and only if, at *t a*'s mind not only produces *P* but also produces itself by producing *P*.

The central idea is that in producing an actual object of thought the mind also produces itself as an actual thing. *Qua* actual, the mind is nothing other than the activity whose content is the particular object of the given episode of thinking. Even this characterization threatens to mislead insofar as it allows for what I call the search light model of mind. Here the mind is thought of as sweeping through noetic space until encountering an object, at which point content materializes. Not only is this less apt as a characterization of a *productive* noetic function but also it counters Aristotle's committment to the thesis of intentionality which requires objects for all *bone fide* psychical functions. Indeed, the prime mover's self-thinking may simply be an advanced expression of the committment. So noetic activity is necessarily activity under a certain objectual mode and a mind producing itself is always a mind producing itself as producing an object.

Of course *M* does not entail any strong thesis about the identity, in general, of *a*'s mind, for such talk is talk about the essential nature of *a*'s mind and this we know to be pure potentiality. Perhaps criteria of identity for such an entity will be forthcoming from facts about the person *a* but in no case will they appeal to facts about an alleged episode-independent entity. And in any case Aristotle displays little interest in the identity conditions for any such entity and so, presumably, no more in its actual existence. This is just what one would expect from a fundamentally functionalist account.

It appears, then, that *M* explains 4, 5, and 6. It also honors 3's causal requirement because actual thinking is produced by the mind's production of the object of thought. The object of thought that produces thinking is, of course, actual and so *M* squares with the requirement that the productive factor in cases of being affected be something actual. The gloss on 1 required that mind's object be internal to it. *M* satisfies this by legislating that the mind itself produces its object.[11] 2 squares with *M* in like manner. The causal role of the object of thought in episodes of actual thinking is no threat to mind's thinking what it wishes precisely because it is the mind itself that produces the object.[12]

Stressing that the analogy between αἰσθάνεσθαι and νοεῖν is directed at the causal aspect of each provides a rather nice account of how the αἰσθητός–αἴσθημα relation is in its turn analogous to the νοητός–νόημα relation. The point of the latter analogy lies, I suggest, in the causal role of the αἰσθητός and νοητός], respectively. Just as the αἰσθητός is the immediate cause of perceiving [αἴσθησις] so also the νοητός is the immediate cause of thinking [νόησις]. And just as the resultant state of the perceiving subject is occurrence of a certain αἴσθημα or perceptual content so also the resultant state of the thinking subject is occurrence of a νόημα. As παθήματα in the soul both αἰσθήματα and νοήματα are, we know from *De Interpretatione* 1, objective and thus the analogy cannot mean to underscore an objective — subjective distinction. And since objects of perception are external but objects of thought not, neither does the analogy mean to distinguish νοητά and νοήματα as external *versus* internal items. Thus the analogy between perceiving and thinking supports *M*'s emphasis on the causal role of certain internal and objective items and, more to the immediate point, it gives a straightforward account of at least part of what the νοητός–νόημα distinction consists in.

It is also worth noting that although the νοητός–νόημα distinction is not that between objective and subjective items of awareness, it does play a role in allowing Aristotle's theory to maintain both the objectivity of objects of thought and their identity with the thinking subject (properly, with the subject's mind). First stress that the thesis of identity is part of Aristotle's theory of νόησις or the activity of thinking. Thus the sense of identity called for will be a function of the role of the thesis in the wider theory. So far as I can see nothing more than formal identity is required. Then, in order to *distinguish x*'s thinking *P* from *y*'s thinking *P* without threat to the objectivity of *P* (the νοητόν that is the same for *x* and *y*) an additional feature is called for. This is just the νόημα which is formally the same for *x* and *y* but which, as a causal result of the νοητόν *P* in distinctive noetic structures, is numerically different.

Some might find it preposterous to propose, as *M* does, that the mind creates itself in producing its objects. Discontent may stem partly from neglecting the fact that the mind thus creates itself only as something actual and partly from failing to appreciate sufficiently Aristotle's committment to functionalism. Given a predominant concern with what noetic agents can do, *M* does provide a partial account of the facts. Conjoined with Aristotle's taste for ontological parsimony, these reminders do something to enhance *M*'s plausibility. What, for instance, would be gained by postulating an episode-independent mind? Certainly not clarity.

There is another line of support for *M* besides its usefulness in explaining 1–7. Numerous passages tout the mind's immunity to error. With respect to its proper objects at least, *noûs* divines truths only. This is a difficult point for the platonist who countenances the independent existence of objects of thought. For here there is logical space to drive a wedge between thinker and object of thought. The usual device for coping with this possibility, namely, endowing *noûs* with an intrepid ability to fathom distinct objects, is notoriously unsatisfactory. Aristotle's solution is more attractive. If the mind not only creates but also is identical with its objects, then there simply is no logical space for the error-enabling wedge.[13]

In any case the infallibility of mind with regard to its objects is problematic at best. Reflexive self-reference, however, fares rather better. One simply does not have any idea how to concoct the required discreditation story, however farfetched. Sincere uses of the reflexive pronoun "I" are *ipso facto* referentially successful uses. So there is a certain sense in which mistakes are not possible here. I raise this point because it suggests a surprising line of support for *M*, a line recently drawn with some elegance by Robert Nozick in a somewhat different context.[14] Nozick suggests that infallibility of genuine reflexive self-reference is among the most salient features of full selfhood and that the notion of the self best explaining the feature is the notion of something that constitutes itself in the very act of reflexively referring to itself. This is surely a close cousin of *M*.

What is more to the point, however, is that there is some evidence that Aristotle himself would welcome such an account. In at least one place, *Nicomachean Ethics* Γ.1, 1111a7–8, Aristotle suggests that while I can be mistaken about almost any other feature of an action I perform, I cannot be mistaken about the fact that *I* am performing the action. Since here, and elsewhere, he displays techniques suitable for handling mistakes in reference *via* descriptions (and, by straightforward extension, proper names), he must have something like reflexive self-reference in mind at 1111a7–8. Aristotle's concerns do not take him deeply enough into problems of reflexive self-reference, nor personal identity for that matter, to find him anticipating Nozick's theory of the self. But they do go far enough to add support for attribution of *M* to Aristotle.

So far I have leaned heavily on Aristotle's usage. This at least has allowed us to explain why productive mind is *productive*. But Γ.5 contains substantive discussion of productive mind and the question that now looms is whether our interpretation can be squared with these remarks. The discussion of productive and receptive mind divides roughly into three stages: 430a10–14 introduces the distinction in terms of a connection between nature and the mind; 430a14–17 explicates the latter term of the connection; 430a17–25 retails the crucial features of productive mind. Suppose, then, we begin with the connection between nature and mind.

The Connection Between Mind and Nature

Since [as=ὥσπερ] in the whole of nature [ἐν ἁπάσῃ τῇ φύσει] there is, on the one hand, something that is the matter for each thing of a given kind [τι. τὸ ὕλη ἑκάστῳ γένει] (and this it is that is potentially all the things of the kind) and, on the other hand, something different that is the cause and that which produces [τὸ αἴτιον καὶ ποιητικόν] by making all of them [τῷ ποιεῖν πάντα] in the way that an art is disposed with respect to its matter, so must these differences hold in the soul.

It is unsurprising that the above passage has been aligned with the view that productive mind is the prime mover, particularly if one retains the bracketed "ὥσπερ" in 430a10. For one might then take the passage to assert an analogy between mind and nature. Just as there is a single thing that makes all things in nature so also in the soul something, presumably the divine mind, makes all things. But two considerations counsel otherwise. First, the conditional structure of the passage urges elimination of "ὥσπερ"[15] and requires that what is said about mind follows from the general rule on whatever is in nature. Second, the fact that Aristotle writes "each *kind* of thing" rather than "each thing" makes it clear that he is not at this point introducing anything as grand as the unmoved mover but merely issuing a rule on any production whatever, namely, that it occur within a genus.[16] This is a standard and familiar requirement.[17]

So there is, initially, no warrant for linking productive mind and the divine mind. What Aristotle does want to focus on is the *way* productive mind produces its objects. Γ.4 approves of those who describe the mind as the place of forms so long as the forms are understood as potentially and not actually present. Thus the general point seems to be that productive mind retrieves, as it were, from receptive mind a given object of thought and thus renders it actual. That this is a rather special sort of production is clear from the fact that it is likened to the way an art produces its result. On the standard account of *On Generation and Corruption* A.7 there are two kinds of productive agent [τὸ ποιοῦν]: Where what produces does not have the same matter as what is affected, it is possible that it move without being moved itself [324a34–35]. Thus, whereas wine produces health by itself being affected the doctor manages this without suffering change. So too for the *art* of medicine [324b5–7] which thus rates as an unaffected producer [τὰ ἀπαθῆ τῶν ποιητικῶν]. Aristotle appears, then, to be urging that productive mind will, like an art, produce thinking or an object of thought without itself being affected. Since this was one of the features of the mind advertised in Γ.4 as well as in Γ.5, it is unsurprising, given our hypothesis, that Γ.5 should explain this in terms of one of the mind's structural features.[18]

It might be objected that although the first part of the passage allows for the above interpretation its sequel does not. So we must square our account first with Aristotle's characterization of the two features of mind and then with the traits assgned to productive mind. 430a14–17 explains the first as follows.

And there is, on the one hand, a mind that is of this sort by becoming all things [ὁ τοιοῦτος νοῦς τῷ πάντα γίνεσθαι] and, on the other hand, one that is of this sort by making all things [ὁ τῷ πάντα ποιεῖν], as a sort of disposition such as light [ὡς ἕξις τις, οἷον τὸ φῶς]. For in some way too light makes colors that are potential [τὰ δυνάμει ὄντα χρώματα] into colors that are actual [ἐνεργείᾳ χρώματα].

Since I am suggesting that the above passage and *De Anima* Γ.5 in general attempt an explanation of individual noetic activity, the mind or minds here introduced must be the same as that discussed in *De Anima* Γ.4. For the latter is clearly the individual mind. In particular the distinction between productive and receptive mind is to be connected with the mind described in 429b5–9 (see page 170 above) and with 429b22–25's puzzle about how the mind can think at all, if thinking is something like being affected.

An additional remark is in order on the passage. The point I wish to stress is that even when the mind is actually$_2$ thinking of, say, a triangle, it is potentially thinking all other objects of thought. This is essential to any account of the activity of thought because the mind cannot actually$_2$ think two objects of thought at one and the same time yet it must be possible for it to actually$_2$ think another object at a later time. If, to reinvoke our puzzle, thinking were *only* being affected, the latter condition could not be met. Since, we have seen, it is clearly the case that thinking is at least something like being affected, there must be some other feature of thought which allows for the actually$_2$ thinking mind to think other thoughts. This additional feature is, I submit, nothing other than the productive mind of *De Anima* Γ.5.

Obviously nothing like productive mind would be required were thought like perception in point of the causal efficacy of its object. For perceivings change at change of object perceived and without agency on the part of the subject. But nothing of this sort is available in the case of thought, where the mind itself creates the object that produces thinking. So if *De Anima* Γ.5 attempts an answer or partial answer to the puzzle raised at 429b22–25, then it is unsurprising that the distinction between productive and receptive mind completes the account of individual mind sketched in *De Anima* Γ.4.

On the interpretation here offered, that productive mind makes all things simply states a condition on the mind actively thinking an object. The mind is productive just in the sense that the object it happens at *t* to actually think was prior to *t* potentially an object of such thought and, in accordance with *M*, in producing the object at *t* it produces itself as well.

The Analogy Between Productive Mind and Light

Aristotle pursues the account by analogizing productive mind to light. The force of the analogy is easily missed, if we neglect certain features of Aristotle's views on light. For our purposes it will be sufficient to consider 418b9–13

> Light is the activity [ἐνέργεια] of this, the transparent *qua* transparent [τοῦ διαφανοῦς ᾗ διαφανές]. Potentially, whenever this is, there is also darkness. Light is a sort of color of the transparent, whenever it is made actually transparent [ὅταν ᾖ ἐντελεχείᾳ διαφανές] by fire or something like the body above (i.e., the sun)

with 418a31–b2

> All color is capable of changing that which is actually transparent [τοῦ κατ' ἐνέργειαν διαφανοῦς] and this is its nature

and 418b18–19

> Since darkness is the privation of this sort of disposition [τοιαύτης ἕξεως] from the transparent, light is the presence of this [ἡ τούτου παρουσία].

On a number of points it is clear that Aristotle means the analogy to be taken seriously and in some detail. 418a31–b2, for instance, supports our linking of the "puzzle" of *De Anima* Γ.4 to the doctrine of productive mind by urging that it is in the *nature* of the actually transparent that it change from one color to another. Analogously, it is in the *nature* of an actually$_2$ contemplating mind that its object change. Indeed this is a requirement productive mind is tailored to fit. Further because what is actually transparent will always be a certain color, the actually transparent does change *qua* red, say, but not *qua* transparent. Similarly, the actually$_2$ contemplating mind does change *qua*, say, triangle because it may contemplate a different object.[19] But *qua* productive mind itself it cannot change.

Note also that while light is the actuality of the transparent *qua* transparent, it does not follow that light is potentially something else but only that the actual *transparent* that is light is potentially dark. This is important for it counters the long received view that productive mind is always active. Light, necessarily active whenever occurrent in any sense, need not be and in fact is not always occurrent. On our interpretation it is unsurprising to find *De Anima* Γ.4 already extending this feature of the analogy to the individual mind. 430a5 states rather plainly that subjects, and here is meant individual contemplating minds, are not always thinking. So unless we assume that the analogy is suddenly not to be taken seriously, additional support is forthcoming for our interpretation.

Even more important, however, is the following. Attention to the account of light in *De Anima* B.7 reveals that light is not a causal factor, in any standard sense, in the account of vision. Rather it is a certain sort of activity that results from the operation of such a standard causal factor on the transparent. Nowadays this causal factor, characterized by Aristotle as fire and the like, would be described as the source of light itself. Nothing of this sort applies to Aristotle whose theory of light excludes any notion of the propagation of light. The fact that Γ.5 omits this factor from the mind — light analogy suggests Aristotle's desire to keep us from taking productive mind to have a *standard* causal role. Certainly, *M* gives it no such role.

What Aristotle *does* want to incorporate from the B.7 account is the characterization of light as both an activity and a disposition [ἕξεως, at 418b19]. What can be made of this, at least *prima facie*, curious dual nature of light? Light, I suggest, is strictly an activity but one that might be called a dispositional activity. While it does not occur without some color, it need not occur as any given color. For Aristotle light does not come in kinds such as blue light, red light, etc. The official view that light is a sort of color of the transparent *qua* transparent does not imply[20] that light is a color that is different from ordinary colors. Rather it is an activity that, while active, is *disposed* to take different manifestations, and must manifest some color.

Indeed just this view seems to be entailed by a pair of B.7's opening theses. The first [418a29–31] asserts that color, the proper domain of the *per se* overlies something else (presumably the transparent) that also is *per se* visible, not in the definitional sense of *per se* but because of having in it the cause of its visibility. The second thesis [418b4–6] asserts that the transparent is visible not in the strict (presumably definitional) sense of *per se* visible but because of the color of something else. Together these theses imply that light never occurs without some color or other and that it itself is visible only insofar as a given color is visible. Analogous points hold for productive mind and its objects.[21]

The picture I want to extract from the analogy is that of productive mind as nothing more than the activity alone of episodes of individual thinking. In terms of *M* it is the inclusion of productive mind as a structural feature that explains how the mind creates its objects simply by its activity. For this just is its activity.[22] It is also productive mind that explains how thinking is possible, thus cashing the second promissory note issued at the outset of Γ.4.

Let me enhance the picture by showing how it accommodates the difficult last sentence of the chapter. Then we shall be in a position to address the more notorious attributes of productive mind. Here is how I read the sentence:

> But we are not mindful [οὐ μνημονεύομεν] because this [τούτο=νοῦς ποιητικός] is not capable of being affected [ἀπαθές] while receptive mind [ὁ παθητικὸς νοῦς] is perishable; and without this there is no thinking [ἀνευ τούτου οὐθεν νοεῖ].

The interpretation I shall suggest fits the lines 430a23–25 into an account of the function and activity of an individual mind. It does so with an ease that is surprising, given the bewildering variety of interpretations so far advanced. First, I assign as the object of "οὐ μνημονεύομεν" productive mind or, more precisely, episodes of productive mind. Whereas we do remember receptive mind or, better, episodes involving receptive mind by virtue of its capacity for images, there is nothing analogous for productive mind. The reason for this is that productive mind is nothing more than the activity, as opposed to the content, of a given episode of thinking. If, for example, I think about something at a given time and later recall the prior thought, what I remember is the object or content of the thought and not the thinking itself. We may be reflexively aware of the activity itself while engaging in it but it drops out as a candidate for memory.[23] It is also clear why Aristotle would raise the issue. Probably only philosophers give attention to the pure activity involved in thinking.[24] Non-reflective thinking, fixed on the object of thought, excludes such thought about thinking and, hence, most thinking subjects are unaware of it. So when Aristotle says that we are not mindful of such thought, he is paying tribute to the fact that we do not notice that aspect of episodes of thinking which is without content.[25]

Second, I take the perishability of receptive mind to amount to the perishability of what occurs in receptive mind rather than to the perishability of receptive mind as a whole (although one *could* talk about *receptive* mind as perishing whenever thinking stops). The passage contrasts what is unaffected [ἀπαθές] with what is perishable [φθαρτός]. So what is perishable is what is capable of being affected and this, I submit, means just that one state of receptive mind gives way as another comes to be. Appropriately, *De Anima* Γ.4. requires precisely this of an account of the operation of the individual mind.

Third, the final clause of the sentence, "καὶ ἄνευ τούτου οὐθὲν νοεῖ," fits the entire chapter into an account of individual minds. Exactly how this works, however, is not completely clear. Hicks took the contrast to be between ἀπαθές and ἀΐδιον and , hence, reasoned that the final clause asserts the dependence of the conditional part of the contrast (receptive mind) on its unconditioned correlate (productive mind). Thus 430a25 would get the reading "and without the productive mind the receptive mind does not think."[26] Interpretation aside, grammar urges reading ἀπαθές in contrast to φθαρτός. This plus the fact that receptive mind is that feature of mind which doesn't answer to the activity itself of thinking requires that the last clause be read "and without the productive mind there is no thinking." Thus the account of *De Anima* Γ.4 is finally completed by explaining that, in addition to the capacity for being affected, the individual mind must also be capable of the sort of activity that goes under the heading of productive mind. I do not, of course, mean to imply by this that the two can proceed independently of one another.

Finally, this interpretation gives productive mind a natural and important place in Aristotle's theory of individual noetic activity. It also explains why productive mind appears exclusively in *De Anima* Γ.5. Since productive mind is required for a conceptually complete analysis of individual *noûs*, the obvious place to introduce the condition is after the discussion of *De Anima* Γ.4 which retailed but did not explain the salient features of individual noetic activity. Where explicit analysis of individual mind is not undertaken, no point would be served by mention of productive mind. On the other hand, were νοῦς ποιητικός God or divine reason, one would expect the notion to occur in other contexts. But it is notoriously absent.[27]

The Attributes of Productive Mind

The theory of productive mind just offered faces a final interpretative task. The remaining lines of *De Anima* Γ.5, namely 430a17–23, attribute a number of characteristics to productive mind, characteristics which most commentators assume hold only of an entity such as God or the prime mover. Even Hamlyn accommodates this intuition by making productive mind an abstract entity required as the metaphysical ground of thought. So it will be necessary to show how these *prima facie* transcendentalistic characteristics can apply to productive mind construed as the activity only of episodes of individual thought. It also turns out that the lines lend support to the hypothesis that Γ.5's distinction between receptive and productive mind is part of a lower level explanation of the sort of mind that Γ.4 discusses.

The characteristics I have in mind are, then, typically located in the following:

> And this mind is (*a*) separate [χωριστός] and (*b*) not capable of being affected [ἀπαθής] and (*c*) unmixed [ἀμιγής] since (*d*) in its being it is activity [τῇ οὐσίᾳ ὢν ἐνέργεια], for (*e*) what acts [τὸ ποιοῦν] is always superior to what is acted upon and the principle to the matter [430a17–19],

and

> And (*f*) actual knowledge [ἡ κατ' ἐνέργειαν ἐπιστήμη] is the same as the thing but (*g*) potential knowledge is prior in time in the individual case [ἐν τῷ ἑνί], although in general it is not prior in time [ὅλως δὲ οὐδὲ χρόνῳ]; but (*h*) it is not the case that it sometimes thinks and sometimes does not [ἀλλ' οὐχ ὁτὲ μὲν νοεῖ ὁτὶ δ' οὐ νοεῖ] [430a19–22],

and, finally,

> (*i*) When separated [χωρισθείς] it is just that very thing that it is [μόνον τοὺθ' ὅπερ ἐστί] and this alone [τοῦτο μόνον] is (*j*) not capable of death [ἀθάνατον] and is (*k*) eternal [ἀΐδιον] [430a22–23].

I shall begin with remarks on the possiblity that productive mind is here accorded some sort of ontological independence. *a* is easily read in this way. Even if this is correct, it is relatively harmless. Given that *a* is grouped with *b* and *c*, the notion of separation here awarded productive mind is too weak to support a Cartesian notion of mind let alone any doctrine of pre- or post-existence. For suppose *b* and *c* referred back to *De Anima* Γ.4's arguments for the independence of mind. Can we now take these arguments to pertain to productive mind? Recall that the mentioned arguments show only that mind is independent of body in the sense that mind is not the actualization$_1$ of any set of physical structures.[28] This is precisely what it means for a faculty to be mixed with body. It also explains the fact that intense objects induce fatigue in faculties that are actualizations$_1$ of physical structures but have no affect on mind. Such objects produce in the physical structures in question durational states that exclude reception of certain inputs.[29] In Aristotle's idiom the faculty ceases to be in a mean with respect ot these inputs (objects). That the mind in general is not subject to such fatigue requires that receptive mind as well as productive mind be independent of any physical structures in the sense that neither is an actualization$_1$ of such structures. Otherwise contemplation of a particularly fine object of thought would after all inhibit the mind's capacity for thought of lesser objects. It follows that the (receptive) mind is dependent on body in the weak sense that it is not the actualization$_1$ of any specific set of physical structures but only requires such structures as a causal condition for its [re]presentational devices (images). Thus in no serious sense does *a* countenance separate existence for productive mind. Hence, we remove one temptation to identify productive mind with a transcendent entity.[30]

The sense just given to separation in *a* is especially plausible in light of *d* which gives as the reason for the separatensess of productive mind that its being is activity. No mention of separate existence need be found here. *b* squares obviously with our interpretation, as does *c*, whether the latter has the sense "unmixed with body" or the sense "unmixed with receptive mind (alternatively, images)."

There is, however, an additional point that demands comment, namely, the fact that Γ.4 has already listed *a*, *b*, and *c* among the mind's characteristics. This has inclined a number of commentators toward the view that both receptive and productive mind are separate, unaffected, and unmixed and that they differ just on the point, mentioned in *d*, that productive mind is, additionally, activity.[31] Unfortunately, problems arise here. First, the view overlooks the fact that *d* is given not just as another, even if distinguishing, feature of productive mind but rather as the *reason* for productive mind being separate, unaffected, and unmixed. Second, Γ.5 certainly appears to deny of receptive mind just these features so how can Γ.4 be supposed to attribute them to it? Finally, there is Γ.5's assertion that receptive mind is perishable. If, as Brentano thought, *a*, *b*, and *c* characterized items are eternal (a somewhat dubious proposition), then Γ.4's mind could hardly be the same as Γ.5's receptive mind. Brentano's rather drastic solution was to deny that the mind that becomes all things, what he called the "*aufnehmende Vernunft*" and took to be the subject of Γ.4, is the same as that which is said to be perishable. Our solution would be to deny that Γ.4 intends in the first place to limit itself to receptive mind. Rather the subject of Γ.4 is simply the individual mind of the ordinary person and Γ.5 provides a (partial) account of how it must be organized to function in the way it does.

So the reason that individual mind is separate, unaffected, and unmixed will be that productive mind has these characteristics. The structure of the passage can be made to support this in the following way. Notice not only that *d* is the reason for productive mind enjoying *a*, *b*, and *c* but also that *e* explains *why d* is the reason. This suggests that Aristotle's reasoning involves the following principles.

8. If (i) what produces is superior to what is affected & (ii) productive mind is the purely active feature in thinking & (iii) what produces in the case of thinking is the purely active feature, then (iv) productive mind is superior to receptive mind

and

9. If (iv) & (v) individual mind has superior qualities *a*, *b*, and *c*, then (vi) productive mind has *a*, *b*, and *c*.

(i) and (ii) are asserted in the passage. (iii) follows from the general principle that what produces must always be something actual. So, if we may assume that separateness, unaffectedness, and unmixedness are superior qualities, at least in the context of Γ.4, then we may assert (vi). Thus, the passage can be construed without requiring that receptive mind also is separate, unaffected, and umixed.

More needs to be said, however, for 9 allows us to say that productive mind has the features in question but not that this explains why *individual* mind has them. And just this is demanded by our interpretation. The following principle, stated at *Metaphysics* 993b24−27 and *Posterior Analytics* 72a29−33,[32] may help:

PT. If *x* has *F* because of *y*, then *a fortiori y* has *F*.

Just as the *Posterior Analytics* ranks as better known those propositions in virtue of which another is known so also might not Aristotle here hold that, if individual mind is separate, unaffected, and unmixed because of productive mind, then *a fortiori* productive mind has these features. And just as the *explanation* of *P*'s knowability involves appeal to the superior knowability of some *Q* so also the *explanation* of individual mind's separability involves appeal to the superior separability of productive mind.

The above interpretation would benefit from a more direct account of how PT applies to the case at hand.[33] Taking the lead from *d*, the suggestion I want to advance is this: It is because of the nature of its *activity* that individual mind must be separate, unaffected, and unmixed; and it is because individual mind enjoys such activity *in virtue of* productive mind that productive mind itself must be separate, unaffected, and unmixed.

Roughly, the idea is that, unlike other kinds of psychical activity, noetic activity is not the activity *of* any set of physical structures. Recall that faculties are distinguished in terms of their objects. Since, for all psychical faculties, the activity, or faculty in activity, is somehow identical with its object, restriction on range of object will entail restriction on range of activity and this in turn, I am suggesting, enforces a corresponding restriction on the physical structures that are the *loci* of the activities in question.[34] It would thus follow that any faculty whose range of objects is limited must itself be unseparate, affected, and mixed *because* its activity just is, in part at least, the activity of certain physical structures. In short, object individuation goes hand in hand with organ individuation.

The fact that mind's objects are unrestricted means, then, that its activity is not the activity *of* any set of physical structures. Now, of course, one might respond that this supports the view that receptive mind also is *a*, *b*, and *c* characterizable. In view of the above paragraph, however, it is unclear why anyone would say this. For, if I am right, ascribing *a*, *b*, and *c* to receptive mind will not, in any case, *explain* why individual mind has these qualities, for such an explanation must proceed in terms of noetic activity and here it is productive mind that plays the crucial role. So it is indeed understandable that Γ.5 would attach *a*, *b*, and *c* to productive mind alone.

We come now to *f*, *g*, and *h*. Although Ross brackets the entire passage,[35] it is in fact quite at home in the chapter. First, Aristotle has just finished a discussion whose interpretation makes use of the thesis of identity between object and active faculty; second, the identity is one of the features (namely, 7 above) Γ.4 attributes to individual mind; third, the solution to how mind can think itself is given in terms of the thesis of identity. So, if nothing else, *f*, *g*, and *h* serve to make exact the sense we are to give to the identity of mind and object.[36] But the passage might also be addressing a more interesting issue. At the end of Γ.4 (430a6) Aristotle says that we must consider why we are not always thinking. Since this query comes on the heels of the assertion that theoretical knowledge is one with its object and since this is just what *f* says, the passage as a whole might be cashing the Γ.4 promissory note. This, at any rate, is the suggestion I want to try out.

Given the identity of actual knowledge and its object (*f*) and the fact that what is known is an eternal feature of the world, there may be some temptation (registered, we may suppose, in the promissory note) to think that individuals are also eternally in possession of actual knowledge. After all the mind in activity is said to be identical with its object. *g* neatly rules this out by emphasizing that in the individual case, which is the case that concerns Aristotle, potential rather than actual knowledge is temporally prior. There are, however, at least two ways to take this. Read with, for example, *Metaphysics* θ.8, 1049b17–1050a3, *g* might mean to assert that actual knowledge is prior in general because for any individual to acquire the knowledge that P is for him to acquire the ability to exercise the knowledge and this presupposes that tokens of such actual knowledge already exist. How else would one learn? Since, however, Γ.4 and 5 show little interest in knowledge acquisition as opposed to its exercise, an alternative account would be preferable.

Metaphysics Z.1 suggests a non-genetic notion of priority in time. Adapted to the case at hand, we get something like the following: *x*'s potential knowledge that *P* is prior in time to *x*'s actual knowledge that $P \equiv$ *x*'s potentially knowing at *t* that *P* does not depend on *x*'s actually knowing at *t* that *P* & *x*'s actually knowing at *t* that *P* does depend on *x*'s potentially knowing at *t* that *P*.[37] In our earlier idiom, being an actual_2 knower entails being an actual_1 knower but not conversely. On this reading, *g* is about episodes of thinking and heads off a potential difficulty for the account of individual thought.

But if this is correct, what role is *h* supposed to play? What else but a transcendent entity could do justice to such an impressive characteristic? Since *we* quite plainly are not always thinking, the subject of "νοεῖ" would appear to be God or an appropriate transcendent analogue. This at least is the received view. But so far from implying, as this requires, that the subject of "νοεῖ" is always thinking, *h* need yield nothing more than what it literally says, namely, that it is not the case that there is something which at one time thinks and at another times does not think.[38] In effect *h* simply denies the existential proposition that there exists a special entity for thinking such that it sometimes thinks and sometimes does not.[39] This is exactly what Aristotle needs to say at this point. For having shown in *g* that his theory does not have the undersirable consequence that individual mind is always thinking, it might then be supposed that individual productive mind must sometimes exist but not think. So Aristotle is quick to point out that productive mind simply isn't an entity of this sort. Socrates may well be such an entity but productive mind surely is not. Indeed, it is not an entity in the usual sense at all: After all it is odd to talk of light as an entity and odder still to countenance the logically difficult locution "Thinking thinks." So *h* can be reconciled with construing productive mind as the activity solely of individual episodes of thinking. It hardly needs pointing out that anyone who held *M* would also be committed to *h*. Aristotle's explicit mention of *h* thus counts in favor of attributing to him *M* or something like it.

What about the admittedly more difficult *j* and *k*? Surely, it would seem, immortality and eternality attach to none but transcendent entities. That appearances may well deceive in this case is, however, suggested by the fact that in *i* Aristotle introduces *j* and *k* by a quite deliberate use of the aorist participle, "χωρισθείς," rather than the "χωριστός" already available from 430a17. Indeed the latter would have Aristotle simply repeating himself.[40] Clearly something new is afoot, but what? Recall that the sense given to "χωριστός" in *u* was not strong enough to yield immortality and eternality. So Aristotle's choice of the aorist participle arguably signals not only that some different sense of separation is meant but also, and in particular, that the separation is sufficient for some sort of ascription of immortality and eternality to productive mind. What is needed is an account of *i* that accommodates *j* and *k* within our general finitistic reading for *De Anima.*

The first of two strategies I shall suggest for accomodation assumes with Zabarella and Hicks that occurrence of the aorist participle in *i* imposes at least one constraint on interpretation, namely, that the separation is to have occurred in past time.[41] But, of course, productive mind may rate either as an individual or as a divine affair. This plus the option of reading "χωρισθείς" either extensionally or notionally yields the following possibilities for *i*:

i 1. when divine productive mind is extensionally separted
i 2. when divine productive mind is notionally separated
i 3. when individual productive mind is extensionally separated
i 4. when individual productive mind is notionally separated.

On *i* 1 it would follow, given the aorist participle, that at some time in the past divine mind, singly or severally, was separated and so that it was previously unseparated. Since to be unseparated from receptive mind is to be unseparated from matter, *i* 1 requires that at one time the divine mind was not separate from matter. But by definition, as well as by proof (of the unmoved mover), it is impossible that divine mind be connected with matter. So *i* 1 will not do.

If we construe notional separation as separation by abstraction, then *i* 2 is no better off. For Aristotle abstraction is always abstraction from something, typically something material. So, again, it can't be divine mind that is separated by abstraction. And this certainly squares with *j* and *k*, for what would be meant, in any case, by attributing, *in abstraction*, immortality and eternality to the divine mind?

i 3 commits Aristotle to some sort of individual immortality, not for all individual mind but at least for individual productive mind. Since here there will be no memory, perception, and the like, proponents of *i* 3 must urge that what thus post-exists the individual is entirely without content. This, of course, raises serious problems about the individuation of such allegedly individual entities. If we finally cannot say what would count as having two as opposed to one such entity on our hands, then either *i* 3 is internally objectionable or it collapses into *i* 1. In either case it fares ill. It would also be rather odd for Aristotle to interject discussion of individual immortality into a context where it serves no apparent purpose. Such objections notwithstanding, for our purposes it is sufficient to provide a credible non-transcendentalistic interpretation.

The remaining candidate is *i* 4. As Hicks indicates there is nothing objectionable about taking the aorist of the active verb to cover separation by abstraction. Less clear, however, and what Hicks doesn't address, is how *j* and *k* can be fit into the abstractionist account. One place to start is with the *Metaphysics* E.1 contrast between the objects of mathematics and the objects of theology. The objects of theology are extensionally separate [χωριστά] and changeless [ἀκίνητα], those of mathematics are not. They may, however, be considered as if they were separable and changeless. Since 1026a10 glosses the latter with "eternal" [ἀΐδιον], we may suppose Aristotle to be saying that mathematics considers its objects *as if* they were separate, changeless, and eternal. *Physics* B.2 takes us a step further, for there such objects are said to be rendered changeless by dint of their being separated in thought [χωριστὰ τῇ νοήσει at 193b34] from motion. So the separating is what yields the changelessness and eternality of the objects of mathematics and, on the assumption that "ἀκίνητον" and "ἀθάνατον" are here interchangeable, we get an exact parallel with *i*, *j*, and *k*. Thus, just as separation in thought is what gives the objects of mathematics apparent transcendental characteristics so also is this the ground for attribution of immortality and eternality to productive mind in *j* and *k*. And just as the objects of mathematics are not extensionally separate so also for productive mind.

In effect, I am suggesting that we relativize ascription of divine properties, counting them as indicators of the most divine thing in us, not of anything absolutely divine. In particluar, I am urging this even for what seems to me to be the most difficult case, namely, ascription of immortality. There is, in fact, independent evidence for relativization. *Nicomachean Ethics* X.7 grants human beings a restricted share in divinity on the basis of an ability to engage in *theoria* and urges, therefore, that we do whatever we can to immortalize ourselves [ἀθανατίζειν]. This can hardly count as making us immotral in the sense of the gods. Hence, it falls short of ascription of any transcendental property to *noüs* (what does the theorizing). This is even clearer from *Protrepticus* B 108 (Düring), which counts *noûs* and *phronesis* alone, of the things in us, as candidates for divine status. For Aristotle says only that this alone *seems* [ἔοικεν] to be immortal and divine. The most this warrants is ascription of a sort of "as if" immortality to human *noûs*, especially in light of the following sentences, B109 and 110, which characterize man as a god *relative* to other animals because of his *noûs*.

So far, then, from demanding a transcendentalistic interpretation, *i*, *j*, and *k* actually support the opposite. The only sense in which apparent transcendental attributes hold of the mind is that in which a certain feature, namely, productive mind, has them and, even then, only in the weak sense of having them in abstraction. It is also clear why, *pace* Aquinas, productive but not receptive mind may be so considered. Not everything is a candidate for abstraction. In particular, nothing whose definition includes or entails matter can be properly considered in abstraction.[42] And, of course, the feature of mind that so requires matter is receptive mind, for an adequate account of receptive mind will require mention of at least general physical structures. In the idiom of *Metaphysics* E.1 it falls to the study of nature to investigate a certain kind of soul, namely, that which is not independent of matter. Only productive mind can be *adequately* considered without mention of matter [note here the "τοῦτον μόνον" at 430a22] and even then only if held separate in its own right from receptive mind. Of course the general account of mind cannot proceed free from such mention because both receptive and productive mind are required for the general account.

A final word is in order regarding how this first strategy makes Aristotle ascribe immortality and eternality to individual productive mind. The parallel with mathematical abstraction is by itself not enough, for we must have some idea of what such an ascription amounts to in its own right. *i* 4 must, in other words, be held to the condition that an object is suitable for consideration in abstraction, only if there is no such actual object but nevertheless we have some idea of what such an object would be like. For objects of mathematics the condition is unproblemtically satisfied, indeed it is worth conjecturing that they would be suspiciously like platonic forms. (In this light Plato's "mistake" was to construe notional separateness as ontological separateness.) Individual productive mind seems rather more recalcitrant. Nonetheless it satisfies the condition in much the same manner. While not actually immortal and eternal, productive mind may, in abstraction, be safely so regarded because not only do we know what such an object would have to be like but also we have actual examples in the divine intelligences and the unmoved mover.

The second promised strategy for accomodation drops Zabarella's assumption that the separation is to have occurred in past time.[43] Thus it is able to read *i* as notionally separating divine productive mind (i.e., *i* 2). How exactly does this work? Assume that productive mind is a generic or determinable notion whose determinations are, or include, individual as well as divine mind. This requires only that in certain general features individual and divine productive mind are alike. Of course, they differ in certain respects as well. In particular only the divine version will satisfy *j* and *k*. In *i*, then, separation is an analytical operation specifying for consideration the divine kind of productive mind. Thus we may paraphrase the entire passage as follows: "When one has specified that type of productive mind that can when all alone be just what it is, then (unlike individual productive mind, which is the type that can be separated only in the weak sense of *a*) we have something that is not capable of death and is eternal (namely, divine mind)." In short the separating in *i* is something *we* do when we consider one rather than another variant of productive mind. This means, of course, that divine mind makes an appearance in *De Anima.* But it does so by way of removing itself from the sort of productive mind that is germane to the *De Anima* account. So that account remains stubbornly finitistic.[44]

I have been arguing that Aristotle's program in *De Anima* does not call for transcendent entities or properties and that this holds for Γ.5's productive mind as well. By way of conclusion let me add a final bit of confirmation for this thesis. In A.1, 403a3–15 Aristotle introduces the topic of mind's separability from body in terms of its dependence, or lack of dependence, on imagination. If thought depends on imagination, then even *noûs* will not be separable extensionally. Γ.8 then appears to establish the point by arguing that all thought requires images. But this may well not be conclusive because the argument for thought's dependence on images employs a premise that Aristotle himself considers false, namely, that there are no actual things apart from extended things. Recognition of the unmoved mover alone falsifies this.

So it may be that Γ.8 is not intended to argue a thesis for all thought (human thought, perhaps?). That this does no harm to my argument is clear from the fact that the final sentence of the preceding chapter raises, but immediately *defers*, the question whether the mind can think separate objects should it itself lack separateness. What is most striking here is Aristotle's staunch neutrality on the extensional separability of *noûs*. Moreover, this is precisely his attitude toward the question at *De Anima* B.2, 413b13–16 and in Γ.4. In light of this, the contention that Γ.5 countenances extensional or *in re* separateness for productive mind loses plausibility, for it asks us to find there an outright assertion of such separateness even though flanking chapters unabashedly promote neutrality. Surely this is asking too much. Besides, if we have succeeded in locating Γ.5 within a fundamentally functionalist account of individual noetic activity, the question need not arise at all.

University of California at Davis

NOTES

* Some twenty years ago Marjorie Grene gave eloquent expression to the puzzles that surround Aristotle's doctrine of *noûs*. In returning to the topic on this occasion I do not pretend to have solved all the puzzles but I do hope to have charmed her, at least a little, by the chase.

[1] Aquinas (1925), §§742–743; Ross (1961), pp. 47–48; Rodier (1900), vol. 2, p. 465; Hyman (1982); possibly also Sorabji (1982).

[2] Clark (1975), pp. 184–186; Brentano (1867); Guthrie (1981), vol. 6 (*Aristotle – An Encounter*), pp. 322–324; Hamlyn (1968), p. 140.

[3] Wilkes (1978), pp. 115–116 and, possibly, Barnes (1971–1972), p. 113.

[4] For more in the way of defense see Wedin, *Mind and Imagination in Aristotle*.

[5] I do not deny the relevance to my topic of *De Generatione Animalium*'s discussion of *noûs* entering from without, of the suggestion at 408b18 that *noûs* comes about in us as a sort of substance, and in general of those passages allegedly attributing divinity to man's *noûs*. Although such passages can, I believe, be brought into line with the present account, considerations of space require that I leave them aside.

[6] One might, for instance, try to argue for mind's dependence on body from the functional role of images in thought.

[7] Notice that the analogy generates two alternatives: either thinking will be a certain being affected [πάσχειν τι] or it will be something different like this [τι τοιοῦτον ἕτερον]. It is clear, as I requires, that Aristotle wants the second alternative. B.2, 417b6–7 announces that perception itself is not an instance of πάσχειν in the strict sense because it does not involve the destruction of something by its contrary. Rather perception involves something (the faculty of perception) developing into itself and into actuality. The latter is called "being affected" by convenience only because "there is no name for the difference" [418a1].

[8] For a persuasive case against emendation see Owens (1976).

[9] See, for instance, 417a17–18: "Everything [that is affected or moved] is affected or moved by something that is able to produce this and is in actuality."

[10] The caveat is simply to hold at bay passages, such as *De Anima* Γ.2, 426b22 & 426b31–427a1, that appear to countenance αἰσθητά as objects of thought.

[11] That *M* does not here commit Aristotle to the absurd thesis that mental content is created from nothing is clear from the fact the *M* is a thesis about fully capable noetic agents, the sort for whom the model of thinking sketched in 429b5–9 holds. For such agents do not simply create objects of thought on their own; rather objects of thought are already part of the public domain to be entertained at a thinker's discretion. Indeed this is at least part of the effect of the, otherwise difficult, lines 430a19–21: "Actual knowledge is identical with its object but potential knowledge is prior in time in the individual but not prior even in time in general." Although Hamlyn brackets the sentence, presumably because the point of their inclusion is unclear, it is quite appropriate as a warning against mistaking the sense in which mind produces objects of though. See also below, page 186.

[12] *M* also seems tailor-made for the dilemma, raised at *Physics* H.3, 247b5–8, concerning how that which knows potentially comes to know actually not by moving itself but by the presence of something else. Since mind's objects are internal, the only way to avoid the outlawed case of the mind moving itself is to have it produce the object that affects it.

[13] Of course the gain is not without cost. For may not the gap be relocated at the point of contact between objects of thought and the world? Subjectivity is not so much the issue, Aristotle seems open to a model of concept acquisition that is globally intersubjective, but rather the details of how the model works are what are in question. Nowhere does Aristotle give us a sustained and adequate account of his views on concept acquisition.

[14] Nozick (1981), esp. "Self-Synthesis," p. 87ff.

[15] Some might retain "ὥσπερ" on the grounds that sometimes it is used to soften or apologize for introduction of a locution that is out of context (see Liddell, Scott, Jones, *A Greek-English Lexicon*, p. 2040). For in the present case a principle from the theory of nature is suddenly enlisted to explicate a point about the mind.

[16] Even had Aristotle written "each thing" rather than "each *kind* of thing," there would be no warrant for identifying the productive factor in nature with the prime mover. Exploiting the scope ambiguity of the quantifier expressions "ἐν ἁπάσῃ τῇ φύσει," "ἑκάστῳ," and "τῷ ποιεῖν πάντα," we could adopt [1A] (y) $(\exists x)$ $(y$ is in nature $\supset$ x produces $y)$ rather than [1B] $(\exists x)$ (y) $(y$ is in nature $\supset$ x produces $y)$ as the correct (partial) reading of the text. Besides this, 1B would, for the case of mind, require that some account be forthcoming of the relation between individual thinking and the master activity of a single global productive mind. But Aristotle nowhere even hints at the need for such an account (in contrast, incidentally, to Averroes and Avicenna) and it is difficult to see what it would consist in. In particular, adoption of 1B, or something like it, would seem to preclude any account of *individual* noetic activity.

[17] See, for instance, *On Generation and Corruption* A.7 and *Metaphysics* Θ.8.

[18] It is clear from *Metaphysics* Θ.8 that producer unaffectability is the intended point of analogy. 1050a30–b1 distinguishes production that results in something apart from the producer from that which does not. In the former, exemplified by building, the activity [ἐνέργεια] is in the result; in the latter, exemplified by *theoria*, the activity is in the producer [ἐν τῷ θεωροῦντι].

[19] This is not as odd as it sounds. I have in mind simply the case where the individual mind is contemplating and so may be said to be identical with the triangle as an object of thought.

[20] *Pace*, for example, Rist (1966), p. 12.

[21] Contrast Ross (1961), p. 296, who finds the characterization of productive mind as both ἕξις and ἐνέργεια to be "the most conspicuous instance" of the carelessness he finds pervasive in Γ.5.

[22] Two passages outside of *De Anima* support our interpretation of what it means to characterize productive mind as a *hexis*. The first, *Metaphysics* Δ.20, lists as the first sort of *hexis* "a sort of *activity* of the haver and of what he has [ενέργειά τις τοῦ ἔχοντας καὶ ἐχουμένου]" and says, in agreement with *M*, that this sort of having cannot be had on pain of infinite regress. The second passage, *Metaphysics* Λ.7, 1072b19–22, explains the thesis that *noûs* thinks itself (in language, incidentally, that closely reflects feature 7 of Part I) on the grounds that for the mind to be actual is just for it possess the object of thought [ἐνεργεῖ δὲ ἔχων]. Since this is just the *hexis* mentioned above, being a *hexis* of a certain sort is nothing other than having an object. So on this point *M* and Γ.5 go hand in hand.

[23] In this sense 430a23–25 plays out the message of *De Memoria* 450a12–13 that even memory of objects of thought requires images.

[24] The analogy with light further supports this line of interpretation. *De Anima* B.7,418b4–6 says that the transparent is not, strictly speaking, visible in itself [οὐ καθ' αὑτὸ δὲ ὁρατὸν ὡς ἁπλῶς εἰπεῖν] but only because of perceiving the color of something else. Analogously, mind is not, strictly speaking, thinkable in its own right. Thus, not only is it the case that the suggestion that we remember the activity of productive mind turns out to be a non-starter but also we have here a view that is obviously of a piece with Aristotle's own account of what is involved in mind thinking itself.
[25] Contrast this with Hamlyn (1968), p. 141, who is unable to find a plausible way to make productive mind the object of "οὐ μνημονεύομεν."
[26] Although Hicks does not say exactly this, it is what he is committed to. See Hicks (1907), pp. 509–510. Hamlyn, on the other hand, explicitly adopts the reading.
[27] And, if productive mind is elsewhere mentioned, it would appear to be in explicit discussions of thinking, such as *Metaphysics* Λ.7, 1072b19–22. In any case this passage agrees nicely with our interpretation (see note 22 above).
[28] Indeed, as the sense given to "χωριστός" at the outset of Aristotle's analysis [*DA* B.1, 413a3–7], it would appear to govern discussion of separation in the balance of the work. Unless, of course, other indications are forthcoming, as in the "χωρισθείς" at 430a22 discussed below.
[29] See, for instance, *DA* B.12, 424a28–32.
[30] It might be urged that *b* and *c* manage to contrast productive and receptive mind, even if *a* is understood not as separate from body but as separate from receptive mind. Here productive mind would not be dependent even in this weak sense. But there is no reason to find in this a claim stronger than that the noetic feature determined by productive mind does not properly involve images. It is sufficient to hold that the purely active feature of an episode of thinking can, for purposes of analysis, be separated from its content. This, incidentally, is quite like the manner in which light is distinct from the transparent and from any given color of the transparent.
[31] For example, Hicks (1907), p. 502 and Rodier (1900), who goes so far as to urge, pp. 460–463, that *a*, *b*, and *c* apply in *different senses* to receptive and to productive mind.
[32] "A thing has an attribute more than other things, if it is in virtue of it that the same quality [τὸ συνώνυμον] belongs also to the other things" and "for that by which an attribute belongs to something always has it more," respectively.
[33] It is, for instance, clear from examples alone how PT applies to the case of scientific knowledge. For *Q* to be more knowable than *P* is, at least, for *Q* to be logically independent of *P* but not conversely. Thus the proposition that planets are proximate heavenly bodies can be known apart from propositions it explains, propositions such as that planets twinkle.
[34] This move needs more argument. Perhaps the discussion in *DA* Γ.1 concerning the exhaustiveness of the five senses contains the seeds of such an argument.
[35] Ross's arguments for bracketing 430a19–21 are curious at best. First, 430a19–20 reappears at 431a1–3. Although Alexander obelizes the latter lines (Brentano and Rodier are comfortable with them at both places), Ross thinks they fit better in Γ.7 because that chapter he regards anyway as a "collection of scraps." This is surely a bizarre sense of fit. Second, the text is alleged to read more smoothly without the lines: "this sort of mind is separable . . . and when it has been separated. . . ." But

notice that the first argument does not justify bracketing 430a22 (our *h*) and the second argument requires bracketing from "ἀεὶ γάρ . . ." in 430a18 (our *e*). Thus Ross's arguments recommend bracketing a full third of the chapter! Surely a more acceptable alternative should be sought.

36 Earlier promises notwithstanding, it is appropriate to say something about the thesis (call it *A*) that mind thinks itself because of its connection with the thesis (*B*) that mind and object of thought are identical. That there is a connection is clear enough: The entailment from *A* to *B* is asserted at 429b5–9 and explained at 430a2–5. What is in question is the meaning of *B*. There are, initially, four ways one might take *B*: (*B* 1) Mind introspects or intuitively grasps itself; (*B* 2) Mind studies itself *qua* mind as just another object of scientific investigation; (*B* 3) Mind can retain its contents and this amounts to mind knowing itself; (*B* 4) Mind thinks itself in thinking any proper object of thought (an object without matter). Since I favor *B* 4 something should be said about the alternatives. *B* 1 is a difficult doctrine in its own right and there is virtually no evidence to support its attribution to Aristotle. Indeed, our discussion so far (see note 24 above) indicates that we are aware of the mind only insofar as we are aware of particular thoughts it has. *B* 2 is true, witness *De Anima* itself, but irrelevant. Γ.4 and 5 do not even pretend to be addressing the *program* of *De Anima* but clearly are treating a specific topic that arises in the course of carrying out the program. Further, Aristotle's explanation of the possibility of *B*, namely, a certain version of *A*, suggests not *B* 2 but *B* 3 or *B* 4. But *B* 3 is difficult because, if to know mind's contents is for mind to know itself, then it seems that mind will have to know some complete set of objects. But this is an implausible reading of *A*, for 429b5–6 says "whenever the mind has become *each* thing [= ἕκαστα and not ἅπαντα which would be required for "all"]. Besides this, *B* 3 denies *B* any interesting theoretical role in the account of noetic activity. *B* 4, on the other hand, squares with the explanation at 430a2–5 and makes *B* part of a theory of thought.

37 1028a33–4 glosses the primacy in time of substance: "of items in the other categories, none can exist separately, but only substance."

38 Most recently, Richard Sorabji takes the unduly strong reading, (1982), p. 305. It is also the view of P. Merlan (1967). In fact, *h* is almost universally held to say that productive mind is always thinking. But, as we have shown, *h* says no such thing at all and, were it to, it would flatly contradict Γ.4,430a5's assertion that we are not always thinking. Hicks (1907), p. 505, registers but does not resolve the "glaring contradiction." Ross leaves *h* without comment but probably bracketed it for the same reason (see note 35 above). Others, for example Brentano, avoid contradition by making divine mind the subject of *h*. But this proliferation of subjects is not needed.

39 Again, the parallel with light is instructive. For there it is obviously mistaken to suppose that light is the sort of thing that is sometimes active and sometimes not. It just is a sort of activity and, hence, is either active or simply non-occurrent.

40 Although I have translated "χωριστός" in *a* as "separate," "separable" might be happier in light of the "χωρισθείς" in *i*. This is acceptable since I have given a weak reading anyway to "χωριστός."

41 But see note 43 below.

42 The requirement is made clear in *Metaphysics* Z.11's claim that because man (unlike circle) is found in a certain sort of matter only, alternatively, is defined as a certain form in a certain matter, it is not possible for us to separate the form [διὰ τὸ

μὴ καὶ ἐπ' ἄλλων ἐπιγίγνεσθαι ἀδυνατοῦμεν χωρίσαι at 1036b7].

43 The first strategy can be pursued even if, as I suspect, Zabarella is wrong to insist that the aorist carries a past time constraint. It is enough, to rule out *i* 1 and *i* 2, that the separation occur at some time. This would, however, not exclude a view on which productive mind, while neither individual nor divine, is nonetheless singular. Here there would be but one productive mind for all individual thinkers (presumably in the way that there is, some argue, just one substantial form *per* species) and differences in individuals will track differences in receptive minds. While I applaud the non-transcendentalistic possiblities of this view, it fails to provide any account of how individual thinking comes about. Moreover, if this single productive mind is always thinking (and, as activity, it would have to be), then we have no account of why we are not always thinking. But we have been promised both accounts.

44 If I am correct in taking *M* as a general thesis on productive mind, then we may want it to cover divine productive mind as well. It turns out that *M* does indeed apply to the unmoved mover as a self-thinker and in a rather surprising way. We know from *Metaphysics* Λ.6 that the unmoved mover must always exist and by *M* we know that, as a purely thinking thing, it can manage this only if it always thinks itself. It will, in short, guarrantee its eternal existence by continuously producing itself through eternally thinking itself. So the unmoved mover is just the limit case of *M*.

BIBLIOGRAPHY

Aquinas, Thomas: (1925), *Sancti Thomae Aquinatis in Aristotelis Librum de Anima Commentarium*, Marietti, Turin.

Armstrong, A. (ed.): (1967), *The Cambridge History of Later Greek and Early Medieval Philosophy*, Cambridge University Press, Cambridge, England.

Barnes, J.: (1971–1972), 'Aristotle's Concept of Mind', in *Proceedings of the Aristotelian Society* **72**, pp. 101–114.

Brentano, F.: (1867), *Die Psychologie des Aristoteles*, Kirchheim, Mainz.

Clark, S.: (1975), *Aristotle's Man*, Clarendon Press, Oxford.

Guthrie, W.: (1962–1981), *A History of Greek Philosophy*, 6 vols., Cambridge University Press, Cambridge, England.

Hamlyn, D.: (1968), *Aristotle: De Anima*, Clarendon Press, Oxford.

Hicks, D.: (1907), *Aristotle: De Anima*, Cambridge University Press, Cambridge, England.

Hyman, A.: (1982), 'Aristotle's Theory of the Intellect and Its Interpretation by Averroes', in O'Meara, 1982, pp. 161–191.

Merlan, P.: (1967), 'Greek Philosophy from Plato to Plotinus', in Armstrong, 1967, pp. 14–132.

Nozick, R.: (1981), *Philosophical Explanations*, Harvard University Press, Cambridge, Massachusetts.

O'Meara, D. (ed.): (1982), *Studies in Aristotle*, Catholic University Press, Washington, D.C.

Owens, J.: (1976), 'A Note on Aristotle's *De Anima* 3.4, 429b9', in *Phoenix* **30**, pp. 107–118.

Rist, J.: (1966), 'Notes on Aristotle's *De Anima* Γ.5', in *Classical Philology* **61**, pp. 8–20.

Rodier, G.: (1900), *Aristotle: Traité de L'Âme*, 2 vols., Vrin, Paris.
Ross, W.: (1961), *Aristotle: De Anima*, Clarendon Press, Oxford.
Schofield, M. and Nussbaum, M. (ed.): (1982), *Language and Logos: Studies in Ancient Greek Philosophy Presented to G. E. L. Owen*, Cambridge University Press, Cambridge, England.
Sorabji, R.: (1982), 'Myths About Non-propositional Thought', in Schofield and Nussbaum, 1982, pp. 295–314.
Wedin, M.: (in progress), *Mind and Imagination in Aristotle.*
Wilkes, K.: (1978), *Physicalism*, Routledge and Kegan Paul, London.

PART III

SCIENCE AND EXPLANATION

NANCY CARTWRIGHT

TWO KINDS OF TELEOLOGICAL EXPLANATION

Majorie Grene has written about teleological explanation in biology, and I follow her lead. If teleological claims were eliminated from our biological theories, the explanatory power of what is left would be critically reduced. Both the *teleology* and the *explanation* in this claim are to be taken seriously. Many important and interesting attempts to save teleology do so only by turning teleology into some kind of efficient causality. Ernest Nagel's[1] classic treatment of functional explanation is a good example. So, too, is Larry Wright's[2] more recent analysis, which claims that an account is teleological in so far as the functional or purposive action is brought about by the fact that the occurrence of that action tends to bring about the specific effect. According to Wright there are no mysterious, unscientific processes involved. The "bringing about" is an ordinary causal process, like most others employed in scientific explanations, a process that can be observed, that can be described, and that can be tested. But from a point of view that takes teleology seriously this very strength in Wright's analysis should be a weakness. Even though the efficient causation involved is of a very specific kind, teleology in the end is reduced to the operation of efficient causes. The teleology is lost.

Real explanatory power may be lost as well. Here is one example: Many biologists and philosophers argue that teleological explanations are superior because they treat the phenomena at a manageable level of description; for many situations, a genetic account or the full biochemical story would be so unwieldy as to be practically useless. Martha Nussbaum[3] gives one such argument in her discussion of Aristotle's *The Motion of Animals*. She contrasts two explanations of a lion's behavior, one that describes the motions of its molecules, and the other that points to its desire for meat. Nussbaum says, "if you were a shepherd in charge of flocks, which account would give you more information that was relevant to your plans and precautions? From eight lines of Homer I can learn more that is general and valuable about the behavior of lions than I would from two volumes of atoms and charts."[4]

A. Donagan, A. N. Perovich, Jr., and M. V. Wedin (eds.), Human Nature and Natural Knowledge, 201–210.

But this common type of defense[5] attends to too narrow a range of human plans and precautions. If I am primarily concerned that my Diffenbachias should not grow lopsided, I may be content to know the general maxim, "Green plants seek light." But if I suspect that the production of carcinogens in tobacco leaves is a threshold effect that depends minutely on the amount of sunlight received, I will need to know exactly what auxins are produced and precisely how they operate. Observations like Nussbaum's, while both true and important, do not give a very persuasive defense of the explanatory power of teleological claims. Even poor tools may perform very well if they are set only a limited number of tasks.

I will propose here an account of teleological explanation that tries to do both jobs that supporters of teleology demand: (i) it does not reduce teleology to efficient causality; and (ii) it makes the power of teleological explanations dependent on the structure of the phenomena, and not on the focus and limits of our interests. A primary obstacle for the first task is the apparent completeness of the efficient causal processes that originate in microstructural events. There seems no room for the operation of purposes. Thus the teleologist retreats. Teleology fills not a gap in nature but a gap in our understanding. But then the second task becomes crucial: to give an account of this understanding that makes it a recognizable part of a legitimate scientific description, and not just a fable.

I am going to argue that the kind of understanding that teleology promises in biology is exactly the same as that provided by a covering-law theory in physics, no more, no less. The idea is from a second argument of Martha Nussbaum's. I aim to set this argument into a framework that makes explicit how teleology can, on the one hand, play the same kind of explanatory role that derivations from fundamental laws play in physics, without, on the other, competing with the causal accounts that tell how, step by step, each separate biological occurrence is brought about. Admittedly, my framework depends on a certain pragmatic view about how physics itself explains. This view is described at length in *How the Laws of Physics Lie*.[6]

1. It is well to start with Aristotle. It is common nowadays to warn that the simple translation of *aitia* as *cause* may be misleading, since modern readers tend to use *cause* in a way most akin to Aristotle's efficient causes. It is better, we are told, to think of *aitia* as explanatory factors. Still, one wants to know, do all four factors play a role in the actual physical production of the explanandum, or not?

John Cooper argues that the answer to this question for Aristotle is undoubtedly, yes. The material and efficient principles for the interaction, for example, of "the watery mobile stuff of the male semen" with the "warm wet" material supplied by the female, are simply not sufficient to produce the complicated structure of the ensuing fetus, argued Aristotle. Indeed Aristotle had good reason for this belief, claims Cooper: "The ancient atomists had no *empirical* reason to think that the powers attributable to matter of different kinds were sufficient to determine any of the actually observed outcomes."[7] But Cooper admits that this is not a reasonable view today "with the hindsight made possible by post-Renaissance experimental physics and chemistry."[8] This kind of interpretation does make a genuine place for teleology, but not in our current scientific view.

What can we learn from the more generous reading of *aitia* as *explanatory factor*? Here our second problem arises. In just what sense do *aitia* explain if they form no part of the causal process that produces the outcome? Julius Moravcsik[9] proposes an answer, and Joyce Kinoshita[10] generalizes this to a complete theory of explanation. Explanations give answers not only to *why* questions, but also to *what* questions. They say *of* something, what it really is. For Aristotle, the world is made of substances, and there are four parts to the concept of a substance. Each of the *aitia* corresponds to one of these. Teleological factors are explanatory for Aristotle, just because they are part of the correct identification of the explanandum. This is the right kind of answer, for it provides a plausible theory, stripped of the concerns or intelligence of the questioner, of what explanation consists in, and shows the role that teleology plays there. But again, this answer does not bring teleology into modern biology, since we do not begin with the assumption that the world is fundamentally composed of Aristotelian substances.

Martha Nussbaum offers yet another proposal. Teleological factors do not supplement material and efficient causes, as Cooper argues. Rather, they compete, and they compete well: "the teleological account . . . is superior in generality and predictive value."[11] Nussbaum contrasts teleological descriptions with certain low level efficient-material descriptions that teleological laws unify. I think phototropism is a good example. If a potted plant falls on its side, the stem bends upwards; if it stands for long on a sill, the leaves turn towards the window; if, experimentally, it is placed in a black box with a hole in the side, it grows towards the hole . . . In short, green plants seek light.

Nussbausm says: "Instead of a conjunction of efficient causal sequences (in C_1, O does x_1, in C_2, O does x_2; etc.), we have the simple law — from which, unlike the conjunction, we can make predicitions about new situations — that in $C_1, \ldots, C_n$, O will "choose" (within natural limits) whatever behavior will bring about y, where y is some component of the creature's *logos*."[12]

This answer promises more relevance for modern science, since the generality and predictive value that Nussbaum claims for teleology are features expected of scientific explanation today. But is her claim plausible? Are we really to believe that a teleological law like "plants seek light" has more generality and predictive power than a causal account that details how light stimulates growth hormones, how these hormones are transmitted, how they differentially affect cells at different parts of the plant, and how this bends the stem? If we had such a scheme at hand, it seems that its predictive power would be complete, or at least as complete as nature itself allows. As with the shepherd and the lion, the predictions may be far finer than we care to bother with, but contrary to Nussbaum's suggestions, the possibility for accurate and for novel predicition from a straightforward biochemical causal account is far better than from a teleological one.

Nussbaum's claim seems to gain its intital plausibility from a conflation in levels of explanation. Following her interpretation of Aristotle, Nussbaum identifies the final cause of an organism's behavior with the formal cause of the organsim: The paradigm healthy adult. These two-as-one she contrasts with the coupled efficient-and-material causes; and she argues for the superiority of the first pair over the second. But we have different categories. We begin with what Wilfrid Sellars calls "the manifest level" — "everyday" facts about persons, animals, and plants, facts that can be confirmed by Mill's method for ordinary observation and induction. Both final and efficient causes can operate at this level.

But for maximal predictive power, we look not to manifest-level causes of either kind, but rather to micro-structural explanations. These, too, are often called 'causes'. Ernan McMullin remarks on this explicitly:

> When the properties or behavior of a complex entity are explained by alluding to the structure of that entity, the resultant explanation may be called a *structural* one. . . . Such explanations are causal, since the structure invoked to explain can also be called the *cause* of the feature being explained. Thus the cause of the sun's emission of radiant energy is the hydrogen fusion going on within the sun.[13]

Nussbaum's defense focuses on efficient-material causes from the manifest level; the "*C*'s" in her scheme are to be filled in with ordinary descriptions of the distance of the plant from the windows, for example.[14] But micro-structural causes might as well be classed as 'efficient-material', and it is these that promise to explain better than teleology. The superiority which Nussbaum finds is a superiority not over micro-structural accounts, but rather over a collection — a short list — of manifest level generalizations, involving manifest level causes. In the philosopher's familiar terminology, she finds her teleological laws to be better than an odd assortment of "low-level phenomenological laws"; but the argument goes no way to undoing the power of chains of efficient causes that operate from the micro-structures upwards.

Yet this way of putting the objection reveals a place for Nussbaum's argument. We may look to physics for a parallel. There two separate and different kinds of explanations are given, side-by-side, for the same phenomenon. First, we may lay out a detailed causal account of exactly how the phenomenon is produced. Secondly, the phenomenon will be fitted into a general explanatory theory: the mathematical laws describing the phenomenon will be derived from a complex of more fundamental laws that bear on a wide variety of different phenomena as well. I will illustrate with a brief summary of an example that I have discussed before,[15] in case readers want to look at the two explanations in detail. Lines in a spectroscope are often not completely sharp, but instead smeared over a small region. How is this line broadening to be explained? Well, the atoms emit photons of a given frequency, they reabsorb them, they remit them, but not always at exactly the same frequency. In physics we are taught that line broadening is due to the emission and reabsorption of real photons. That is the causal story of how the smear gets produced.

Besides the causal story, I can also show you a theoretical treatment of line broadening. The treatment begins with the Schroedinger equation and a variety of auxiliary assumptions. We fill in the forces or energies for an atom in interaction with the electromagnetic field; simplify; approximate, and eventually derive a phenomenological law that describes broadening. This second kind of explanation is in no way causal. We write down some laws that we think of as fundamental and show how others, more complex and less general, can be dervied from them. But we do not trace the causal process by which the phenomena described in the more complex law are produced. The derivation provides explanations of a different kind, the kind that theoretical subsumption in general provides; unification of diverse phenomena into one systematic scheme.

It is just this kind of explanatory power that Nussbaum points to in teleological claims. In this view teleological and micro-structural explanations are not in competition, nor do they separately contribute, each their own bit, to the same explanatory story. Rather they provide explanations of very different types. The first lays out how bits of regularly occurring behavior are produced; the second takes the complicated laws that describe this behavior and derives them from a set of laws more simple, more abstract, and more general. If the second kind of explanation is a good one, the basic explanatory scheme should have breadth, precision, and the possibility of novel prediction. All of this Nussbaum claims for teleological schemes.

We may ask again, "Are Nussbaum's claims correct?" To answer the question properly one needs to look carefully at a lot of examples to see just how essential a role the teleological claims play in the kind of explanatory schemes she points to. But I think that, put in the context of the second kind of explanation, her defense of the explanatory power of schemes using teleological claims has a great deal of initial plausibility.

The examples given so far are very simple in structure, however; the force of the plausibility may not be apparent. I shall close by describing one example in the recent history of biology that shares some of the richness of structure that explanatory theories in physics have. The example is that of biological clocks, which came to be seen *as* biological clocks in large part due to the work of Colin Pittendrigh. Pittendrigh himself argues for the importance of teleological thinking in his research so we need only to follow his own arguments.

2. We have long known that plants and animals have rhythms that repeat approximately every 24 hours. These are called "circadian oscillations". In the early 1950's these oscillations came to be seen as manifestations of an internal clock. The first impetus came from studies that showed that honey bees orient themselves by the poistion of the sun. This raised a problem. Information about the locations of the honey source is communicated to other bees by the wagging dance. Yet the dance occurs in a dark hive, and sometimes considerably later than when the source is originally discovered. So the dancer must somehow change the orientation of the dance as the day goes on,

and to do to so she must have some internally accurate time-keeping mechanism — the biological clock.

We say, then, of the mechanism that produces circadian oscillations in the bee: this mechanism functions as a biological clock. On the other hand we might just say: certain very felicitous behavior in the bee runs in parallel with, and is partly caused by, the cycles of this mechanism. The teleological claim is richer in its implications, and in ways borne out in Pittendrigh's own research on fruit flies. For example, fruit flies eclose only at dawn, when their moist wings will not be dried out by the heat of the day. Careful experimentation shows that this is true even if they are ready before. This provides structure to the biological clock in the fruit fly: ciradian oscillations in fruit flies work "like a master clock on a synchronous computer."[16] When something is ready early, a gate is shut to keep it waiting for the slowest component.

A second consequence is with respect to temperature changes. With no further information, one would expect a rhythmical mechanism in a biological organism to be highly sensitive to temperature changes in the external environment. Exactly the opposite is to be expected if the mechanism is to serve as a clock. In fact, the oscillations should be resilient under any normally occurring variations in the environment. Again, experiment strikingly supports the consequences of the functional description. Here is what Pittendrigh himself says about these experiments:

> It was the hypothesis that circadian oscillators were indeed the biological "clocks" implicated in the discovery of time-compensated sun-orientation that led Pittendrigh to look for and find the invariance of their frequency in the face of temperature change. The temperature-compensation of [the period] has been stressed ever since 1854 as one of the most characteristic features of circadian oscillation generally. But in that emphasis, we seem to have lost sight of a more general propostion that the same teleonomic (functional) logic invites: that to function as a reliable clock, circadian should be homeostatically protected against frequency change by *any* variation in the cellular milieu.[17]

A structure much like an explanatory theory in physics emerges. So long as we restrict attention just to the time-compensation of the bee dance, the teleological description and the non-teleological one amount to much the same. So too with the Schroedinger equation and its phenomenological consequences, when we consider just a single quantum system, like the radiating atom. But the Schroedinger equation can be coupled with other pieces of knowledge to yield new consequences for new situations. The phenomenological laws, that before stood separate and isolated, are fitted with the aid of the Schroedinger equation into one coherent scheme. The functional descripition of the circadian oscillation provides the same kind of unification.

One may still puzzle, "How can the fact that the oscillator functions as a clock make it temperature resilient?" But this is an entirely wrong kind of question for theoretical style explantion. What makes it temperature resilient is some happening in the distant evolutionary history of the species. The teleological claim does not give the cause of the temperature resilience. Rather, it organizes it into a single theory in which time-gating and compensation for other environmental variations figure as well.

When will such functional ascriptions be correct? As with any theoretical claim, when enough of its consequences are borne out. The theory is better the wider the variety of consequences, the greater their precision, the more roads for the further investigation opened, the more problems solved, and so forth. There is one useful side-effect to seeing teleological explanation as theoretical rather than causal, that I would like to point out in conclusion: the theoretical point of view allows us to separate teleology from evolution. Evolution is supposed to provide a ground, in efficient-material terms, for teleological causation. How can the purpose or end of a piece of behavior serve as its cause? Here is the answer: Organisms that were able to serve that purpose were selected for; serving that purpose is that part of the chain of events that causes organisms that do behave in that way to survive and develop. This idea has led biologists to identify the function of a piece of regular behavior with the consequence of that behavior (if any) that was the historical basis for natural selection. But this identification is unfortunate. Very often a new and very useful function can piggyback on a process whose selection value was entirely different. Again, Pittendrigh sees this in his work in biological clocks: "It is clear, however, that this spectacular clock fucntion is recent in the history of life; it has exploited already existing circadian oscillations; it does not account for their initial evolution."[18]

Yet the connection between selection and teleology is a natural one so long as functional claims are supposed to describe causes. We want to know, "What are the grounds in the empirical corpus that warrant a teleological claim?" The evolutionary answer is: *p is the purpose of behavior b* is empirically warranted when *p* is a consequence of *b* that served as the primary basis for natural selection. When functions need no longer be causal, we can instead give a different answer modelled on conventional views about theory acceptability in natural science: *p is the purpose of behavior b* is empirically warranted when it is an essential claim of a total explanatory theory that best saves the phenomena. (Use your own favorite account of what constitutes "saving the phenomena best" — or even well enough.) Here is kind of teleology that is no efficient causality in disguise. Nor does its ability to explain depend in any important way on purely local interests. Its explanatory power is coextensive with its empirical warrantability, and how great that is depends on what kind of theoretical organization works best in practical biology.

Stanford University

NOTES

* This paper was written while I was doing research at the Center for Interdisciplinary Research in Bielefeld, West Germany, and I would like to thank the Center for its support and also Lorenz Krueger, who made this possible.

1 Nagel (1961), chapter 12.

2 Wright (1976).

3 Nussbaum (1978).

4 *Ibid.*, p. 70.

5 Cf. also Sorabji (1980), chapter 10.

6 Cartwright (1983).

7 Cooper (1982), p. 211.

8 *Ibid.*

9 Moravcsik (1974).

10 Kinoshita (1983).

11 Nussbaum (1978), note 4, p. 80.

12 *Ibid.*, p. 80.

13 McMullin (1978).

14 We might try instead to exploit Nussbaum's idea by filling in the *C*'s with micro-structural descriptions rather than manifest level descriptions, and sometimes this may be what Nussbaum has in mind. But in that case her claim for superiority of the teleological claims loses its force, for a general understanding of how the casual mechanisms work will also show us "how to go on" in producing new entries in our short list of laws. But, as I am trying to argue, I think Nussbaum is right that teleology may provide a superior organization of some of these laws, whether at the manifest or at the micro-structural level, than does a pure chemical theory. It is one of the

significant features of theoretical explanations, as opposed to causal explanations, that the very same phenomena may be a derived consequence in two equally acceptable, and accepted, theoretical structures.

15 Cf. Cartwright (1983), essay 4.

16 Pittendrigh (1965).

17 Pittendrigh, Caldarole, and Cosby (1973).

18 Pittendrigh (1965), note 17, p. 107.

BIBLIOGRAPHY

Arm, D. (ed.): (1965), *Science in the Sixties*, Tenth Anniversary Air Force Office of Scientific Research Scientific Seminar, University of New Mexico Office of Publications, Albuquerque.

Cartwright, N.: (1983), *How the Laws of Physics Lie*, Oxford University Press, Oxford.

Cooper, J.: (1982), 'Aristotle on Natural Teleology', in Schofield and Nussbaum, 1982, pp. 197–222.

Kinoshita, J.: (1983), 'Explanatory Realms: Theory and Illustrations,' Ph. D. Dissertation, Stanford University.

McMullin, E.: (1978), 'Structural Explanation', *American Philosophical Quarterly* **15**, pp. 139–147.

Moravcsik, J.: (1974), 'Aristotelian Adequate Explanations', *Synthese* **28**, pp. 3–17.

Nagel, E.: (1961), *The Structure of Science*, Harcourt, Brace, and World, New York.

Nussbaum, M.: (1978), *Aristotle's "De Motu Animalium"*, Princeton University, Press, Princeton.

Pittendrigh, C.: (1965), 'Biological Clocks: The Functions, Ancient and Modern, of Circadian Oscillations', in Arm, 1965, pp. 96–111.

Pittendrigh, C., Caldarole, E., and Cosby, S.: (1973), 'A Differential Effect of Heavy Water on Temperature-Dependent and Temperature-Compensated Aspects of the Circadian Systems of Drosophila Pseudoobscura', *Proceedings of the National Academy of Science* **70**, pp. 2037–3041.

Schofield, M. and Nussbaum, M. (ed.): (1982), *Language and Logos: Studies in Ancient Greek Philosophy Presented to G. E. L. Owen*, Cambridge University Press, Cambridge, England.

Sorabji, R.: (1980), *Necessity, Cause, and Blame*, Duckworth, London.

Wright, L. (1976), *Teleological Explanations: an Etiological Analysis of Goals and Functions*, University of California Press, Berkeley.

MICHAEL FREDE

PHILOSOPHY AND MEDICINE IN ANTIQUITY

Throughout antiquity the relation between philosophy and medicine was very close. The author of "Decorum" gave the advice to carry philosophy into medicine and medicine into philosophy (chap. 5). Obviously the advice was heeded by many doctors and philosophers. Burnet (*Early Greek Philosophy*, p. 201, n. 4) went so far as to claim that from the times of Empedocles onwards "it is impossible to understand the history of philosophy . . . without keeping the history of medicine constantly in view." And as far as medicine is concerned, it is generally agreed, and indeed obvious, that ancient medical authors, from the Hippocratic writers onwards, heavily relied on philosophers, not just for their views on physiology, but also for their conception of their art and their moral precepts for the doctor. But often they also formed fairly detailed philosophical views of their own. In fact, there is a whole tradition of philosophical thought in ancient medicine, particularly concerning the nature of medical knowledge, which is fairly independent of the thought of the philosophers, and which was substantial enough to at times even influence the views of the philosophers.

Ancient philosophers reveal a surprising amount of interest in medical questions. Part of this, no doubt, is due to the fact that, ever since medicine in the 5th century B.C. became an intellectually respectable discipline, educated men quite generally took an interest in medicine. To a good extent this interest was practical. The health-system required it. Often, especially in small places, no doctors were available, and one had to care for oneself as well as one could. Since access to the medical profession was completely unregulated, and since doctors tended to be itinerant, and hence hardly subject to the ordinary social sanctions against abuse of trust, there were many quacks and incompetent doctors, and one had to try to make some informed judgment as to whether one should trust the competence of the doctor who offered his help. The relationship between the doctor and the patient was different. A good deal of the responsibility for the treatment was carried by the patient. *Epid.* I, 11 tells us that it is the

A. Donagan, A. N. Perovich, Jr., and M. V. Wedin (eds.), Human Nature and Natural Knowledge, 211–232.

patient who must combat the disease with the help of the doctor. Especially if the patient was educated and, hence, as a rule, socially superior to the doctor who, again as a rule, was regarded as a simple artisan. The doctor offered explanation, advice, and help. The choice of treatment and, hence, the primary responsibility, was the patient's. Thus, naturally, he tried to be as informed as possible. The author of *Regimen for Health* (chap. 9) advises the wise man to learn to take care of his ills by his own judgment. At least as early as the 4th century, and possibly even earlier, there were people who had a good knowledge of medicine (Aristotle, *Politica*, 1282^{a}4; Plato, *Politicus*, 259A), though they were not practicing doctors, at least not professionally, presumably in part because their social status would make it very difficult and, in any case, unnecessary for them to join the medical profession. How knowledgeable laymen could be we can see in the case of Celsus, the author of a large hand-book on medicine in the reign of Tiberius which is still extant. Historians of medicine have had great difficulties in determining whether Celsus was a doctor or not. His vast knowledge of medicine seemed to suggest the former. But we also know that Celsus, like Varro before him, was an encyclopedic author who had covered various arts and who wrote for an audience which regarded medicine as one of the liberal arts, a good knowledge of which was fitting for, if not to be expected from, an educated free man. Gellius tells us a revealing story about how he once was visited by Calvisius Taurus, the leading Platonist of his day, and a group of followers of Taurus, when he was just being attended upon by his doctor since he lay ill with a fever (XVIII, 10). The doctor claimed that Gellius was getting better and that Taurus could see this for himself by feeling his vein. Taurus' following was consternated by so much ignorance; for the doctor meant the artery. Gellius decided that it was shameful "not just for the doctor, but for all free men with a liberal education, to be ignorant of even those things which are relevant for knowledge of our body which are not too deep and too hidden." And so he started to read up on the medical literature.

But the philosopher's interest in medical questions was not just the more or less passive interest he, as an uncommonly educated man, would have been supposed to take in medicine anyway. Philosophers from a very early point onwards showed an active concern for medicine, in particular medical theory. We may find this surprising, but we have to keep in mind that the first philosophers were trying to give a general, unified account of nature in terms of which one could explain the most conspicuous, interesting, important, curious natural phenomena. The persuasiveness of such general accounts was in good part measured by their ability to explain particular phenomena. Thus it was natural that philosophers should try to show the strength of their account by trying to explain the most complicated phenomenon of nature: the constitution of man, the way he functions and behaves, or fails to function. And their interest in a physiological explanation of human beings got stronger in the 5th century when man and his behavior became more and more the centre of attention on the part of philosophers, so much so that it took men like Plato and Aristotle to assure the continued study of natural philosophy in general. Thus, the physiological study of human beings was an important part of the very enterprise of natural philosophers almost right from the beginning. And in pursuing these studies the philosophers did not have to have the feeling that they were intruding on somebody else's territory. For when they started to concern themselves with medical questions, doctors did not yet rely on any medical theories, or at least not on anything the philosophers would have regarded as such. There was, of course, the magical view that illness is a matter of the intrusion of a foreign entity, spiritual or grossly bodily, into the body; and there were more specific views which correlated different diseases with different spirits, demons, or gods (cf. *Sacred Disease* 1 ff; *Airs, Waters, Places* 3; Celsus, *De medicina, prooemium*, 4). But for the most part medical practice just relied on the tradition of an accumulated experience with wounds, fractures, dislocations, and some vaguely diagnosed internal diseases. Thus it was the philosophers who started medical theory, and at least in principle they never gave up their claim to the subject. As late an account of natural philosophy as the post-Posidonian account of the Stoic philosophy of nature in Diogenes Laertius tells us that of the three subdivisions of natural philosophy one, namely etiology, has two parts, of which one is the common concern of both the physician and the philosopher (VII, 132).

One could have imagined a division of labour, e.g. one according to which the philosopher deals with the soul, whereas the doctor takes care of the body. But this presupposes a conception of the dualism between body and soul which was not widely shared. Both philosophers and physicians tended to assume that an interest in the soul was an interest in the various life-functions, like procreation, growth and nutrition, respiration, perception, thought, and both had an interest in all these functions. Thus it was not uncommon for doctors to write treatises on the soul. Asclepiades of Bithynia wrote one, and so did Soranus and Sextus Empiricus. Moreover doctors tended to concern themselves with all kinds of disturbances as long as they clearly also involved bodily disturbances, whether it be insanity (mania), effeminacy, lethargy, morbid hunger, or melancholy. Caelius Aurelianus devotes a chapter to the rejection of the claim that hydrophobia is a disease of the soul and not the body (*De morbis acutis*, III, 13), and hence, it seems to have been suggested, not of concern to the doctor. To dismiss the suggestion he points to the bodily symptoms, the physical cause, and the sympathetic accord of body and soul. Doctors attended not just to the bodily effects of mental disturbances, but also to the mental effects of what they regarded as bodily disturbances. Thus Soranus in the case of insanity (Caelius Aurelianus, *Chron.* I, 166–167) in addition to complex dietary measures, prescribes attendance of philosophical discussions for those who are willing: "for by their words they (sc. the philosophers) diminish fear, sadness, and anger, and no small amount of good arises from them for the body." Thus, given that most philosophers and doctors did not accept a simple, sharp dualism between body and soul, a division of labour along these lines was not possible, even if it had been desired.

It is for the reasons given, then, that philosophers took a very active interest in medical theory and continued to do so even after the rise of a science of medicine in the medical schools. Aristotle claimed: "It is also the task of the natural philosopher to have a view of the first principles of health and disease. Hence most natural philosophers . . . finish discussing the matters which fall under medicine (*De sensu*, 436^{a}17–22), and again "the most refined philosophers of nature end up by discussing the principles of medicine" (*De respiratione*, 480^{b}28–30). These Aristotelian testimonies are fully born out by the evidence we still have. The first Presocratic of whom we are explicitly told that he took an interest in medicine is Pythagoras (Diogenes Laertius, VIII, 12). It is difficult for us to determine the core of truth in the legend which grew up around him and which the Pythagoreans themselves seem to have fostered. Given what else we know about Pythagoras it seems reasonable to believe that, as Diogenes Laertius (VIII, 33, 35) claims, health for Pythagoras consisted in the harmony of the bodily constituents and, correspondingly, disease in an upset of this harmony. He seems to have thought that this harmony or balance primarily is upset by a lack of proportion between bodily exertion, intake of nourishment, and rest (Iamblichus, *Vita Pythagorica*, 163; cf. 244), that diseases arise from excess in any of these respects (Iamblichus, *Vita Pythagorica*, 218; Diodori Siculi, X, 7), that such excess leads to indigestion, which in turn is the cause of most disease (Diodori Siculi, X, 7). These views are extremely general, but they did provide an extremely useful framework for thought about internal diseases and suggested a way of dealing with them, namely by regimen. Internal diseases caused the greatest difficulties for traditional medicine. Here was a promising approach to them. It was taken up by Iccus of Tarentum and Herodicus of Selymbria, a gymnastic trainer a generation before Hippocrates, whom tradition made one of Hippocrates' teachers, presumably because already the Hippocratic Corpus in so many of its parts reflects the dietary approach, which changed medical thought and medical practice thoroughly. The author of 'On Ancient Medicine', e.g., in his account of the rise of medicine talks as if medicine were a matter of regimen and does not even mention surgery and pharmacology, though it had been surgery and drugs which traditional doctors had mainly relied on. As we can see from Celsus' *De medicina prooemium*, (9), already in antiquity the rise of scientific medicine was associated with the rise of dietetics, as opposed to surgery and pharmacology. Dietetics seemed to presuppose some kind of theory about the constitution of the human body and its exchange with the environment. And it was this kind of theory which the physicians of the 5th and 4th centuries tried to provide. In Alcmaeon of Croton at the beginning of the 5th century we find a philosopher whose primary concern seems to have been medical theory. Diogenes Laertius, anyway, tells us that Alcmaeon wrote mainly on medicine (VIII, 83); and most of what we know about him does, in fact, concern medical theory. Hence historians have often talked of Alcmaeon as a philosopher and a doctor, but it should be clear from what was said above that there is no reason to think that he was a doctor just because he took such an interest in medicine. Like the Pythagoreans Alcmaeon starts out from a very general schematic view about the constitution of the human body and the nature of health and disease. The body is constituted by a large number of pairs of opposites, the moist and the dry, the hot and the cold, the bitter and the sweet, etc.

Health obtains if we have a harmonious blending (*krasis*) of these constituents, if each of them does not play a larger role than is appropriate for it (if there is *isonomia*); disease is produced by an undue preponderance, by a domination (*monarchia*), of one of the constituents which is destructive of its opposite and hence prevents it from playing its role (*Aetii de placitis*, V, 30, 1).

But Alcmaeon also clearly tries to construct this schema in such a way that it fits experience and can be increasingly adapted to it. For Aristotle explicitly points out (*Metaphysica*, 986^{a}22 ff.) that Alcmaeon, unlike the Pythagoreans, assumed concrete opposites which could actually be observed, and that he did not settle for a definite list of them. Moreover, Alcmaeon seems to have been intent to fill out this schema with as much concrete detail as possible. Thus, he tried to locate the origin of disease in particular parts of the body, the blood, the marrow, or the brain (*Aetii de placitis*, V, 30, 1). He seems to have been the first to recognize the central importance of the brain, in part as a result of anatomical studies which revealed the connection between the sense-organs, in particular the eyes, and the brain (Theophrastus, *De sensu*, 26). There is no reason to suppose that he was the first to do dissections, but his work will have made it very clear that enormous progress could be made by a closer and systematic study of anatomy.

Empedocles, it seems, plays a bigger role in the history of ancient medicine than any other philosopher. Among his students were doctors, e.g., Pausanias and Acron of Acragas, whom later tradition sometimes took to be the first Empiricist doctor (cf. Galen, *De subfiguratione empirica*, p. 43, 1 D). Empedocles tries to give an account of how living beings, including men, did come into existence, and of how they are constituted by the four elements blended according to different ratios which will account for the different kinds of constituents of the body. This approach also allows him to account for the difference between different human beings. For the mixture in different men may vary to some extent. Thus, he can assume individual natures (B 110, 5), a notion of great importance for Hippocratic medicine because it goes to justify the view that treatment has to be individualized to the individual patient. He apparently attempts to develop a physiological doctrine of all the major vital functions, reproduction and the growth of the embryo, presumably growth and nutrition quite generally, respiration, sleep and death, perception and thought. However inadequate Empedocles' physiology seems, and must have seemed then, it will have served the important function of persuading people that it was important and possible to arrive at an adequate and comprehensive physiological account of human beings.

And thus from Empedocles onwards all major Presocratics, in particular Anaxagoras, Diogenes of Apollonia, and Democritus, tried to develop a more or less detailed physiology. Diogenes even wrote a treatise on the nature of man. Democritus wrote various medical treatises, among them one on the nature of man and one on the causes of diseases (Diogenes Laertius, IX, 46).

Thus the philosophers in the course of the fifth century developed increasingly elaborate medical theories. When we come to the fourth century the situation has hardly changed, as we can see from the detail with which Plato's *Timaeus* discusses the constitution of the human body, health and disease, and from Aristotle's remarks referred to above. The situation did change significantly in Hellenistic times. Philosophers now are primarily concerned with ethics, the best of them exercized their ingenuity on the theory of knowledge. The school which we might expect to contribute most to medical theory, namely the Stoic school, scrupled by profound scepticism, was very hesitant to engage in detailed etiological research. Ironically enough, the defenders of the theoretical possibility of certain knowledge took a rather dim view on our ability in practice to gain a scientific understanding of the concrete phenomena of nature. It also must have made a difference that by now physicians had developed quite intricate medical theories and clearly had taken the lead in a subject which one could no longer master without specializing in it. Nevertheless, as we see from the passage in Diogenes Laertius referred to above, philosophers did not give up their interest in medical theory. It surely is no accident that Posidonius, who — unlike his fellows Stoics — did pursue etiological research in great detail (Strabo, II, 3, 8), also studied medicine in Alexandria under Zopyrus (Apollonius Citiensis, *De articulis*, p. 12, 5).

It would also be a mistake to assume that the interest of the philosophers in medical theory was marginal to their philosophical interests. Work on the physiology of the senses had a considerable influence on the theory of perception. The theory of growth and nutrition has considerable metaphysical ramifications, as we can see from Aristotle's *De generatione et corruptione* and the history of the

"Growing Argument" from Epicharmus onwards; it raises questions concerning matter and the continued identity of objects. The question of the central organ, heart or brain, was closely connected with the much discussed question of the seat of the ruling part of the soul, the distinction of nerves, and that of different kinds of nerves, was of obvious relevance for the explanation of motion and action. Moreover speculation concerning human physiology had an effect on the general physical theory. The theory of pores, e.g., seems to have its origin in the physiology of Alcmaeon and Empedocles, but it was used by Empedocles and later authors to account for mixtures of substances, magnetism and other natural phenomena outside the human body.

The interest of the philosophers in medicine was not restricted to the theoretical level, though. We will hardly give much credence to Celsus' claim (*De medicina, prooemium*, 6–7) "At first the science of healing was supposed to be part of philosophy such that both the cure of diseases and the contemplation of the nature of things originated with the same authorities, i.e., with those who most needed the cure, since they had weakened the strength of their body by restless thinking and nocturnal vigil. Thus we find that many teachers of philosophy were experts in medicine, the most famous among them being Pythagoras, Empedocles, and Democritus." But it does seem that in the 5th century philosophers became more and more convinced that the new knowledge should be put to practical purpose to improve the lot of mankind. Thus Empedocles is said to have brought to an end a plague at Selinus which was caused by water pollution; he diverted two rivers into the river which ran through Selinus to clear its water (Diogenes Laertius, VIII, 70). Empedocles himself promises to impart knowledge of all drugs against illness and old age (B 111, 1) and boasts that people come to him to hear the word of healing for all sorts of diseases (B 112, 9–10). Similarly Democritus must have taken a practical interest in medicine, if indeed he is the author of a treatise on regimen (Diogenes Laertius, IX, 46). But the interest of philosophers in the practice of medicine presumably also was due to a moral concern for a rational, healthy life. This seems fairly clear in the case of Pythagoras and the Pythagoreans who laid so much emphasis on a proper regimen (Diodori Siculi, X, 7; Iamblichus, *Vita Pythagorica* 97; 163f; Porphyry, *Vita Pythagorae* 34; Diogenes Laertius, VIII, 19; cf. Plato, *Respublica*, 660 A).

Such, then, was the concern of philosophers for medical questions.

But, to get a complete picture of the role medical questions and the art of medicine played in philosophy, one would also have to talk about the influence which particular medical views and more general philsophical views developed by doctors had on philosophers, and about the role medicine played in philosophy as an example of an art. It is fairly clear, e.g., that Aristotle's views in ethics about the difficulty to find the mean appropriate to one have their paradigm in the difficulty of the doctor to find the balanced regimen appropriate for the individual patient, an analogy which extends far into the details. More about the influence the views of doctors had on philosophers will be said below.

To understand the attitude of doctors towards philosophy we have to distinguish two kinds of doctors, traditional healers and physicians. There were the traditional healers who had learned their craft, case by case, as assistants of a doctor. Their status was low, their education negligible; they were just following their craft uncritically, as it had been handed down to them, with no understanding of what they were doing. But, at least from the 5th century onwards, there also were carefully trained, sometimes highly educated doctors, who took a critical attitude towards tradition, were interested in medical theory, tried to find new ways of healing, and who, because of their competence and their learning, might enjoy a considerable reputation (cf. Plato, *Leges*, 720A-C; 857-D; Aristotle *Politica*, 1282^{a}3–4). This distinction is a rather crude simplification. For there will have been physicians with very little learning who relied largely on traditional practice, and there will have been traditional practitioners to whom some of the learning and the practice of physicians will, in one way or another, have filtered down. Still, the distinction is clear and marked enough. And what we will be concerned with in the following are the physicians.

Their interest in philosophy was very much part of the very origin of the art of the physician, or scientific medicine. Traditional medicine had suffered from severe limitations. Doctors knew how to handle fractures and dislocations, they also could help with wounds, but they were largely powerless in the case of internal diseases. We rightly admire the 42 case histories of the *Epidemics*, but the fact is that 26 of them ended with the death of the patient.

Yet these limitations on the doctor's art for a long time were accepted by the public. There were a large number of defenses, which made the limitations seem natural, a given about which nothing could be done; e.g., that once the illness has gained mastery over the body there is nothing anybody could do; that one can do nothing against the will of the gods; that many diseases are hidden from the understanding of human beings; that patients are unreliable and weak, etc. In the course of the 5th century, though, the educated public became less and less willing to accept these defenses. It now was so obvious that the limitations on the art of the doctor were not just due to the nature of things, but also in large measure to the fact that doctors did not know things which could be known, that doctors had not made sufficient efforts to overcome the extrinsic limitations of their art. Some people went so far as to argue that there is no such thing as the art of medicine, that the doctor can do no more than what an educated, thoughtful layman could do, and that the rest was mere chance (*Art* 1ff.; *Reg. in morb. acut. 8; vet. med.* 4 *et passim*). In this situation educated doctors, at least, naturally felt very threatened. Doctors were considered with a good amount of distrust and suspicion anyway. Even a good doctor had to make an enormous effort to win the confidence and trust of a clientele, often to no avail. For if his patients died the public might hold it against him, even if it was not his fault. Thus, ancient doctors were very hesitant in the first place to take on obviously dangerous cases. This, of course, all the more raised the question what one needed doctors for. For the harmless cases one could also take care of oneself, or perhaps rather they took care of themselves. Nevertheless, the good doctor still had been able to live with the thought that at least he knew that he had mastered a complicated and important art, though the public might find it difficult to recognize this and to distinguish him from the incompetent doctor and the quack. But now his very art came under criticism, and it was called into question whether he really had any special expertise. His art had lost the intellectual respect on the part of those on whom he depended for the kind of living he was accustomed to, the educated elite. For the educated doctor must have been upwardly mobile in social background. Otherwise he would not have had an education, and certainly would not have been able to pay the fees asked by the schools of medicine which arose towards the middle of the 5th century. On the other hand, he cannot by background have been part of the upper class. For in this case he would not have practiced medicine, certainly not for a living. Thus, it was exactly the educated doctors who must have felt particularly threatened, whereas the ordinary, traditional healer could continue to rely on an unbroken consensus with the vast uneducated majority of the population, especially in backward areas, the kind of consensus between doctors and the public which any functioning health-system presupposes. This threat provided the motivation for the attempt to reform medicine systematically and, thus, led to the rise of Western medicine.

What made it easier for the educated doctor to acknowledge the fact that his art was severely limited and stood in need of reform was a view characteristic of the enlightenment of the later fifth century. On this view the various arts and crafts had evolved in the course of the history of mankind to enable human beings, who originally had been exposed rather helplessly to a hostile environment, to gain a secure life, and, thus, if one just systematically applied one's reason, instead of mindlessly following traditional belief and practice, ever further areas of life would be brought under rational control and human beings would flourish. Socratic ethics is perhaps the most splendid and extreme reflection of this attitude. Thus, medicine was naturally conceived of as an art which had arisen rather late, but which could be set on the road of firm progress, if one merely put one's mind to it, instead of just following the traditional practices and beliefs, as the ordinary doctors did.

Given the conflicting practices and views of doctors, which now became all the more apparent with the rise of a medical literature, it was clear that what was needed was reliable knowledge rather than mere opinion. And given the limitations of the art due to ignorance it was also clear that this knowledge had to be extended systematically, so that in the end it would form an organized body of knowledge which would put one into a position to deal with all disease to the extent that this is humanly possible. The need for such a systematic body of knowledge must have become particularly apparent when, in the new schools of medicine, doctors started to give systematical lectures instead of just teaching their students case by case in the manner of the traditional healers.

The question, then, was "what is to count as knowledge, and by which method does one manage to systematically arrive at medical knowledge?" Thus, a concern for epistemological and methodological questions stood at the very beginning of the new art of the physician. And it continued to characterize physicians throughout antiquity. For answers to these questions the physicians naturally turned to the philosophers who in the course of the fifth century had themselves become increasingly concerned with epistemological questions and who by the fourth century had developed rather elaborate views as to what an organized body of knowledge had to look like and how it would be arrived at.

But right from the beginning there also were physicians who objected to the way in which their fellow-physicians tended to follow the philosophers' methods. Thus the author of "On Ancient Medicine" criticizes physicians who proceed by the hypothetical method, i.e., who postulate certain basic constituents of the body and try to derive the appropriate treatment from the assumption, e.g., of an imbalance between the postulated basic constituents. Against them he claims that medicine has long found its own method, the method of error and trial, that many discoveries have been made by this method, and that on the basis of these discoveries medical knowledge will grow further, if one just follows the method by which medicine has progressed so far. The new philosophical methods, on the other hand, do not, he claims, lead anywhere (chap. 2). Thus, there developed in medicine right from the start a tradition of independent thought concerning the origin, nature, and scope of medical knowledge and of knowledge in general. Part of the reason for this was the special situation of medicine. It conceived of itself as a growing subject. For this was the only way in which one could acknowledge the inability of the art in its present state to deal with a large variety of cases, and yet maintain that it was a true and worthy art. Hence, physicians, much more than philosophers, took an interest in methods of discovery. Moreover, whereas the philosophers were mainly concerned with theoretical knowledge, the physicians' concern was eminently practical knowledge, on whose reliability much depended in a very obvious and concrete way. Much more than the philosophers they were concerned with the application of general knowledge to particular cases, especially since most of them took the view that the individuality of each case and of each patient has to be taken into consideration. They reflected on the inherent limits of practical arts; no matter how knowlegeable and skillful a doctor is, he may not achieve his aim for reasons which do not reflect badly on his competence or the state of the art. For the most part, though, they were concerned with the nature of the general knowledge which the doctor applies to the particular cases. And on the nature of this knowledge they disagreed radically.

Most physicians down to Herophilus and Eriasistratus, especially those who followed the philosophers, tended to adopt a Rationalist position, i.e., they thought that reason allowed us to determine the nature of a disease, to find out its cause, and, on the basis of this, to find the kind of treatment which would remove the cause of the disease. Thus, they thought that the practical knowledge the doctor relies on was based on some theoretical knowlege which would enable the doctor to account for his practice. In the extreme case they looked for a theory from first principles from which these results could be obtained. But many Rationalists, it seems, took a more cautious position. Diocles, perhaps the most distinguished physician of the 4th century, warned against the temptation to look for this kind of theoretical causal account for everything (Galen, VI, 455; Wellmann, frg. 112). For, he said, such knowledge is rarely of practical use; moreover many facts of nature are primitive, hence, have to be accepted as such and do not allow for derivation from a theory; moreover such theories often depend on questionable assumptions. One should rather trust what one has learned from experience. Similarly Herophilus, the first of the great Alexandrian physicians, pointed out that in medicine one often has to take as a given or a principle what in the true nature of things is not a principle (*Anonymous Londinensis*, xxi, 21; Galen *De methodo medendi*, X, 107, 15). Galen even chides Herophilus for refusing to pronounce himself on the nature of the primary bodies (Galen, *De methodo medendi*, X, 461, 17). That Herophilus developed a list of arguments against causality which closely resemble those Diogenes Laertius attributes to the Pyrrhonean sceptics suggests, too, that he had strong doubts about theories of the extreme Rationalist type (Galen, *De causis procatarcticis*, 197ff.). Perhaps the kinds of doubts we find in Diocles and in Herophilus were what had already moved the author of "Nature of Man" (chap. 1) to say that though we need some kind of account of human physiology, we do not in medicine need the kind of questionable account from ultimate principles which the philosophers give.

Nevertheless the Rationalists were all agreed that the mere experience of the practitioners did not suffice to provide the doctor with the practical knowledge he needed. Herophilus and Erasistratus must have argued the point explicity (Galen, *De sectis ingredientibus*, 9, 14; *De methodo medendi*, X, 184.; Pliny, *Historia naturalis*, xxix, 5, 6). This does not necessarily mean that this view had been explicitly defended by any physicians. Some sophists and rhetoricians had taken the view that rhetoric is a matter of mere experience (e.g., Polus in Plato's *Gorgias*). There was, early in the 3rd century B.C., the view that political expertise may be a matter of experience and practice (Philodemus, *Rhetorica*, BI, 27–28). Plato had characterized the lowly practitioner of medicine as somebody who just relied on experience (*Leges*, 720A-C; 857C-D). And both Plato and Aristotle had argued quite generally that knowledge, properly speaking, i.e., scientific knowledge, had to go beyond mere experience and to advance, by means of reason, to a theoretical understanding of the phenomena.

But the 4th century saw a bewildering proliferation of such medical theories. What was worse, they all seemed to be in conflict with each other, and yet there seemed to be no way to decide between them. All had arguments in favour of them, and all had their drawbacks. Moreover Rationalist medicine had a tendency to become academic and scholastic, a term actually used by Galen to refer to the Alexandrian physicians who no longer cared to look at patients (*In Hipp. De nat. hom., CMG* V, 9, 1, p. 88, 1ff.). As we know from another passage in Galen (*De dogm. Hipp. et Platon*, p. 598 M.), Erasistratus had stopped practicing medicine altogether. Presumably he and Herophilus devoted all their efforts to the development of their physiological and anatomical theories and, thus, turned medicine into an academic subject of great subtlety and complexity (cf. Pliny, *Historia naturalis*, XI, 219; XXVI, 11). Pliny (*Historia naturalis*, XXVI, 11) puts the matter this way: "This celebrated rational discipline, though actual practice is the most efficient teacher of all things, but in particular of medicine, little by little degenerated into mere words and garrulity. For it was more pleasant to sit in school and devote one's time to lectures than to walk in the wild to look for the different herbs at the different times of the year."

It was in reaction to the proliferation of competing theories and the neglect of practice and the experience which accrues from the practice that in the first part of the 3rd century a new school of thought arose, the Empiricists, as they called themselves. The Empiricists[1] took the position that all knowledge, and in particular all medical knowledge, is a matter of mere experience which one only acquires in actual practice; experience in the sense that through long observation we come to know what is harmful and what is beneficial. Thus the Empiricists got themselves into the position where they had to show (a) how experience, as a matter of fact, would account for all medical knowledge, (b) that the Rationalist arguments to the contrary were inconclusive, and (c) that Rationalism did not provide an alternative way to gain medical knowledge.

According to the Empiricist all the doctor has to know is what is beneficial and what is harmful to a patient. For his aim is to heal, and not to understand. Hence he does not need a theory which allows him to gain a theoretical understanding of his practice. But what is harmful or beneficial can be known by mere observation of what turns out to harm or benefit people. What is more, the Empiricist, relying on sceptical arguments, denies that the hidden natures and hidden causes, that anything hidden to normal observation can be known. Hence, he rejects all medical theory, not just the physiologies of the Rationalist, but also the anatomy of Herophilus and Erasistratus. For though the facts of descriptive anatomy in some sense are open to observation, the observation is not normal in that it presupposes an artificial interference with the object to be observed, namely the cutting up of the body which may affect it in ways we have no control over (cf. Celsus, *De medicina, prooemium*, 40ff.).

Whereas the Rationalists, before the rise of Empiricism, had not seen themselves as a group, and, in fact, continued to disagree among themselves, they were united by being attacked by the Empiricists and by having to rebut the Empiricist arguments against them. Thus, one came to talk of a Rationalist and of an Empiricist school of medicine.

An interesting feature of the debate between the two schools was that it turned out to be rather academic. The Empiricists, apparently more ready to rely on traditional forms of treatment, made more use of drugs, whereas Rationalists tended more to a dietary treatment, but there was no real basic disagreement about the actual practice of medicine (Galen, *De sectis ingredientibus*, 1, 12ff.; 7, 16ff; 12, 12ff.). The Empiricists did not deny that good Rationalist doctors were good doctors, and they were ready to adopt whatever practice showed promise.

The situation changed when late in the first century B.C. a new school arose, that of the Methodists[2]. The Methodists challenged the practice of Rationalists and Empiricists alike, claiming that neither had understood by which method one comes to know the right form of treatment. They claimed that all internal diseases were forms of constriction or dilation or both and, hence, required the corresponding forms of dilation or constriction or both. To know that constriction requires dilation is obvious to reason and is not something which we just know by experience. But nor is it a matter of theoretical knowledge. With some training we can learn to directly recognize states of dilation or constriction. They are not theoretical states whose existence we have to infer from the observable symptoms. They themselves are observable. In this way the Methodists tried to find a position of their own between Rationalism and Empiricism. The practical knowledge the doctor relies on is a matter of reason, rather than experience; but it is not a matter of reason in the sense that it is based on some theoretical knowledge which we have *a priori* or which we acquire by inference from the observable.

Rationalism, Empiricism and Methodism were the major positions adopted by Hellenistic doctors. Yet Galen, in the second half of the second century A.D., found none of them acceptable. He took the position that experience did suffice to gain knowledge which would make one a competent doctor. But he also thought that if the practical knowledge of the doctor was to be complete, if he was to know all that was practically relevant, he also had to master the theory of medicine (Galen, *De methodo medendi*, 10, 29; 31; 122; 159; 272). For all his stress on, and obvious love for, medical theory, Galen also seems to have lost the naive trust in theory construction which characterized the Rationalists. He suggests that one should clearly separate the body of knowledge one has acquired by theory-construction from the body of knowledge acquired by experience because in this way theory can be confirmed by experience (*De methodo medendi*, 10, 31; 127; 159; 246).

Thus physicians developed elaborate accounts of the nature of medical knowledge. But Rationalists, Empiricists, and Methodists also thought that their accounts applied *mutatis mutandis*, to human knowledge quite generally. And, thus, philosophers had to take notice of the epistemological views of the physicians. The first reflection of this are Plato's famous remarks in the *Phaedrus* (270C) about Hippocrates' method. Later it was primarily the Empiricists whom the philosophers took an interest in. The Stoic view of the nature of an art like divination, as it is represented in Cicero's *De divinatione*, seems to be Empiricist. Demetrius Laco (fr. 12, *De Falco*), Philodemus, (*De signis*, xxxviii, 25) Cicero (*Academica priora*, 122), and Sextus Empiricus, of course, (*Pyrrh. hypotyposeis*, I, 236ff.; A.M. VIII, 327) refer to the Empiricists. And the version of Phyrrhonean scepticism which we find in Sextus in some respects, e.g., in the use of the notion of historia, i.e., reports of one's own experience, clearly is influenced by Empiricism.

The first physicians, then, by the very nature of their enterprise started a rich tradition of epistemological thought within the schools of medicine. But because most of them saw the solution to their problems in the development of a medical theory, of a science of medicine, they also got engaged in the philosophy of nature. And again, naturally enough, they turned to the philosophers for an appropriate physical and physiological theory. It has been pointed out often enough and in sufficient detail how much physicians throughout antiquity depended on philosophers for the elements of their physical theory. Some of the authors of the Hippocratic Corpus clearly are influenced by Heraclitus, others by Diogenes of Apollonia. There may be an influence of the Pythagoreans and Democritus. The author of "On Ancient Medicine" (20) himself points to the influence of Empedocles. Those medical theories which are based on the assumption of 4 elements clearly are Empedoclean in inspiration; but so presumably are doctrines based on the assumption of 4 basic powers and the doctrine of 4 humours. Diocles and Erasistratus are influenced by the Peripatus, other Rationalists are indebted to Stoicism, Asclepiades relies on Atomism, the Pneumatists are influenced by the Stoics and in particular by Posidonius. But, again, also here there was an almost immediate reaction against the uncritical acceptance of philosophical views. The author of "On Ancient Medicine" went so far as to say that one had to be a doctor to be able to develop an adequate physiology (20). And the author of "Nature of Man" takes the view that medicine cannot rely on the philosophers' philosophy, but has to develop its own. It is obvious how much physicians contributed to physiology and anatomy. But it may well be the case that even in the theory of nature in general they had something to contribute and took an influence on the history of philosophy. Thus, Asclepiades of Bithynia may well be the first to use the notion of a law of nature (*Anonymous Londinensis* 36, 47ff; 39, 5) – not the law of the nature of a particular kind of thing, but exactly a law which guarantees the uniformity of nature, and, hence, allows us to make inferences from what has been observed to what has not been observed, from what can be observed to what cannot in principle be observed. If it is true, as I think it is, that a passage in Cicero's *De oratore* (I, 62) forces us to assume that his floruit is well into the second century B.C., then Asclepiades' great reputation in Rome may well be responsible for the widespread acceptance of Epicureanism in Rome in the first century B.C. It also may be the case that some physicians not only wrote their own physiologies in which they developed their own notions concerning the theory of nature in general, but that they felt confident enough to write their own physical treatises. Thus, Galen refers to a treatise "Physica" in several books by Praxagoras (XVII, 838).

For our purposes, though, it may be more interesting to briefly consider the role of these theories than to look at their details. On the Rationalist view these theories provided the foundation for the practical knowledge the doctor relied on. But the role medical theories actually played is much more complicated. Given that ancient natural theory was by no means sufficiently developed to allow for the construction of a medical theory which came anywhere near to fullfilling the role it was supposed to play we have to suspect that the great interest in medical theory can hardly have been due just to the expectation that it would greatly improve medical practice. There clearly was a social motivation. As we saw earlier, traditional medicine had been no longer intellectually acceptable. And this threatened the position of the educated doctor. Had he earlier been looked at basically just as an artisan, now even his artisanship was called into question. Thus, educated doctors had a strong social motivation to turn medicine into a scientific subject. It helped them to regain intellectual respectability. At the same time this learning helped them to set themselves off from the ordinary, lowly practitioner. And it helped them to distinguish themselves from the class of artisans quite generally. In fact, as we have seen, some of the Alexandrian physicians removed themselves from practice altogether. Moreover, theory played a role in the practice of medicine quite independently of the role ascribed to it by Rationalism. The doctor was supposed to give the patient an account of the illness and the proposed treatment. In this way he tried to convince the patient that he was a competent doctor who knew what he was doing. But just as he tried to give an accurate prognosis, mainly to win the confidence of a clientele, and not necessarily for any practical medical purpose, thus he was under considerable pressure to produce a theoretical account, whether or not his treatment actually was based on this account. Already the author of "On Ancient Medicine" (chap. 5) suspected that there was no actual connection between the account the doctor gave and what he then did; if he were a good doctor he would do what good doctors would do in a similar case, except that he vacuously added a story. The Empiricists similarly claimed that medical theories had been constructed in hindsight after the cures had already been found by experience (Celsus, *De medicina, prooemium,* 36) and that it was this experience which good doctors, even if Rationalists, were actually relying on. Similarly Methodists like Soranus (Gyn. p. 4, 7; 6, 6ff.) maintained that one had to know medical theory, not because it provided a reliable basis for sound treatment, but because otherwise patients might think that one was incompetent and because it was a matter of good education. Thus, theory did play, and was recognized to play, a social role. If one reads Galen, especially his criticism of the Methodists, the social function of the physician's learning becomes very apparent. Galen, like many Rationalists, does not just think that the doctor should know the intricacies of medical theory and philosophy, he also thinks that for medical purposes, the doctor should have some knowledge of those liberal arts, such as geometry and astronomy, that are supposed to have some application in medicine. He is quite aware of the fact that the kind of training and education which he received and which he demands from the best doctors is only available to the few who can afford it. It is very clear from his discussion of the Methodists that he deeply resents the fact that they promise to provide an adequate medical training in a short time which does not require any great learning (*De sectis ingredientibus*, 15, 6; 24, 22; *De Methodo medendi*, X, 5, 2) and, thus, opens up the profession to what Galen regards as the rabble (*De methodo medendi*, X, 4, 17ff; 5, 9ff.). Again and again he refers to the humble social background of Thessalus, who is the main proponent of Methodism, as if this disqualified Thessalus from taking a critical attitide towards established medicine (*De methodo medendi*, X, 8, 17ff; 10, 5ff; 11, 7). Soranus, on the other hand, a man perhaps of no less education and learning than Galen, is treated by him with respect, though he is a Methodist, too. One can see here very clearly how education is used to draw social boundaries and to protect social status.

In any case, the physicians, for the reasons given, took a great professional interest in epistemology and methodology, or logic as it was then called, and in physics. But they were also concerned with ethics. The distrust towards doctors was not just due to the fact that so many of them were incompetent quacks; doctors also had to fight against a good amount of suspicion against them as human beings. Part of this may just have been the primitive fear of the powers of the medicine man. But part of it was no doubt due to the callous way in which some doctors misused the trust placed in them. There will be good reason why the Hippocratic Oath asks the doctor to promise: "I will use treatment which will benefit the sick to the best of my ability and knowledge, and refrain from injury and injustice . . . Into whichever house I go, I will enter for the benefit of the sick; I will abstain from any willing injustice or harm, in all other respects and in particular in sexual matters, in the case of female and male bodies, whether they be slaves or free." The doctors defended themselves against such suspicion by adopting a manner and bearing which would instill trust and respect. But they also strove to be rigorously reliable in their moral behavior, or at least to give the public this impression. In this effort they tended to adopt rather conservative ethical views. Thus the Oath enjoins the doctor not to perform any abortions, though throughout antiquity this was standard practice and was accepted even by philosophers like Plato (*Respublica*, 460E–461C) and Aristotle (*Politica*, 1335^{b} 22ff.), although Aristole in this context also discusses the question up to which point of the development of the foetus abortion should be legal. Doctors were suspected of greed, and the *Corpus Hippocraticum* contains a good deal of advice for the doctor on how to deal with the problem of fees. Galen found it very upsetting that the Empricist Menodotus had claimed that doctors were practicing for reputation and money (*De dogm. Plat. et Hipp.* 764 M.). Doctors liked to see themselves as motivated by philanthropy (*Precepts*, 6; *On the Physician*, 1: Galen, *De dogm. Plat. et Hipp.*, 765 M.).

Given their somewhat insecure status it was natural that doctors would try to convince the public of the elevated worth and dignity of their subject. Thus the author of *Lex* (1; cf. *Vet. med.*, 14) claims that medicine is the most important art. Similarly the Methodist Thessalus proclaimed the superiority of medicine over all other disciplines (Galen, *De methodo medendi*, X, 11, 8). What made this possible was the great value antiquity placed on health. Sextus Empiricus can attribute even to Herophilus the view that health is the highest good; for without health all the other goods are of no avail (*Adversus Mathematicos*, XI, 49–50). It may well have been the case that Hreodicus of Selymbria not only saw in his dietetics a way to maintain and regain health, but the way to lead a good life. Otherwise it is difficult to understand why Plato in the *Republic* would criticize Herodicus so sharply. Hence, it may very well be true that some physicians and some philosophers saw medicine and philosophical ethics as rival arts of life. In any case, both their subject and their social situation forced educated doctors to develop moral views concerning their role and practice.

Given all this, it is not surprising that Galen was able to write a whole treatise to show that the best physician will be a philosopher. Already the author of "Decorum" had written (5) that a doctor who is also a philosopher will be godlike, and he had gone on to explain how medicine has all the marks of wisdom. Galen himself, in fact, in his own work did go much beyond a discussion of those topics philosophically minded physicians traditionally had concerned themselves with. He obviously had the ambition to be recognized as a philosopher as well. And this recognition he was almost immediately granted, though perhaps somewhat grudgingly. His contemporary, Alexander of Aphrodisias, the greatest of the Aristotelian commentators, devoted monographs to the criticism of some of Galen's philosophical views. Similarly, the later commentators refer to some of Galen's views on a variety of topics. Galen was not the only physician who managed to acquire a reputation as a general philosopher. Sextus Empiricus, too, was a doctor and a medical author, but is best known for his two extant philosophical works, the *Outlines of Pyrrhonism* and *Adversus Mathematicos*, our main sources for Pyrrhonean scepticism. In fact, unlike Galen, Sextus Empiricus does not seem to have made any significant contribution to medicine properly speaking. His distinction as an Empiricist doctor relied entirely on his elaboration of empiricist doctrine (Ps. Galen, *Med.*, XIV, 683, 11ff.) It is clear from his remarks in *Outlines of Pyrrhonism* (I, 236ff.) that he was dissatisfied with the dogmatism of the traditional Empiricist doctrine of the nature of medical knowledge. But the evidence also clearly indicates that he never gave up Empiricism. Thus the contribution to Empiricism Ps. Galen refers to must have consisted in his attempt to bring Empiricism further into line with Pyrrhonism. In Sextus Empiricus, then, we have a doctor whose reputation rests entirely on his philosophical work which secures him a place in the history of philosophy, whereas he is little more than a name in the history of medicine.

Philosophy and medicine in antiquity were so intertwined that we find philosophers like Alcmaeon of Croton whose fame rests entirely on his contributions to medical theory and doctors like Sextus who were primarily philosophers. As a result of the close connection between the two subjects there evolved in medicine a rather independent tradition of philosophical thought, mainly concerning the nature of medical knowledge. Unfortunately, this tradition has not received the attention it deserves. Historians of medicine tend to leave it to the historian of philosophy. But historians of philosophy tend to overlook it because it is not part of the history of the philosophy of the philosophers. This is most unfortunate. For in this way a large amount of material which is both of intrinsic interest and of considerable use to historians of philosophy and of medicine lies waste.

Princeton University

NOTES

[1] For an account of Empiricism the most important sources are: Galen, *De sectis ingredientibus, De subfiguratione empirica, On Medical Experience, De methodo medendi*; Ps. Galen, *De optima secta*; and Celsus, *De medicina, prooemium.*

[2] For a convenient summary of the Methodist position cf. Galen, *De sectis ingredientibus*, p. 12, 9ff.

GEORGE GALE

ANTHROPOCENTRISM RECONSIDERED

The history of science may be usefully divided into two phases, classical and modern. The modernization of natural science is accurately gauged by the progressive elimination of anthropocentric concepts from scientific thinking. Indeed, many historians and philosophers of science place the origin of modern science in Copernicus's astronomical scheme, which eliminated the ultimate example of literal classic anthropocentrism from the science of astronomy. It is somehow ironically fitting that it is this very same science, astronomy, which today proposes that a contemporary version of anthropocentric thinking might provide a valuable tool for a new understanding of the natural world.

Anthropocentric thinking in classical and early modern science was not limited, however, to the literal sort of case we see in geocentric astronomy. Insofar as they relied upon scientific concepts discovered by Aristotle and his successors, all natural sciences were more or less implicitly anthropocentric. Thus, it is the progressive elimination of Aristotelian concepts, first from physics, then chemistry, and finally from biology, which tells the story of science's coming of age in the modern world. However, even as a new style of literal anthropocentric thinking is reappearing in today's astronomy, it can be fairly argued that an analogous implicit anthropocentrism is reappearing in other modern sciences, such as physics and biology.

In what follows, I will first briefly describe the anthropocentric elements which were eliminated during the rise of modern scientific thought. Then I will review the contemporary situation in both astronomy and quantum physics, in an attempt to make the case as sharply as possible that a new, post-modern anthropocentric perspective — anthropocentric both literally and implicitly — has come into being. Moreover, I will argue that the new perspective, while risky, offers at least several methodological advantages to the scientific enterprise.

Reflection of the literal and explicit anthropocentric perspective which pervaded classical science is found in the dictum "Man is the

A. Donagan, A. N. Perovich, Jr., and M. V. Wedin (eds.), Human Nature and Natural Knowledge, 233–241.

measure of all things." But equally important for the implicit anthropocentrism of classical science is the notion that "Man is the measur*er* of all things." The former dictum seems, in great degree, to once again be at work, this time in contemporary astronomy, while the latter dictum certainly operates in much of today's quantum physics.

Aristotle's thought, and that of his followers, exhibited characteristics derived from both dicta, explicit and implicit. Aristotle was a biologist, and a good one. From his activities in biology, he extended ideas into the other natural sciences, and, perhaps more importantly, into his logical and metaphysical thinking.[1] Since these two philosophical disciplines were ultimately responsible for laying the methodological foundations for all subsequent development in the classical sciences, the "biologizing" of these sciences came as an unavoidable matter of course.

The supreme example of this intellectual fertilization is to be found in Aristotle's thoroughgoing teleology. For the classical biologist, the ultimate reality is the concrete, individual specimen, and, derivatively, the species it exemplifies.[2] Teleological activity, that is, functional or goal-directed behavior, is evident to even the most naive observers of biological entities. Animals of all kinds, and their organ systems, appear obviously to behave "in order to" attain particular goals, or carry out particular functions. Aristotle believed, thus, that one of the main tasks for a biologist was to discern the patterns of teleological activity of his specimen animals, plants, and their component systems. He further generalized this task to that of a central analytical desideratum in logic and metaphysics. Consequently, when the Aristotelian chemist or physicist carried out his specific tasks, he in part investigated functional or goal-directed behavior as a matter of straightforward scientific methodology. It must be pointed out, however, that Aristotle himself at no time presented the discovery of *conscious* purposes as a requirement in the analysis of inanimate systems.[3] Rather, the scientist would presume that systems exhibited functional or other goal-directed behavior in the absence of conscious purpose. It was only the later Christian Aristotelians who instituted the requirement that not just function, but rather *purpose* was to be found in nature, in particular, the conscious purpose of the Creator. It is fair to say that Aristotelian teleology is based simply on the fact that Aristotle was both a human and a biologist; he thereby necessarily saw teleological activity both in his own behavior and that of his subjects.

His biological analyses then served as the explicit basis for extensions into all other natural sciences and philosophy.

Other features of Aristotelian thought provided the implicit bases for an anthropocentric perspective in the various sciences. In astronomy, for example, Aristotle, just like all other ancients, relied implicitly and uncritically upon the information of his human sense organs functioning in a human-centered world. Thus, scientific observations were not far removed from common sense. An example of the ramifications of this point concerns the structure of the universe itself. When the heavenly bodies are carefully observed, the fact of motion is most plain. First the star is directly above the oak tree; later, some distance has opened up between oak and star. The scientific question in the case comes down to this: when one observes the sky, *something* is moving; now, either the objects in the sky are moving, or the astronomer himself is moving; but, by the plain testimony of his own senses, the astronomer is not moving. The conclusion for him is inescapable: the heavenly objects are moving. Equally important, this plain common sense reasoning squares well with the already ingrained idea that humans are the pinnacle of the biological realm. Earth-centered astronomy follows directly from these ideas. "Man is the measure of all things" conjoins with "Man is the measurer of all things" and produced directly and reasonably the earth-centered Ptolemaic system of the universe.

One final aspect of Aristotle's human-centered science is particularly noteworthy. It concerns his physics. Aristotle's physics takes the natural world as it is actually given to human observation, and generalizes these observatons, into scientific principles. In this sense, his physics is not just a *human*-centered physics, it is a human *observer*-centered physics. His physics of motion provides an excellent example. According to Aristotle, velocity is proportional to force. Thus, the cart moves at a speed directly proportional to the effort put into it by horse or man. Even today, an automobile is observed to move at a constant velocity when a constant engine speed is maintained. The modern physics account, of course, notes that, in the absence of friction, it is acceleration that is proportional to force; but for Aristotle, *equally* of course, his human world had no objects speeding in the absence of friction. It is the actual, experienced world of the observer, the observer's world, which is described by the classical natural sciences.

In all cases, the actual standpoint of the human observer, coupled to a conceptual system developed by a biologist, sets the foundational standards for classical scientific discovery and investigation. But modern science, as is well-known, progressively eliminated this standard, and, through this rejection itself, it acquired the very essence of what we call "modern." Thus, "modern" in science came to mean "objective," *not* human-centered, "universal." What I would like to suggest is that the period of "modern" science is perhaps at an end. And this *precisely* because the once-rejected, human-centered scientific standard appears to be once again on the rise.

Contemporary astrophysicists are investigating the possibility that the existence of life, in particular human life, may set constraints upon the allowable conditions of the early universe. Arguments such as this involve what has come to be called the "Anthropic Principle."[4] Although these efforts differ somewhat from one another in their underlying philosophical spirits, they all converge upon one fact: in order for life to exist today, an incredibly restrictive set of demands must have been met in the early universe. These demands include such seemingly independent factors as the gravitational constant, the expansion rate of the universe, the strong force constant, the amount of matter in the universe, the isotropic distribution of matter and radiation, the ratio of matter to antimatter, the presence of stable elements heavier than helium, and so on. When all these demands are conjoined together, a "window" of incredibly small dimensions is created, through which the possibility of life must pass.[5] Yet as the Anthropic argument goes, necessarily we conscious beings must observe a universe which contains life, since, if it didn't, we wouldn't be doing any observing! Or as Dicke has marvelously put the case, "it is well known that carbon is required to make physicists."[6]

Although the narrow boundaries of each of these requirements must be investigated separately, it will suffice for us now to examine just one such case. It would seem that the presence of planets is necessary to life as we know it. But the possibility of planets depends upon the star in question, and, ultimately, upon the existence of galaxies. Galaxies depend not only upon the distribution of matter, and the force of gravity, but also upon the currents and eddies within the initial fireball, as well as the velocity of universal expansion following the original big bang. Present-day measurements show that matter and radiation are uniformly distributed — isotropic — to an extremely high degree, and that galaxies represent clumpings within that overall uniform distribution. Mathematical models of the early universe have shown a remarkable resistance to the formation of galaxies within isotropic conditions. Indeed, the class of models which allow such evolution is vanishingly small. Thus, the existence of galaxies, and thereby life, is highly improbable and in desperate need of explanation.

One possible explanation of the remarkable coincidence demonstrated by life's existence, might be provided by reference to that life itself. According to British astrophysicists Collins and Hawking, "since it would seem that the existence of galaxies is a necessary condition for the development of intelligent life, the answer to the question 'why is the universe isotropic?' is 'because we are here'."[7] So goes the now classic Anthropic Argument.

Taking Collins and Hawking as a typical case, let us now look at the logic and methodology exhibited by Anthropic arguments. We must be extremely careful here to accurately represent what is, and what is not, at work in the various examples of Anthropic Principle reasoning. In the first place, only some Anthropic accounts (Wheeler's,[8] for instance) evidence an explicit use of teleological thinking. On this view, one might reason that, for example, "The universe is isotropic in order to produce intelligent (human) life." This view does not *necessarily* involve conscious purpose, but rather, might instead reflect only an Aristotelian-style functional analysis of the actual observed situation.

Secondly, on the other hand, most accounts using the Anthropic Principle certainly contain at least implicit teleological reasoning. Thus, it is not a long distance to travel getting from "Because we are here, the universe is isotropic" (Collins and Hawking) to "The universe is isotropic in order for us to be here." Since many investigators would perforce deny such implicit Aristotelianism, extreme care must be taken in order to avoid it while formulating Anthropic Principle explanations.[9]

Thirdly, it is clear that there is a purely methodological use to which the Anthropic Principle may be put. On this view, the explanations are formulated in the form, for example, "We exist only if the universe is isotropic." In this formulation, the various requirements function as a set of necessary conditions, and not sufficient causes, of the existence of life. Analogous to this logic would be something like "Offsprings have blue eyes only if both parents have blue eyes." The logic here is perfectly typical, and quite "harmless" from the modern scientific viewpoint. However, even though teleological thinking is hereby avoided, it is incontestable that the fact of human existence is *the* central fact of all such explanations. Thus, although non-teleological, such explanations are inescapably anthropocentric.

A final point is required. Although it is usually not stated, most anthropic explanations depend upon *human* life as the central point. This excludes, for example, life based upon various other proposed schemes such as ammonia instead of water or silicon instead of carbon. This seems to be *too* anthropocentric. Although the very nature of the anthropic logic centers on the fact that we humans do, indeed, exist, it seems to me plausible to believe that *other* intelligent beings, should there be any, would justifiably have their own versions of anthropic logic. Thus, even though necessarily only *we* can use the *Anthropic* Principle, other beings might use their own version of the same thing. We cannot rule out the possibility, say, of an Andromeda-thropic Principle.

Here, in the Anthropic Principle, we see evidence of the return of a literal sort of anthropocentric perspective into contemporary science. In those uses of it which involve implicit or explicit teleology, we see additionally a revival of analogues to Aristotelian biological-observer-based schemes. But whether man-centered or biological-observer-centered, anthropic principle thinking most certainly represents the return of a classical mode of thought.

A return to implicit anthropocentrism is also evident in contemporary quantum physics. If we divide the world into three dimensions, namely, the very small, the very large, and the middle dimensions, then quantum physics describes the world of the very small; cosmology and astronomy, which we have just glimpsed, describes the very large. Mankind, of course, inhabits the middle dimensions. Modern science self-consciously (and self-critically) rejected the idea that the middle dimensions were the central, most significant dimensions. But as we have just seen, the Anthropic Principle reflects the penetration of the very large by the human scale, and, next, in quantum physics, we shall see the exact same penetration by the human dimension of the very small.

In quantum physics, a main principle has come to be held by almost all practitioners, to wit, that an event or an object is not real unless and until it is observed. This quantum observation principle, or as some call it, the quantum reality principle, has been necessitated by the fact that quantum systems behave somewhat paradoxically when compared to ordinary systems such as cells or automobiles. The reality principle, for example, avoids some of the excruciating problems involved with light's behaving at one time wave-like and at another time particle-like. Thus, rather than trying to answer the question "What is light *really*, wave or particle?", the quantum reality principle advises us that it is what we observe it to be, when we observe it, and only when we observe it. In an experiment which relies upon observations of particular wave-like properties of light, light is observed to be wave-like. Similarly for experiments which measure essentially particle-like properties. To put the case most sharply, the nature of most quantum phenomena inevitably depends upon the interests, intents, and activities of the observer of that phenomenon. It is now widely agreed that it is simply impossible to factor the observer out of the quantum physics equation. No longer does it seem plausible to believe in the old-style, objective world existing "out there" independently of the scientific observer. The observer, via his very observation, participates in the nature and creation of the world he observes.[10] One sees clearly here analogues to the anthropocentrism implicit in Aristotle's biological-observer-centered science.

Quantum reality has been linked by some to the Anthropic Principle. The result of this linkage is an explicit and literal anthropocentric metaphysics, that is, a philosophy of reality grounded in human beliefs.

John Archibald Wheeler has argued for a cosmology that he calls "self-referential." In this universe, the observer confers reality upon the phenomena that he observes, while simultaneously, the universe confers reality upon the observer. As Wheeler states, "quantum mechanics must lead us to take seriously the view that the observer is as essential to the creation of the universe as the universe is to the creation of the observer."[11] Not only does Wheeler present us with a thoroughgoing and obvious teleology, but also an evident and thoroughgoing observer-centered perspective, an anthropocentrism of the first order.

Again, as in the case of cosmology, a certain amount of care is needed to sort out what is, and what is not, going on here. Clearly, not many quantum physicists would wish to extend the quantum reality principle to the limits exhibited by Wheeler. However, it is becoming ever clearer that physicists have despaired of finding an objective, observer-free world to observe. The essential coupling of the observer to his observed subject lends necessarily an anthropocentric orientation to the observation itself. The quantum world is as it is observed to be, in large part because it is observed to be that way. Since we cannot under any circumstance say what it would be like without observing it, it must be as we observe it. Thus, the contemporary world of the physicist is inevitably observer-centered, and implicitly anthropocentric precisely to that degree. Just as Aristotle the physicist, we must take the world as we actually find it, since an observer-free observation is just as impossible for us as was a friction-free motion for Aristotle. In the end, we must return, sadder perhaps, but certainly wiser, to a scientific method deeply rooted, centered in fact, in the biological nature of the observer. Aristotle would not be surprised.

University of Missouri-Kansas City

NOTES

[1] This thesis, although not radical, is certainly still somewhat controversial. I came finally to accept it because Marjorie Grene argued it so well. See her superb (1963), pp. 32–33.

[2] Aristotle's *Categories*, prolegomenon to everything else he does, exhibits precisely this scheme of reality. See especially chapter 5.

[3] Ross (1964), p. 179, is quite explicit on this point.

[4] See Gale (1981).

[5] The best reviews of these myriad restrictions are Demaret and Barbier (1981) and Leslie (1982).

[6] Dicke (1961), p. 440.

[7] Collins and Hawking (1973), p. 317.

[8] Wheeler (1977).

[9] On the other hand, Wieland's idea ((1975), pp. 146–147) that Aristotle used an "as if" teleology, that is, one in which telos functions as "a concept of reflection, and not a cosmic or metaphysical principle" seems to me here to present an obvious way to guarantee a "safe" use of the "in order to" at this level of analysis. That is, of course, if one in fact *desired* to adopt such language. Some do; cf. Leslie (1982).

[10] Wheeler (1977) states this case most strongly. However, he most certainly is not alone in his perspective; see, e.g., DeWitt and Graham (1973).

[11] Wheeler (1977).

BIBLIOGRAPHY

Barnes, J., Schofield, M., and Sorabji, R. (eds.): (1975), *Articles on Aristotle: 1. Science*, Duckworth, London.

Butts, R. and Hintikka, J. (eds.): (1977), *Foundational Problems in the Special Sciences,* D. Reidel, Dordrecht, Holland.

Collins, C. and Hawking, S.: (1973), 'Why Is the Universe Isotropic?', *Astrophysical Journal* **180**, pp. 317–334.

Demaret, J. and Barbier, C.: (1981), 'Le Principe Anthropique en Cosmologie (Deuxième Partie)', *Revue des Questions Scientifiques* **152**, no. 4, pp. 461–509.

DeWitt, B. and Graham, R. (eds.): (1973), *The Many-Worlds Interpretation of Quantum Mechanics,* Princeton University Press, Princeton.

Dicke, R.: (1961), 'Dirac, Cosmology, and Mach's Principle', *Nature* **192**, pp. 440–441.

Gale, G.: (1981), 'The Anthropic Principle,' *Scientific American* **245**, no. 6, pp. 154–171.

Grene, M.: (1963), *A Portrait of Aristotle,* University of Chicago Press, Chicago.

Leslie, J.: (1982), 'Anthropic Principle, World Ensemble, Design', *American Philosophical Quarterly* **19**, no. 2, pp. 141–152.

Ross, D.: (1964), *Aristotle,* Barnes and Noble, New York.

Wheeler, J.: (1977), 'Genesis of Observership', in Butts and Hintikka, 1975, pp. 3–33.

Wieland, W.: (1975), 'The Problem of Teleology', in Barnes, Schofield, and Sorabji, 1975, pp. 141–160.

ROM HARRÉ

LOCATION AND EXISTENCE

THE GENESIS OF A SPATIO-TEMPORAL MANIFOLD

The first step in the genesis of the most primitive space and time concepts will be to show that certain elementary principles of logic generate the basic topography for both space and time, that is the basic structure of the set of relations we call "space" and "time," with the help of some very simple empirical assumptions. As a first step we are to suppose that we are given an individual called "*a*" and a property ascribed by the use of the predicate "*F*." Let us consider all possible cases of qualitative and numerical identity. Let us suppose that our wordly experience requires us to assert a proposition which has the form '*Fa* and *not-Fa*'. If we adhere to the principle of non-contradiction, this assertion requires to be remedied. (Does this imply that space-times are differently constituted for intuitionist or quantum logic? I think that it must do so, though I shall not explore that issue here.) It can be remedied in two ways.

(i) If *a* is an *extended* thing, the acts of predication "*Fa*" and "*not-Fa*" could be about different parts of it separated by an "interval of space," provided those acts of predication were simultaneous. This is the minimal spatial concept.

(ii) Alternatively, if *a* is an *enduring* thing, the acts of predication "*Fa*" and "*not-Fa*" are about different moments in its history, that is they are separated by an interval of time, provided that *F* and *not-F* are co-located.

This is the minimum temporal concept. By assuming the distinction between qualitative and numerical identity and requiring that contradictory predications be remedied, we deduce *a priori* a minimum topology based on the bare concept of interval, noticing that the interpretation of (i) as "*a* is extended" requires the assumption of simultaneity, while the interpretation of (ii) "*a* has endured" requires the assumption of co-location. By negation, simultaneity and co-location engender endurance and extension.

I hope to substantiate the intuition in the coming argument that no

A. Donagan, A. N. Perovich, Jr., and M. V. Wedin (eds.), Human Nature and Natural Knowledge, 243–257.

further structure can be imposed on space, time, or space-time, by *a priori* argument independent of all empirical considerations. In the *a priori* deduction above, that which I have called "spatial" and "temporal" are perfectly symmetrical. Further structure must be introduced to develop the formal distinction between spatial and temporal intervals. Further, the possibility of any metrical tests for comparative judgments of interval will require no less than a physics, a theory of the cause of change.

The next strongest principle to the logical law of non-contradiction governing predication, is Leibniz's law of the identity of indiscernibles, governing reference. If two names refer to two putative individuals all of whose properties are the same, then there is one, not two such individuals. We can use this law to demarcate spatio-temporal from all other properties. Most classes of physical beings are such that at a given degree of specificity of description there are many individuals which differ only numerically. Suppose now, relative to the preservation of Leibniz's law we wish to remedy the assertion

Fa and Fb,

where "*a*" and "*b*" are distinct names, we can do so by claiming that *a* and *b* are numerically distinct beings, differing only in spatial or temporal location. They could be simultaneously co-existing things, or co-located but successive events, or various other weaker cases. It is clear that Leibniz's law is also inadequate to distinguish spatial and temporal relations formally.

I conclude that some further principles, in particular one which specifies the way causal influence is propagated in the world, would be required to make a distinction. With the help of the principle that the propagation of causal influences has a maximum but finite velocity, and the idea of the light cone, space-like separations can be distinguished from those which are time-like and by reference to the requirement that time-like separations on a worldline are within a light cone, while space-like separations are on worldlines intersecting the surface of a light cone. Reichenbach (1958) has shown that the choice of invariants is related to considerations of economy in the introduction of non-uniform fields.

To introduce metrical tests which would enable us to rank intervals according to some such distinctions as longer or shorter, earlier or later, we require a principle of invariance, leading to a metrical test which would enable us to compare spatial intervals, or temporal intervals, one with another, and this would require certain assumptions about the invariance of standard lengths and standard durations. The transport of a rod or the reproduction of a temporal sequence is assumed to be invariant with respect to the metrical properties of space and time, if each can be isolated from extraneous causal intervals. Following Reichenbach's principle, we note that, in the end, whether such isolation is achieved is a matter of fiat or the laying down of a convention.

The upshot of this argument is the identification of a system of relations between material individuals and their having of properties which derive from general constraints on criteria of identity for those individuals. The relations are a-causal by virtue of the invariance assumptions which are needed to ensure the coherence of those quantitative comparisons defining a metric. It follows from these considerations that any thing or event meeting the criteria of identity and individuation above, must stand in spatial, temporal, or spatio-temporal relations to one another and, furthermore, to every other.

The next step in developing a universal reference scheme is to consider whether it is possible to abstract and generalize a system of spatio-temporal relations so we could free it from dependence on actual material bodies, to provide a universal and unique system for locating things and events with respect to which individuals of various kinds can be unambiguously located and for which successful location will entail existence in the same mode as that of the material beings from which the abstract system of location has been derived. I shall speak of the abstract system as the "space-time manifold" and the system of spatio-temporal relations between material bodies and their changing properties I shall call the "system of spatio-temporal relations." Exploration and comparison of the properties of the abstract spatio-temporal manifold and the concrete system of spatio-temporal relations, will have to answer the question of their realities.

With the abolition of the ether, the case for the reality of the abstract system seemed hopeless. However, considerations deriving from physical theories more general than Special Relativity suggest that the question might be reopened, with the consequence that we are once again free to examine the relation between the abstract and the concrete relational systems. Does the latter exist only because it is embedded in the former, or is the former a mere abstract mathematical fiction?

At this point we must turn to an investigation of the relationship between the spatio-temporal system and the metaphysical theories within which a general physics that gives sense to these spatio-temporal systems is embedded. Physics in general shows a progression in that the metaphysics embedded in the physics of successive periods changes. I propose to argue that the solution to the possibility of the reality of the abstract space-time manifold will differ, depending on the particular general physics with respect to which the argument is pursued.

The central issue is how far the general physics of a period is committed to the principle that space, time and matter are in mutual interaction. We can compare the three forms that the metric formula or law of the invariance of the line element takes in three progressive general physics. The line element formula represents a way in which the geometry of the spatio-temporal system works. A line element is the general arc length, so the line element law is the law of calculation of arc length from whatever other parameters of the system are required. Consider three line element laws; beginning with

$$\textit{Newtonian} \quad ds^2 = dx^2 + dy^2 = dz^2.$$

This formula represents a metaphysics for physics in which space-time and the system of material bodies are independent of one another. The metaphysics or principle of dependence is reflected in the fact that the line element law contains no terms for time or the material system.

Consider next:

$$\textit{Minkowski} \quad ds^2 = dt^2 - (1/c^2)(dx^2 + dy^2 + dz^2).$$

The metaphysics of the physics in which this is the line element law clearly requires the non-independence of space and time. It is the metaphysics of Special Relativity in which the independent systems of separations and locations in Newtonian and commonsense physics are treated as one system, the system of space-time. It is notable that the calculation of the length of an arc requires both a temporal parameter and spatial coordinates. However, in the metaphysics of space-time there is no relation between the system of material bodies and the spatio-temporal system. The system could be concrete relational in a very strong sense in that not only would empty places be defined counterfactually with respect to the reference frame of real thing-events, but also every observer would have to apply modal predicates to every frame other than his own. So, with respect to relations of non-existence and succession, what was real for one would necessarily not be real for another. Reality, like spatio-temporal location, would become indexical and so 'exists' would join 'here' and 'now' in a peculiar logical category.

Finally, we have:

Robertson-Walker $ds^2 = dt^2 - (1/c^2)\, r^2\,(t)\,(dx^2 + dy^2 + dz^2)$.

The metaphysics of the physics of which this, or something similar, is the line element law is radically different from either the Newtonian or the Minkowskian laws in that the calculation of arc length now requires measurement of the density of the system of material bodies at some moment. At least the local geometry of the universe for which this was the line element law would be influenced by the distribution and quantity of the material system. The most elaborate line element law entails the following:

> Measures of space, time and matter/energy are functionally related.

This sub-conclusion is susceptible of a variety of different explanations not all of which require that an independent reality be assigned to the space-time manifold. It could be argued that the inter-relation of measures of space-time and matter/energy is a natural consequence of a physical interaction between the space-time manifold as real and the system of material bodies. On this interpretation they are independent substances mutually interacting. In support of this proposal one could cite the fact that there are solutions for the differential equations expressing line-element and other general laws defining specific universe models even in the absence of any matter, solutions which imply a structure for space-time (as 'the vacuum'). There seems to be a case for the view that the spatio-temporal manifold has properties which are independent of the material system. An argument of this form was proposed by Nerlich (1976), based on the idea of treating some geometrical properties of the manifold as 'shape'. I will return to consider his argument in more detail below.

Alternatively, vacuum solutions could be interpreted as mathematical accidents, comparable to the physically implausible solutions for kinematic parameters in elementary mechanics and the many deleted solutions throughout physics. Vacuum solutions could be declared physically meaningless on metaphysical grounds. It might be held that in the absence of the material system the spatio-temporal manifold would not exist, perhaps because no concrete spatio-temporal relations could exist in those conditions and *a fortiori* the manifold would have no properties.

A more radical explanation of the appearance in physics of relationships such as those expressed in the Robertson-Walker metric would be to deny that they reflect mutually interacting but distinct substances. They could be experientially differentiated aspects of the development of a single substance. Adopting this explanation of the appearance of mass-energy terms in geometry allows us the comfort of Occam's Razor. We can take the appearance of line-element laws like the Robertson-Walker metric as evidence of the existence of matching systems of properties of a deep system one of whose manifestations is the array of interacting material bodies. The spatio-temporal manifold is rather more than just the relational properties of the system of material bodies. But it is not an independent existent. This seems a solution most in keeping with the considerations we have so far advanced. In short, there is no need for a second substance.

THE REALITY QUESTION ITSELF

There is an air of paradox about this question in the context of the defence of a realist theory of science. The space-time system of relations are being offered as the foundations of a technique by which the establishment of referential relations to an item located in the system determines the reality of that item. To what then might the reality of the space-time system be referred? Bracketing for the moment the question of the relation of the abstract and concrete systems to one another, the paradoxical air can be dispelled by looking more closely at the reasoning which would be involved in fully spelling out the claim to reality based upon successful location. Whatever may be the status of the abstract system it is the fact that the relations of the concrete system obtain between things and events that serves as the reality donating premise of the reasoning that uses as minor premise the establishment of spatial and/or temporal relations to undoubted existents.

The abstract system which I have called the space-time manifold provides for a location whether or not a concrete thing or event is in the vicinity. It provides an actual place where, relative to the system of concrete spatio-temporal relations, there are only possible places. The concrete system on the other hand, must follow Leibniz's specification of relational space and time: "space denotes, in terms of possibility, an order of things which exist at the same time . . . " (*Clarke-Leibniz Correspondence*, Leibniz' Third Letter, para. 4). So for the concrete system to comprehend all of space and time, as we might wish to use it as an ultimate foundation for reality proofs *via* establishment of reference, it must allow counterfactual conceptual components such as "where nothing is, but something could be." It looks as if a non-paradoxical grounding in reality of the concrete system of actual relations amongst things and events requires inclusion of counter-factuals in its specification. With all the problems inherent in any claims we may need to make as to their truth (and the use of empty places 'in space' to locate reference in non-existence proofs seems to require them), the concrete system becomes paradoxical in reality proofs. On the other hand, the abstract system is defined without recourse to counterfactuals since all its locations are equally part of the space-time manifold; it is created as an abstract system of relations 'lifted off' any particular grounding in things and events.

The status of the problem I am considering can be made a little clearer by recasting it in the form it took in classical physics. Within that scheme there were two material modes of being, the mechanical and the electromagnetic. While the mechanical was defined for things and events that were discrete and in many respects like those of the general material system on which I have suggested we define the concrete system of spatio-temporal relations, the electromagnetic mode of being was defined for a universal aether, providing for a realist interpretation of the abstract spatio-temporal manifold, actual at every point. The abstract system can still be treated relationally but it is now a relational geography of a uniform and pervasive medium. It parallels exactly the use of lines of the latitude and longitude for locating 'places' at sea.

ARGUMENTS FOR THE REALITY OF SPACE

Two recent studies of space have offered arguments for the independent reality of space (Hinckfuss, 1975, and Nerlich, 1976). I shall concentrate on Nerlich's argument in this paper. His project is to defend the position that space is absolute, as opposed to relational. So far as I can see the only meaning that can be given to the thesis that an entity is "absolute" (saving the attribution of it to God) is that it is a substance. As we shall see Nerlich does seem to be committed to the view that space is an individual substance (*op. cit.* p. 206). His argument depends on two main premises:

1. Space has a shape, that is it has geometrical properties;
2. The shape of space has independent non-geometrical effects.

Underlying these premises, which are elaborated in detail in the study, there seem to be two deeper metaphysical principles upon which the argument finally rests.

For the demonstration that space has a shape to license an inference to its absolute existence, a principle something like the following seems to be needed:

> Anything which has a property is a substance, at least it is not a relation or system of relations.

Unless some such principle is admitted it is open to the relationist to claim that Nerlich's demonstrations of the geometrical properties of space only go to show that a relational system, namely that we call space, has a property, since the above principle seems to be obviously false. Relational properties can have properties. Thus the interval between a missile and a plane can become shorter, a journey between two specified places can be accomplished more quickly, the height of a tower may be awe-inspiring, and so on.

The second phase of the argument occurs in the context of general relativity in which Nerlich argues that a material thing propelled through a shaped space will change the relations between its parts, and so will experience forces, real dynamical effects, caused by the spatial distortions in its path. But for this argument to go through it must not be possible to refer the shape of space to the presence of material bodies, which engender a force field which is ultimately responsible for the dynamic effect in the test body. Admittedly there are mathematical fantasies in which cylindrical or other spaces are defined and through which imaginary things pass, but the physics of space-time does not admit intrinsically curved spaces, unless vacuum solutions are taken seriously. The curvature of space is an effect of the presence of material bodies. In the thorough-going mathematical reciprocity between the metric tensor and mass-energy tensor in General Relativity there is no direction built in, leaving us free to draw on our general metaphysics which assigns a causal direction from masses to masses *via* space-time distortions.

GLOBAL MANIFOLDS AND LOCAL SPACE-TIME

It is easy to show by example that for any given local geometry there are many global topologies, the simplest and most easily visualized case is the embedment of a local Minkowsky metrical geometry into global Space-Times. For instance, the global structure of cylindrical Space-Time and the global structure of flat Space-Time are both compatible with the metric $ds^2 = c^2dt^2 - dx^2$ as a local constraint on the line element. The difference between these two global structures is readily seen in Figure 1. In cylindrical Space-Time there are closed

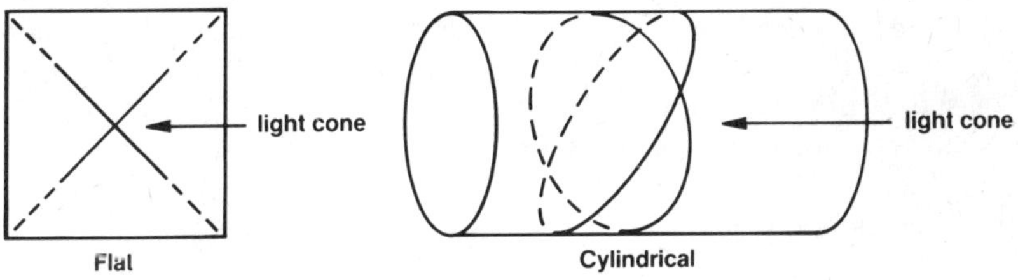

Fig. 1

time-like world lines, while in flat Space-time world lines are time like, since they must lie within the light cones, but cannot close on themselves. Now these examples might be taken to show that the global topology could be treated just as any other theoretical entity and used in the explanation of observable effects. I wish to dispute the suggestion that the abstracted space-time manifold could be real, and that we could argue for the differentiation of one such manifold from another on empirical grounds.

The logical independence of metrics and global geometries gives us the options for claiming on the basis of a restricted form of Nerlich's principle that the metrical geometry is a real adverbial property of the relational properties of the material system, in particular those properties having to do with ordering, and that the global geometry is a formal fantasy arbitrarily chosen from a limited range of compatible global systems. The fact that we can assign different properties to global geometries is no ground at all for saying that any of them exist. By parity of reasoning one could claim that this is a general feature of physical systems in that in any physics the abstracted space-time manifold is to be treated as a fiction. Clearly the absolute space and time of Newtonian physics is quite naturally so treated.

This makes it essential to insist on non-realist interpretations. If the metrics I have been using as examples are related to the presence of actual matter in the universe, then if there were locally discernible effects on the spatio-temporal system in the absence of matter, one would not be so easily able to reject the abstracted manifold as real.

The arguments I have proposed against taking vacuum solutions seriously can be summed up simply. If the abstracted space-time manifold were substantial it *would* have a structure as expressed in the vacuum solutions. This seems to be indisputable. It does not follow from this that we have any good reason for thinking that the abstracted manifold exists, i.e., that any of those counter-factuals are true. The material system and its relational properties provide the basis for a spatio-temporal system co-extensive with the material system, and therefore adequate for reference. The most economical hypothesis is to suppose that the one is a relational property of the other.

INDEXICALITY AND REALITY: THE NECESSARY UNREALITY OF THE GLOBAL SYSTEM

This argument can be sharpened if we consider the existential problem in a less general way. The existence of the abstracted spatio-temporal manifold as a real substance encourages a tenseless view of change. On that view all events are equally real and change is impossible. If we take the ideal of a relativistic space-time manifold seriously, the tenseless view seems attractive on the grounds that there is no way of identifying a unique space-like hyper-surface orthogonal to the world lines of material things. Thus the very idea of events coming into existence is equivocal to say the least, since it becomes impossible to define a global concept of the present. However, there are immense difficulties in the tenseless view, not least the implausibility of the existential proposal that past, present and future are all three real.

On the other hand, a naive tensed view in which events are

supposed to acquire different properties, changing from past to present to future, and in which the direction of time becomes the fundamental explanatory concept of the temporal relations of past and present seems to involve a contradiction: For events successively to have the three properties of pastness, presentness and futurity, they must have some form of sempiternal existence to be the bearers of those properties. Events on the tensed view will have to change. However, such a view is naive. A more satisfactory tensed view can be constructed on the basis of a careful consideraton of what existential claims are required to be sustained in each view and what has to be explained by any theory of time.

It seems to me that the following features of experience and the world must be satisfactorily accounted for. (1) *We* are able to order our knowledge of change (events) and things in a rather particular way, with respect to the possibility of causal intervention, in that our own activities, or the activities of other agents, are only able to be exerted in the present but we can conceive of them as having future effects but not as having effects upon the past. Even this, as we shall see, is a careless way of speaking. (2) Our consciousness is narrowly focused, comprising only the present. (3) Our language contains indexical sentences, whose referential meaning is a function of their relation to the immediate context of their utterance. Typically such sentences include the indexicals 'here' and 'now'. Should we treat 'exists' as another indexical?

The simplest and most elegant solution to all these problems is to suppose that the universe exists 'always' and only in the present. Tenses order presents: Pasts and futures are human intellectual fictions. This assumption explains our relative causal impotence. We can act only upon that which is real. Neither the past nor the future are ever real, that is, at no present is either real. The effects of our present interventions do not ramify into the future now, but rather constrain and modify the way the universe is at each successive present. These are asymmetrically ordered with respect to those we are potent to effect, since we cannot act on another present from that at which we act now because *a fortiori* only that which is now is real. Restricting reality to the present accounts neatly for the narrow focus of our consciousness since it seems reasonable to argue that we can focus only upon that which exists. Our inability to focus on the past or the future, other than through some present trace or anticipatory condition or representation, follows naturally. Finally, the indexicality of certain sentences and our intuition that the content of these sentences is not expressible in a tenseless language of temporality, is also readily accounted for since it is no more than an expression of the limitation of existence to the present. Both the naive tensed view, that events change from past to present and become future, and the tenseless or block universe view are fictions reifying representations.

There remains a final refinement. If that which is real exists only in the present and Special Relativity forbids us to identify a unique spacelike hyper-surface upon which a universal moment of simultaneous events could be mapped, then there can be no universal present. Does this count against the theory that the reality of each existent is defined only for its present? It would seem that while each element or component of the universe was real by virtue of existing at its present, the totality of such elements, having no universal present, could not exist. But what is it that is most frequently offered as the one true universal being? It is some form of spatio-temporal manifold proposed as absolute and hence real. Global space-times must then retreat to the status of fictions. Since no space-like hyper-surface can be defined, the indexicality of existence favors the relational theory.

FURTHER CONSIDERATIONS CONCERNING REFERENCE AND EXISTENCE

If space-time is real only as an attribute of the system of material things, then the most fundamental ontological hypotheses must concern the existence of a material system. In contrast, the reality of the past, present and future demanded by any form of block universe theory, particularly one which holds that the complete space-time manifold is real, multiplies the material system indefinitely. A world line consists of an indefinitely large number of manifestations of the individual material thing which exists at each temporal location on that world line. This should be enough of a difficulty by itself to dispose of the idea that the spatio-temporal manifold is real as a whole. The intuition behind this argument will not become fully clear until the relation of indexicals to their own and other world lines has been defined below.

I hypothesize that the material system exists only in the present, that only the present is real, and that there is only one material system.

There are, of course, representations or traces of other presents in *the* present and the anticipations of yet others. These traces and these anticipations, though, are part of the existing reality, of the spatio-temporal system as it exists. That is, they are present existences. In order to adjudicate between the two theories we must examine them in the context of the Special Theory of Relativity.

(1) A 'block universe' theory of space-time can act as a reference-system for everything that can be located by the use of 'placed' and tensed language, which is the logico-linguistic counterpart of the theory that the universe exists only once and all at once. But this theory runs into difficulties in trying to cope with indexical statements such as "I am writing now" and "You are reading here," statements whose sense is determined by their spatial and temporal context. This sense cannot be expressed by any calendrical indexing, since it is clear that the statement "I am writing now" contains more information than the statement that I am writing at five past twelve on the twenty-second of June, nineteen hundred and eighty two. It is the additional information relating my awareness (and existence) to an event that the block universe theory with its tenseless discourse is unable to incorporate.

(2) The hypothesis that the universe exists only at each present explains everything the block universe theory explains and it makes possible the extra content of indexical statements because that content depends upon the special status of human awareness as indexical of the fact that I only exist *now*. The additional information that is present in the indexical statement is just that. On this view, the distinction between past, present and future is a psychological distinction, having to do with memory and imagination. If we adopt the relational indexical view with its tensed discourse and its ineliminable claims about the particularity of the present, we should no longer be tempted to reify the space-time framework. Most of that framework, namely all of the past and all of the future, is created only in the imagination and the memory of people. In *rerum natura* there are only contemporaneous traces and anticipations. The space-time system of relations must be the basic referential grid, since it is to that, with its anchorage in things, that indexical reference connects us.

However, the statement that the universe as 'each and every existent' exists only now has to be qualified rather carefully in view of the presently accepted truth of Special Relativity. There is no one hyper-surface orthogonal to the world lines of all particles. So the ontological hypothesis which I am arguing for has to be relativized to each observer on his own world line and based upon only his indexicals. That is, though there are a set of true indexicals they cannot be lined up, as it were, into a single orthogonal hypersurface. Thus, Figure 2. In the future, which still contains the fiction of a

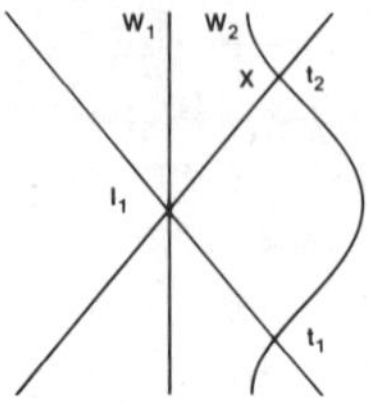

Fig. 2

world line, we distinguish between the indexical I_1, which specifies the existence of the world, and a *t*-statement, which speaks of the existence of the world with respect to the world line W_2. Which *t*-statement is contemporaneous with I_1? With respect to a light cone with origins at the uttering of I_1, the *t*-statement is no more than a hypothesis *indexically* undertermined between t_1 and t_2 on W_2, since *a fortiori* the observer has no place on world line W_2. Should the observer who utters I_1 come to utter I_2 later, our picture must be drawn very carefully indeed to avoid the hypostatization that we are trying to eliminate. Compare in Figure 3 representation RA with

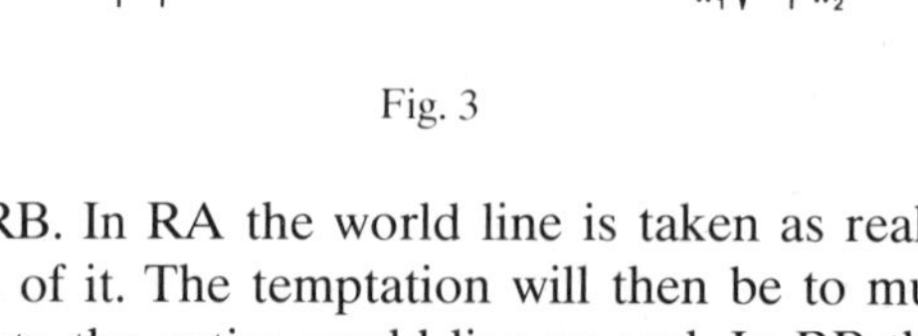

Fig. 3

representation RB. In RA the world line is taken as real and I_1 and I_2 are components of it. The temptation will then be to multiply realities so as to constitute the entire world line as real. In RB the world exists for the observer only if I_2 and he remembers having said I_1 and the balloon represents his remembering it. The world line through I_2 is no longer an existent, but the projection onto a representation space of the traces of the past and anticipations of the future that exist only with the uttering of I_2. A world line on this view is a series of iterated and embedded utterances, not a track through a higher order universal geography.

It seems, then, that it is possible to acknowledge, and indeed emphasize, the truth of special relativity without at the same time being forced into a reification of the spatio-temporal manifold *via* the world line mode of representation. On the tensed indexical ontology advocated in this paper it can be no more than a map of the sets of true indexicals which have been uttered by all possible observers *as* all possible world lines. It is not a map of the universe but a representation of a possible discourse about it.

Oxford University

BIBLIOGRAPHY

Hinckfuss, I.: (1975), *The Existence of Space and Time*, Clarendon Press, Oxford.
Nerlich, G.: (1976), *The Shape of Space*, Cambridge University Press, Cambridge, England.
Reichenbach, H.: (1958), *The Philosophy of Space and Time*, Dover Books, New York.

WILLIAM C. WIMSATT

FORMS OF AGGREGATIVITY

At least since Plato and Aristotle, philosophers and biologists have puzzled over the differences between systems which are mere "aggregates" of their parts (such as a heap of stones) and systems which are "complex organized wholes" (such as biological organisms). The rallying cry for many of these differences has been the aphorism that "an organism is more than the sum of its parts" — whereas a heap of stones, presumably, is "nothing more" than the "sum" of its parts.

This puzzle has also become inextricably connected with a number of other issues, most strikingly with concepts of reduction and emergence. Thus, traditional reductionists have held that for a property to be emergent is for it not to be reducible to properties of the parts and their relations, and denied that for *any* system is "the whole more than the sum of its parts". Traditional holists have embraced the view that for organic systems "the whole is more than the sum of its parts" and, accepting the reductionists' formulation of the issue, seen this as committing them to denying the reducibility of properties of that system to properties of the parts and their relations. The analysis that I will propose of aggregativity and failures of aggregativity gives a clear sense in which a system property may be reducible to properties of its parts and their relations and still be spoken of as emergent. Probably more importantly, the concept of aggregativity and the variety of ways in which system properties may fail to meet its conditions provides a rich variety of tools for describing organized systems in biology, in engineering, and throughout the behavioral, cognitive, and social sciences.

A total analysis of what differentiates stones from organisms (even considered synchronically, thus ignoring developmental and evolutionary dimensions) involves at least distinguishing and separately analyzing the concepts of aggregativity and emergence, functional organization, and organizational complexity. The last two topics are dealt with in other papers: functional organization in Wimsatt (1972) and complexity and organization in Wimsatt (1974). I will not

A. Donagan, A. N. Perovich, Jr., and M. V. Wedin (eds.), Human Nature and Natural Knowledge, 259–291.

reproduce those analyses here, but I will make some comments on how the topics of this paper relate to those issues.

1. AGGREGATIVITY AND INTERSUBSTITUTABILITY

It seems clear that there is no point in speaking of systems *in toto* as being or not being aggregates of their parts. An organism is a paradigm case of a system which is "more" than the "sum" of its parts, but the mass of an organism is nonetheless equal to the sum of masses of its parts under any exhaustive and mutually exclusive decomposition of it into parts. Similarly, though a pile of rocks would be readily counted as an aggregate of the individual rocks, the stability and shape of the pile depends in very different ways on the stability and shape of the individual rocks. Both properties of the pile are obviously functions of the particular arrangement of the rocks — a primitive criterion for non-aggregativity. Both paradigm aggregates and paradigm wholes are aggregates for some of their properties and not for others. For this reason, it is better to talk about properties of systems and their parts, and to analyze aggregativity as a kind of relation between these properties.

The common intuition involved in saying that a property is not an aggregative property is that it depends in some way on the mode of organization of the parts. But there are many possible modes of organization, so it makes more sense to analyze what is involved in saying that a system property is *not* a product of the way in which the system is organized. A partial explication of this is to require that the parts are *intersubstitutable* without affecting the system property. Failures of intersubstitutability suggest both that the parts have different properties and that the system is organized in such a way as to make special use of the different properties of the parts to produce the system property in question.

This suggests the following condition: Suppose that system and parts' properties are related in a theory by a function, F:

$$(1)\qquad P_j(S) = F[p_1, \ldots, p_n(s_1), \ldots, p_1, \ldots, p_n(s_m)],$$

where $P_j(S)$ is a system property, s_1 through s_m are parts of system S under an exhaustive and exclusive decomposition, and p_1 through p_n are properties possessed by these parts. For $P_j(S)$ to be aggregative relative to p_1 through p_n of s_1 through s_m of S, the "composition function" F for $P_j(S)$ must be such that its value is *invariant* with respect to intersubstitutions (or permutations) of the terms '$p_1, \ldots, p_n(s_j)$'.

Several things should be noted about this condition:

(1) It is relativized to given decomposition of the system into parts. (See Kauffman (1971) and Wimsatt (1974) for discussions of the significance and ramifications of being able to decompose systems into parts in different ways.) It is possible that under different decompositions, the parts $\{s_j\}$ might not exhibit some of the properties $p_1, \ldots, p_n$. This would be expected if some of the parts' properties depended upon the internal organization of the parts, and the different decompositions into parts cut across parts boundaries of the first decomposition. (This situation, which I have called "descriptive complexity" is analyzed further in Wimsatt (1974)). It is also possible that although all of the parts would exhibit all of the relevant properties under another decomposition, the value of $p_j(S)$ in the composition function would exhibit variation under intersubstitution from one decomposition to the next.

(2) This suggests some stronger possible conditions, namely, invariance of the composition function not only for intersubstitutions within a given decomposition, but also for intersubstitutions of parts within *each* decomposition, and equivalence of the value of the decomposition function for each decomposition. A system property meeting these stronger conditions could be called *continuously aggregative*, since the conditions would then hold for all possible divisions of the system into parts. There are, as it were, no atoms or smallest units which when further decomposed lose or change any of their properties. Some properties such as mass meet this very strenuous condition, for decompositions at least down to the level of elementary particles, but the condition is too strong for some properties that we might wish to regard as aggregative. A weaker condition relating to different parts' decompositions will be discussed in section 3 below.

(3) It is possible to imagine a system composed of parts with properties showing these kinds of invariance relations in some of the system properties as a result of some rather special features of the parts which might not be had by any other parts. Aggregativity would then turn out to be a kind of accidental property of the peculiar character of the parts of the system, which would break down if any

other parts were substituted in. A plausible way of ruling this out is to talk about a reference class of relevantly similar parts and to require invariance under intersubstitution of system parts for those in this reference class. Some further instances of related phenomena which occur when a part is added to or substracted from a system will be discussed in section 2 below.

(4) The discussion of invariance with respect to intersubstitution of terms in the composition function might seem to be rendered more precise if we simply required that the function be commutative with respect to the terms, or more intuitive if we talked simply about how the value of the composition function changed under physical intersubstitution of parts. My own partiality to scientific realism suggests this latter strategy, and I will present an example later which raises problems for any purely formal account of aggregativity. Taking the realist line of talking about the parts and operations on them, it seems plausible to presume however that the interpretation of physical intersubstitution of parts within the system would be commutation of the terms '$p_1, \ldots, p_n(s_j)$' in the composition function.

Many of these problems will be solved by the addition of further conditions which will be discussed below. I will assume that the proper statement of this condition now is:

CONDITION IS. $P_j(S)$ is invariant under intersubstitution of the parts of S with one another or under substitution of one or more of the parts with other parts from a domain of relevantly similar parts. The notion of relevant similarity would presumably involve equivalence of the properties p_1 through p_n of the substituted parts.

It is worth noting that position effect, in which the location of a gene determines when or whether it is expressed, is an example of failure of aggregativity on the first criterion. This effect was not explicable on the beads-on-a-string model of the classical gene, (a model of gene effect which treated it as an aggregative property, since gene-expression was supposed to be identical over permutations in location of the genes in the chromosome), but became explicable by postulating the existence of operon-like control elements determining gene-expression in molecular genetics. In the modern theories, if a gene were moved from the domain of control of one operon-like unit to another, it might be turned on or off at different times, thus causing changes in the phenotypic expression of a trait. A standard complaint among those sceptical of the "beads-on-a-string" model was the atomistic view of genes it represented. Goldschmidt, for example, argued that the chromosomes or even the whole genome was an integrated whole. (See Carlson (1967) for a discussion of this case.)

2. AGGREGATIVITY AND SIZE

Another plausible condition which seems appropriate for a notion of aggregativity is that the system properties should remain qualitatively similar (though possibly differing in value) under addition or removal of a part from the system, and indeed for any number of such additions or subtractions. This would rule out a variety of threshold phenomena as aggregative properties. Consider the following cases:

Suppose that we have a semicircular arch made up of congruent blocks, each of which is a truncated isosceles triangle in cross section and of equivalent depth. Suppose that the system property in question is the stability or integrity of the arch. It does not matter to the stability of the arch which of the blocks from a much larger pile of congruent blocks we use in the construction, or whether (if we do it in the appropriate manner) we interchange two blocks in the arch or change one for one in the pile — which we might do, for example, for aesthetic reasons, if their surface grain or figuring differed. It seems clear that we would not want to regard stability as an aggregative property of those blocks, for it does clearly depend upon having all of the blocks in place. (Each of the ones near the top of the arch functions as a keystone: removing one of these blocks would lead to the collapse of the arch.) Stability of the arch meets the first criterion, but is diagnosed as a non-aggregative system property by this condition. Similarly, adding a block to the arch would distend it, and either lead to collapse, or at least substantially reduce its stability (by an amount which was a function of the base angles in the blocks and the static coefficient of friction between blocks.) This kind of case has obvious application in structural anatomy.

A similar situation arises where production of a gene product must fall in a certain range of concentrations to produce a certain macroscopic trait of an organism. Suppose that two active genes are required to do this, so that an organism which was homozygous for it, or one which was homozygous for a tandem duplication of that gene (and therefore had 4 genes producing the product) would not. The theory of sex-determination provides another example, if we treat sex as a dichotomous variable. Here, the ratio of X chromosomes to the number of each type of autosomal chromosomes determines sex, with males having a ratio of one-half (in XY-male and XX-female determination systems), females of 1, and various abberrant forms, showing intermediacy or hypersexual features having ratios falling in between or outside of these numbers. (Because of the involvement of autosomal genes in determining this ratio, there are other reasons why it is not aggregative, but it has at least this feature.)

This condition could presumably be rendered formally by requiring qualitative similarity of the relevant composition functions for different numbers of parts. Presumably, this could be further explicated in terms of the recursive generability of the set of composition functions which would hold for that system property for different numbers of parts. I note this possibility for those who would prefer to proceed in this way, but will not take it in the statement of the condition:

CONDITION QS. The system property remains qualitatively similar (differing only in value) under addition or subtraction of parts.

3. AGGREGATIVITY, ASSOCIATIVITY, AND SUB-SYSTEM REAGGREGATION

I have already noted that a system property may be aggregative with respect to one decomposition of the system into parts, but not for others, or even aggregative for each, but not invariant across different decompositions. A partial handle on this can be gotten by interpreting the mathematical property of associativity as describing the physical aggregation (or in reverse, the disaggregation) of diverse subsystems into large (or smaller, or different) parts, and then requiring that the composition function be invariant under associative operations on its terms. Here we are not talking about additions to or subtractions from the system as a whole (as in (2) above), but about redistributions of the material of the system into different parts, with some individual sub-systems being added to and others being subtracted from. (This would cover many (but not all) of the alternative decompositions of a system into parts envisioned in Wimsatt (1974).)

A simple example illustrating failure of this condition can be derived by reconsidering the example of the arch discussed in the last section. Suppose that we are allowed to fuse some of the blocks in the arch together (e.g., by using mortar). Since the blocks thus fused are no longer free to slide over one another, we have in effect created larger blocks with a greater angle between the faces over which sliding can occur. Since the stability of the arch is a function of the angle between these faces, we have changed (in fact, increasing) the stability of the arch through reaggregation of the parts without either adding to it or subtracting from it.

For more complicated but more interesting and important examples, consider the organization of genes into chromosomes or linkage groups. Recombination, transposition, inversion, and translocation can move blocks of genes from one chromosome to its homologous pair, attach it to another non-homologous chromosome, or move it within the chromosome to a different place, or, keeping it in the same place, reverse the order of its component genes. Since biologists, for a variety of reasons, regard chromosomes as natural units of organization of the genome, these operations produce the kinds of reaggregation of parts envisioned in the statement of this condition.

On the beads-on-a-string model, in which genes behave as individual units which are strung together to make up the chromosomes, these operations can produce only the kinds of reaggregations described above, in which changes in location of the genes does not affect the character of the gene-products or when or whether they are produced. In this case, the expression of traits is an aggregative property of the genes in the genome.

In the modern molecular models which recognize the fine structure of the gene and the existence of translation frames and control units, these operations can also produce "frame-shift mutations" (in which reading may start in a different place and the same sequence may produce different proteins) or "position effects" on gene expression like those described in the first section. Since the character of a protein or when it is expressed can affect the presence or absence of a phenotypic trait, these traits are not, on the models of molecular genetics, aggregative properties of the genes of the organism.

Even if these effects are ignored (as they are on the classical model of genetic organization), the phenomenon of linkage and the existence of linkage groups produce effects which fail to meet this criterion. These effects arise because different genes may have different contributions (either positive or negative) to fitness, and because the closeness of a given gene in the linkage map to others affects the frequency of recombination events which will separate it from these various neighbors. Genes which are particularly closely linked on the chromosome will be separated relatively infrequently from each other by recombination. This can allow a deleterious or neutral allele to be carried along with a nearby neighbor which makes a significant positive contribution to fitness. (The term normally used for this is "hitchhiking".) It is in fact one of the major mechanisms used to explain the continued persistence of neutral and even deleterious traits.

There is continuing debate over the consequences of linkage-selection interactions and the exact conditions under which they are manifested (see, e.g., Lewontin (1974), ch. 6; and Maynard-Smith (1978), ch. 5), but it is clear that they can have significant consequences over times spanning several to many generations both for what genes survive, and for how they are organized in the genome (see, e.g., Templeton (1982)). These cases involve selection, which raises special complications and a particularly important example of apparent aggregativity which I will discuss further below, in section 8.

There is a very important phenomenon involving linkage which is found in the classical model even in the absence of selection. The arrangement of genes into linkage groups affects the frequency with which gametes of different types are produced, and thus the frequencies of different genotypes produced by mating. If these are regarded as system properties, then genotypes organized as Mendelian systems assorting in ways constrained by linkage and recombination mechanisms are not aggregates of their genes with respect to these properties. These cases are rather different than those discussed so far, since the system properties in question are historical or trans-generational ones, but they are nonetheless important properties of these systems for the purposes of population genetics.

In summary then, this condition could be stated as follows:

CONDITION RA. The composition function for $P_j(S)$ is invariant under operations involving decomposition or reaggregation of parts.

In the extreme case, when any decompositions or reaggregations are allowed, this reduces to the concept of continuous aggregativity discussed in the first section. The condition is more interesting and more widely applicable if it is understood as relative to given decomposition and aggregation operations.

4. AGGREGATIVITY, COOPERATIVITY, AND MUTUAL INHIBITION

The fourth condition is necessitated by the widespread occurrence of mutual cooperativity or mutual inhibition among groups of similar elements in natural phenomena. Thus, the hemoglobin molecule is a tetramer, composed of 4 like subunits (2 alpha chains and 2 beta chains) which interact cooperatively to reduce the amount of energy required for oxygen uptake in the lungs and oxygen release in the capillaries. The energy required to bind the first of four globin (oxygen-containing) complexes is high, comparable to that required in the myoglobin molecule, which is similar to one of the hemoglobin subunits, and presumably to the primitive monomeric hemoglobin from which the tetrameric form evolved. Due to conformational changes caused by interactions among subunits after the first globin is bound, subsequent globins require much less binding energy, producing a more energetically efficient loading process. In the capillaries, higher partial pressures of carbon dioxide lead to release of globin complexes, with similar facilitation among subunits. (See Edsall (1972) for further discussion of this phenomenon and its historical and adaptive importance.) Here we have an almost paradigmatic example of a case in which the behavior of the whole (in regards to thermodynamic efficiency) is greater than that of the sum of the parts (the subunits studied in isolation).

A parallel case is found for the property of "critical mass" in nuclear physics. Spontaneous decay of nuclei of any of a variety of radioactive isotopes of uranium and transuranic elements produces particles ("slow" neutrons) capable of causing the fissioning of other atoms of the element, a breakdown which produces further "slow" neutrons. If enough mass of this isotope is arranged in a sufficiently compact form, the fissioning of atoms is a self-amplifying or "runaway" process, leading in bombs to an explosion or in reactor vessels to production of power at a (hopefully controlled but) rapid rate. The possibility of such a reaction requires both a fission reaction in which more slow neutrons are released in an atomic decay than are required to produce that decay, and an arrangement of atoms of that type in which a sufficient number of the released slow neutrons do in fact cause fissioning of other atoms of the material. If enough of the material is present in an arrangement which is sufficient to produce such a "runaway" process, the mass of substance is described as "critical" (or "super-critical" if there is more than enough).

The notion of "critical mass" is a function not only of how much stuff is present (a kind of threshold phenomenon like that considered in the discussion of the second condition), but also of how it is arranged. Thus a long wire or rod of uranium isotope U-235 may be sub-critical if it is strung out in a line in which most of the slow neutrons emitted as decay products shoot off without colliding with another atom of U-235, but become super-critical if the mass is brought together in one place. Early alternative designs for a bomb involved bringing together two sub-critical masses which were to be kept separate until the appropriate time, or (apparently more practically) setting off explosives around a sub-critical mass to compress it, increasing the proportion of slow neutrons that would be captured and making the mass super-critical. (The metaphor of "critical mass" has become common now to describe a variety of "emergent" threshold phenomena in human interaction, though conferences and other groups usually produce much less spectacular and (thankfully) safer results. In these cases, although we speak of them as "threshold" phenomena, the threshold is obviously not only a function of mass, but also of the arrangement or organization of parts.)

In chemistry, combination of volumes of different substances sometimes leads to interaction among the molecules of these substances such that the volumes of the mixtures is less when they are combined and mixed than the sum of their volumes before combination. This is common in solvent-solute interactions. Lateral inhibition networks of neurons lead to lower mean firing levels than would be expected for the neurons in isolation. These are both cases where a property of the whole is in some sense less than the sum of properties of the parts.

I am much more certain that these cases need to be ruled out (they after all represent paradigmatic failures of the standard maxim that the whole is equal to the sum of the parts) than I am sure how to describe the condition which seems to be required, or even if this condition is indeed independent of the first three. It might seem natural to express this condition by interpreting "sum" literally as addition, but there is a class of cases which seem to meet the other requirements and in which the mode of aggregation in the "composition function" is not addition, but multiplication. I will discuss one such case in the next section.

What seems intuitively to be required is some sort of "normalization" condition which preserves the right kind of scale relations between the system property and the various parts' properties. Perhaps the best that could be done would be to simply rule out cooperative and inhibitory effects, but this is to leave these effects unanalyzed. Nonetheless, this is what I must do for the time being.

One thing that will not do is to require that the system property is identical with the complex of parts' properties given in the composition function. This gives a kind of normalization of properties and property values — that implicit in the converse of Leibniz's law — of the indiscernibility of identicals. But this would tie aggregativity too closely to reduction, and would fail to distinguish between "normal" cases and those in which effects of cooperative or inhibitory interactions were present. I would want to require that the "composition function" expressed a synthetic identity in any case, but that fact can, it appears, do no further work here.

CONDITION CI. There are no cooperative or inhibitory interactions among the parts of the system.

With this fourth condition, we have all conditions which seem relevant to distinguishing between aggregative and non-aggregative phenomena. Among non-aggregative phenomena, however, there are a number of cases with some kinds of partial or restricted aggregativity which are worth discussing. I will do this in a series of examples in the next section.

5. VARIOUS KINDS OF RESTRICTED AGGREGATIVITY

Consider a set of linear amplifiers which might be put together in different ways. (Figures 1a–1d.) If each shows a linear response throughout the whole range of signal intensities being considered, then they can be strung together in series in any order without effecting the amplification ratio of the system, which is the product of the amplification ratios of all the amplifiers in the system. (Figures 1a, b.) Adding or

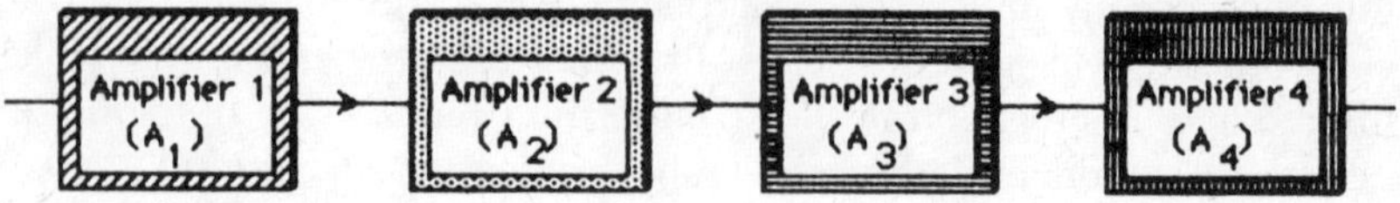

1a: Total Amplification Ratio, A_t, is the product of the amplification ratios of the individual amplifiers: $A_t = A_1 \times A_2 \times A_3 \times A_4$.

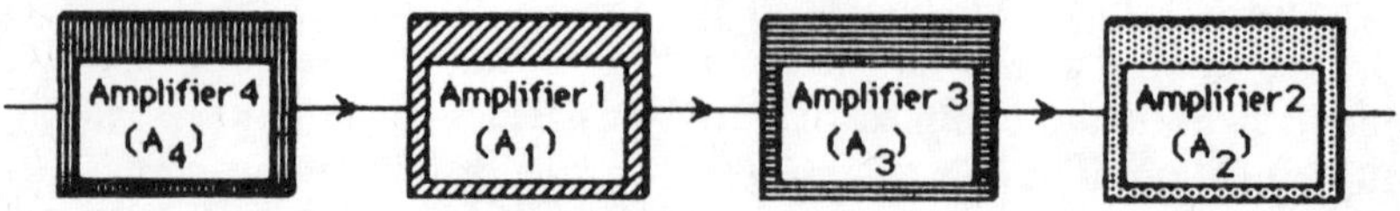

1b: Total Amplification Ratio, $A_t = A_4 \times A_1 \times A_3 \times A_2$ remains unchanged under intersubstitution of the amplifiers or commutation of the A's.

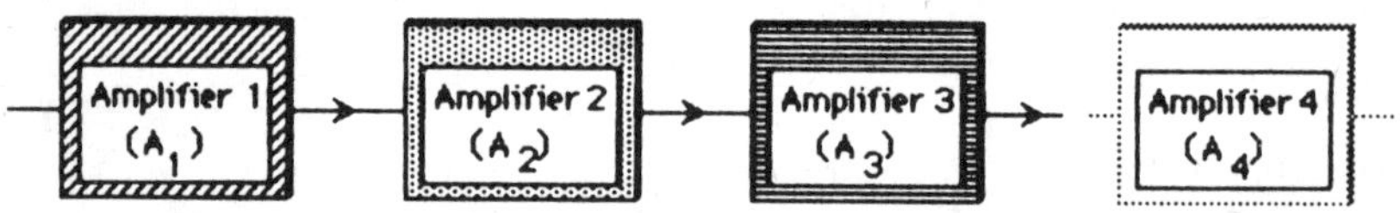

1c: The formula for amplification ratio remains qualitatively similar when adding or subtracting parts. $A_t(n) = A_t(n-1) \times A_n$.

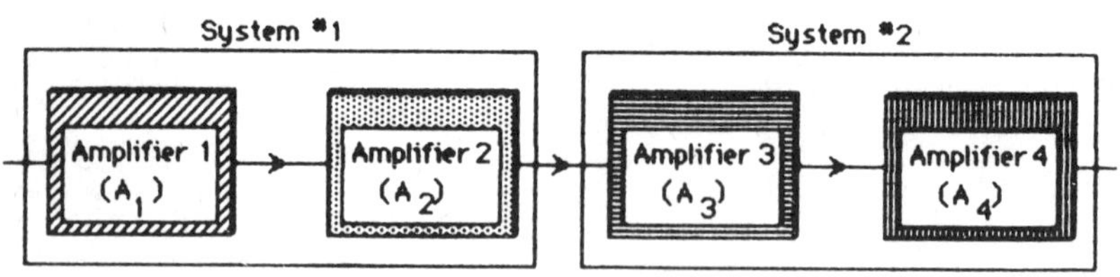

1d: Total Amplification Ratio is invariant under subsystem aggregation--it is associative: $A_1 \times A_2 \times A_3 \times A_4 = (A_1 \times A_2) \times (A_3 \times A_4)$.

Fig. 1 Conditions of Aggregativity illustrated with idealized linear unbounded amplifiers.

subtracting an amplifier would change the amplification ratio of the system, but would not change the fact that it has an amplification ratio. (Figure 1c.) In this situation, the amplifiers can also be aggregated in subassemblies in any fashion, with, for example, the first two amplifiers grouped in one subsystem and the last two in another, or the first three in one subsystem and the last in another, or all four in separate subsystems or all four together without affecting the amplification ratio. (Figure 1d.) (One could suppose that subsystem grouping might be desirable if there were no single power supply capable of handling all four simultaneously.) Finally, unless multiplicative signal growth is regarded as a kind of cooperative interaction (see the discussion of multiplicative fitness models in section 8 below), the amplifiers do not interact cooperatively or inhibitorily, so the amplification ratio of the system should be regarded as an aggregative property of the amplification ratios of its parts. Figures 1a through 1d depict this situation, in which all amplifiers are represented by blocks of the same size, indicating that they all have the same maximum power output. (In subsequent figures, block size indicates maximum power output.)

Real amplifiers are not quite this ideal, however, and thus cannot be intersubstituted freely. Very powerful amplifiers are not very sensitive to very weak signals, and highly sensitive but low power amplifiers introduce increasing distortion at higher power levels. (Indeed, this is the whole point of constructing amplification systems as series of amplifiers, rather than as one big amplifier which does all of the amplification in one step — the distortion is lower (the system response is more linear) if the amplification is done in stages.) In most real multi-stage amplification systems, the power range through which amplifiers are acceptably linear is sufficiently narrow that no intersubstitution is possible without introducing distortion or burning out overloaded circuits. (See Figure 2a.)

If we have better or more general purpose amplifiers which are linear through a broader range, or if larger deviations from linearity are tolerable, we may have situations in which restricted intersubstitution is possible. Thus it might be possible that the first two stages of amplification could be accomplished by one type of amplifier, and the last two stages by another. If the first two comprise one subsystem and the last two another, then there would be limited intersubstitutability of components within subsystems, but none across subsystems. (This is diagrammed in Figure 2b.) I will call this property *sub-aggregativity*, for aggregativity within subsystems.

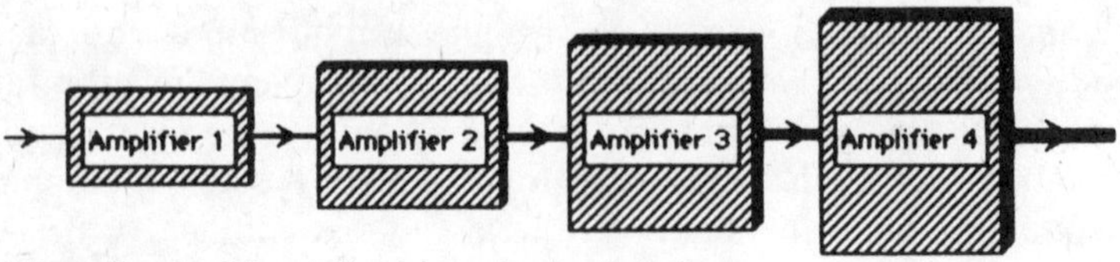

Figure 2a: Amplifiers with minimum input and maximum output signal strengths, indicated by their relative heights. Their order isn't interchangeable without introducing non-linear distortion.

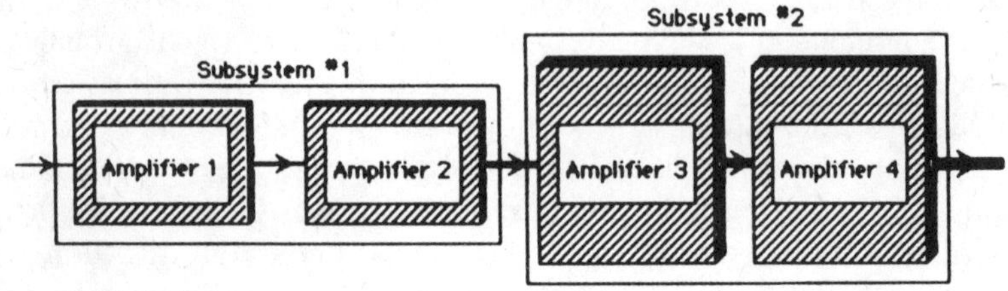

Figure 2b: Partial or sub-aggregativity:
The amplifiers in each of the two subsystems are interchangeable with each other without introducing deviations from linearity in the overall system, but inter-substitutions between sybsystems introduces non-linear distortions.

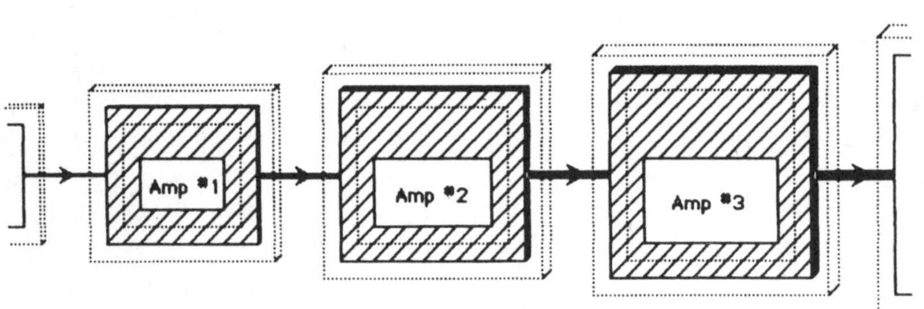

Figure 2c: Neighborhood aggregativity:
Any amplifier is intersubstitutable with its nearest neighbor to the left or to the right, but not further without introducing non-linear distortions.
(Range of intersubstitution is indicated by dotted boxes inside and outside each amplifier.)

Fig. 2 Varieties of Partial Aggregativity illustrated with amplifiers linear only within a bounded range.

Alternatively, we might have a system with four amplifiers of different sizes, each of which has a very broad linear range such that it could be substituted for the amplifer either just below or just above it. Here there would be no well-defined sub-system boundaries, but aggregativity for intersubstitutions within a neighborhood around each point. (See Figure 2c.) This mode of partial aggregativity has many instances. It arises whenever there are discriminability thresholds. Thus, for example, if the visible spectrum is divided into a large number of intervals, individuals may be unable to discriminate colors *k* or fewer intervals apart, but able to do so if the separation is greater than *k*. Here perceived color is invariant under local intersubstitutions but not under global ones. Parallel phenomena may arise for auditory tones, or for judgments of utility, which results in the oft noted intransitivity of the indifference relation. I will call this property *local* or *neighborhood aggregativity*.

Both sub- and local aggregativity suggest a further possible restriction, since the system property may be invariant for substitutions in one (or some) direction(s) but not in (the) other(s). It might be for example that the more powerful amplifiers have the same sensitivity as the less powerful ones, but that the less powerful ones would cause undue distortion or burn out if substituted for the more powerful ones. This situation could be called *semi-aggregativity*.

With sub-aggregativity, subsystems are implicitly defined, if at all, in terms of the sets of components which are intersubstitutable without changing the value of the system property. But it may be that subsystems are individuated on other grounds. If the parts of the subsystems are differentiated, it is possible that these parts are not at all intersubstitutable for other parts in the same subsystem, but are intersubstitutable for corresponding parts (presumably parts having the same values for p_1 through p_n), in other subsystems. This is a particularly important form of intersubstitutability, since it characterizes the situation in many of our designed artifacts which contain multiple copies of component parts which themselves have further parts, be they bolts, spark plugs, or memory or CPU chips in multi-processor parallel computer systems. This kind of situation will be called *modular aggregativity*.

All of the preceding cases implicitly involved standards of accuracy or tolerable errors in judging invariance of the system property. "Linear" amplifiers are seldom exactly linear, but are just "nearly" so, where how near is near is a judgment which depends upon context. In general if $P_j(S)$ is a metrical property, one can talk about aggregativity within limits of approximation for all of the above types. That is, intersubstitutions of parts leaves $P_j(S)$ invariant within bounded limits (within $+/-\ \varepsilon$ for some ε). This property could be called *near-*

aggregativity. Whether a property is nearly aggregative is obviously a function of the choice of a tolerable error ε, since something may be nearly aggregative for a large value of ε, but not so for a smaller value. We can extend this in obvious ways to talk about near (sub-, local, modular, semi-) aggregativity.

6. OTHER KINDS OF AGGREGATIVITY

There are two other interesting kinds of aggregative organization which can be characterized in terms of the conditions or distinctions I have introduced so far. The first and simpler of these occurs if we consider whether the composition function involves only one property of the system's parts or more than one. Thus, for example, the mass of a system is simply the sum of masses of its parts (under any decomposition in which the parts are mutually exclusive and exhaustive.) If we look however at the concept of the "expected utility" of an action, it is defined as follows:

$$(2) \qquad U_e = P_1u_1 + p_2u_2 + \ldots + p_nu_n \quad \left(\sum_{i=1}^{n} p_i = 1 \right)$$

where p_j and u_i are, respectively, the probability of getting the *i*-th alternative outcome and the utility of that outcome. These probabilities are assumed to be mutually exclusive and exhaustive. Here we have a situation in which the value of U_e specified by the composition function is invariant under commutation of the terms 'p_iu_i' but not if just the p_i's or u_i's are commuted independently. That is, we can interchange both the probabilities and utilities of two alternatives but we cannot interchange just their probabilities or just their utilities without changing the expected utility of the action. This kind of situation can be called *complex aggregativity.*

The concept of "mean Darwinian fitness" in population genetics also has exactly this form, in which the p_i's are the relative frequencies or the different genotypes in the population and the u_i's are their fitnesses. Any weighted average would have this form. But weighted averages in which the weights represent frequencies in a population are only apparently cases of complex aggregativity, which can be converted into simple cases by noting that the relative frequencies are the absolute numbers of the various types divided by the total number in the population. When the case is redescribed in this way, the fitnesses assigned to individuals can be permuted in any way without affecting the mean fitness of the population. Viewed in this light, the request to permute p_i's amounts to a request to consider a population of different proportions of types — i.e., a different population. Similarly, a request to permute the fitnesses of the different genotypes involves an implicit assumption that the environment has changed — a change for which we should not expect the property of the fitness of the population to remain invariant. The new population has been constructed from the old by substituting parts which are not relevantly similar, or which are considered in a different environmental context.

The case of expected utility cannot be handled in this way, however, since the probabilities are not relative frequencies (or needn't be interpreted as such) and thus cannot be broken down as incidence of cases with various values of utilitics. (A similar case would occur for the expected mean Darwinian fitness in an array of mutually exclusive possible environments, in which the fitnesses of each type are represented as vectors with the fitnesses of the different possible environments as components. The composition functions of this type are aggregative in a manner which is ineliminably complex.)

A second interesting kind of aggregativity is a species of restricted aggregativity. Suppose that a system property is sub-aggregative — i.e., aggregative for intersubstitutions within subsets of parts but not aggregative for intersubstitutions of parts in different subsets. Now consider these subsets as parts, and consider whether the system property is (1) aggregative, (2) non-aggregative in any sense, or (3) sub-aggregative for some new (higher-level) subsets. If case (3) holds, then the same operation can be applied again on the second order subsets as parts. Recursive application of these steps can generate an organizational hierarchy of levels.

This kind of situation appears to characterize functional hierarchies with a great deal of redundancy of functional components for tasks at various levels. If at the lowest level, each function can be accomplished in a variety of ways (but not in any way whatsoever), these functions comprise functional systems for accomplishing higher-level tasks. If these higher level tasks are themselves at least partially redundant functional means for accomplishing still higher-level tasks we again have sub-aggregativity at this higher level, and so on. Computers with built-in hardware redundancy, software with a variety of ways of accomplishing various tasks at various levels, and the backup redundancy of excess capacity of cells within organs, and bilaterally symmetric redundant organs within the organism are all examples of this type of organization. For more on functional hierarchies and functional organization, see Wimsatt (1972). For more on the importance of redundancy in system design, see Glassman and Wimsatt (1984) and Wimsatt (1981a).

Hierarchial organization can be produced in other ways which involve restricted aggregativity. I will discuss one of these ways in the next section.

7. AGGREGATIVITY AND NEAR-DECOMPOSABILITY

H. A. Simon (1962) has introduced a concept which he calls "Near-decomposability" which has widespread applications in the simplification and analysis of complex systems. Suppose that a system is decomposed into parts or state variables, and that it is possible to order the relative strengths of interaction among these components of the system. Suppose also that strengths of interaction are not continuously distributed, but can be divided into two groups, with the weakest of the stronger interactions significantly stronger than the strongest of the weaker ones. Suppose also that the components can be divided into subsets such that no strong interactions exist between components in different subsets. Then, viewing the subsets as subsystems, all of the strong interactions are contained within these subsystems. (These subsystems are defined dynamically, in that being in the same subsystem implies nothing about relative spatial location or continuity. It is possible that all of the subsystems would be spatially interdigitated. See Wimsatt (1974) for further discussion.)

If a system is organized in this fashion, then the analysis of its behavior can be simplified in two different and complementary ways by using two different approximations. In the first approximation, the weaker interactions are ignored, and the various subsystems are treated as if they were isolated from each other. (In general, if the ratio of the strongest weak force to the weakest strong force is denoted by ε_c, then ε_c is a measure of the maximum error introduced by ignoring the weaker forces. (See Wimsatt (1974).) If we assume, as seems generally true, that forces go to equilibrium at rates proportional to their strengths, then approximate intrasystemic equilibria will be reached in roughly ε_c proportion of the times it takes for the weaker intersystemic forces to equilibrate. (The approximate equilibrium of the subsystems will be what Lotka (1924) called a "moving equilibrium" in which the weaker intersystemic forces act as relatively slowly changing parameters which determine the local equilibrium values of the subsystems.)

In the longer run, with the subsystems nearly at equilibrium, a second approximation is possible. The internal complexity of the subsystems can then be ignored, with the many variables of the various subsystems replaced by a single index for each, and the problem solved as an interaction of these index variables. Given that the complexity of equations for analyzing many-variable systems grows as some power of the number of components (see, e.g., Wimsatt (1979), pp. 362–363 for further examples, one of which is discussed in greater detail in Wimsatt (1980b), pp. 222–230), the reduction in complexity that these two approximations provide can be enormous, as long as ε_c is tolerably small for the purposes at hand and we do not need to know the behavior of the system over time periods between the short and long runs. These approximations in effect break the system down into levels of dynamical organization, with the stronger interactions at the lower level and the weaker inter-systemic interactions at the higher level.

If the weaker interactions between subsystems can be further broken down into two groups with a significant gap in relative strengths, such that the medium strength interactions are bounded within subsets of the components, this process can be repeated, with the weakest of the three classes of interaction defining a third still higher level of organization. n repetitions of this process would generate $n+1$ levels.

In this case we see that the tolerable error imposes a kind of hierarchial organization on the system. Since the ratio of the weakest forces to successively more powerful groups of forces gets smaller and smaller, a prediction of the behavior of the system at a given level must go down a level only when the ratio of forces at the two levels is not such as to exceed the tolerable error. Smaller tolerable errors or less difference between the magnitude of forces at different levels requires investigation of variables at lower levels to achieve acceptable predictions at the given level. More accurate predictions or stronger coupling between levels requires a deeper analysis of the system.

This kind of property is essential in the analysis of almost all complex systems. Consider the simplest binary switching element of an integrated circuit. It does not switch from one state to the other instantaneously, but if the clock cycle time from one machine instruction to the next is significantly longer than the time it takes to do so, this time (and the detailed physical characteristics of the switching element) can be ignored in the analysis of the logical characteristics of the circuit. This is exactly the kind of condition required by functionalist theories of mind, which seek to analyze functional organization at one level which is invariant over variations in the microstructural details at lower levels of organization.

The problem-solving heuristic of assuming near-decomposability of the parts of a system is in effect the meta-heuristic of reductionistic analysis. The system is broken down into parts, whose characteristics are studied in isolation. These parts are then put together or articulated to explain system properties through their interactions. (See Kauffmann (1971) for a general account of this strategy). This is an effective strategy because in it the complex problem of analyzing the behavior of a system is broken down into a bunch of simpler problems — of analyzing separately the behavior of a number of parts of the system, and then analyzing as a separate problem the behavior of these parts taken together. Simon's metaphor of the combination lock (Simon, 1962, Wimsatt, 1980b, 1984) illustrates clearly the computational advantages of this strategy.

This strategy is interesting because it produces a kind of aggregativity not of the system studied *but of the problem of analyzing the system!* The parts of the system may not be intersubstitutable at all, but if it can be broken down into parts which can, with acceptable errors of approximation, be treated as isolated, then *the parts can be studied or analyzed in any order.* Within each part, its sub-parts may need to be analyzed in a particular order imposed by the paths of interaction among its components, and the same holds true when the subsystems are analytically reaggregated to study the system properties, but the strategy yields a new degree of freedom in analyzing the parts.

8. A CASE STUDY — AGGREGATIVITY AND DECOMPOSABILITY IN POPULATION BIOLOGY

The concepts of aggregativity provide useful tools in analyzing a current controversy in population biology — that of the existence of higher-level units of selection. I will not attempt to survey all of this now voluminous literature. (Sober (1984) has collected many of the earlier relevant papers.) I will focus on my views (Wimsatt, 1980b, 1981b) and those of biologists whose work has been central to them, for it is here that issues of aggregativity are most clearly expressed. Here also, the appearance of aggregativity is used as a criterion for asserting that the problem of analyzing the system is decomposable in Simon's sense, and that the units of this decomposition are the units of selection (see Wimsatt, 1980b, pp. 222–225).

On my analysis, any entity which shows heritable variance of fitness can (and will) function as a unit of selection if a further qualifying condition is met. This condition is that there is an additive (or heritable) component of variance in fitness of entities at that level which does not appear as an additive component of variance in fitness of entities at any lower level of analysis.

The requirement that there be a component of variance in fitness which is additive is centrally anchored in the conceptual structure of evolutionary theory by Fisher's "fundamental theorem" of natural selection, which states that the rate of evolution is proportional to the additive variance in fitness. With no additive variance in fitness, there is no evolutionary change. If there is an additive component of fitness at a given level which also appears at some lower level, then evolutionary processes at that lower level are sufficient to explain evolutionary changes at the higher level. If the higher level component does not appear as additive at the lower level, then there can be no evolution at the lower level, and the higher level units must be invoked to explain the evolutionary change. Conversely, if all of the variance in fitness is additive at a given level, there are no higher-level units of selection, since there is no remaining (non-additive) component of fitness which could be converted to an additive component at higher levels by appropriate chunking of the lower-level units.

This analysis was first presented in my 1980b, and is further elaborated and defended in my 1981b. To summarize the relevant discussions there, the centrality of additivity to evolutionary discussions is explained by connections of that notion with two other concepts — that of heritability and that of context-independence. All three of these are naturally related to the idea of aggregativity.

To say that variance in fitness is totally additive at the level of individual genes is to say that the fitness increase in genotype or other higher level unit is a linear function of the number of genes of a given type present in it, and nothing else. (This claim is of course made relative to a given environment.) But this entails that the contribution to fitness of one of these genes is independent of the genetic background in which it occurs, which is to say that its contribution is independent of (genetic) context. Additivity is thus a special case of *context-independence.*

In this case, the gene has the same effect on fitness no matter how genotypes are reassorted in successive generations by Mendelian segregation and recombination. But this is to say that its fitness contribution is *heritable* under such rearrangements. Conversely, fitness contributions which are non-additive, and which are thus context-dependent, will not be heritable under such rearrangements of genetic context. Additivity is thus important because it is equivalent to the third of Darwin's principles (Lewontin, 1970), that variance in fitness be heritable!

Does total additivity imply that the fitness of a system is an aggregate of the fitness contributions of its parts? It would imply that the problem of analyzing the evolution of the system can be decomposed into the problems of analyzing the selection forces on its parts in isolation — exactly the condition which Williams envisions in arguing for his genetic reductionism. The system (in Simon's terms) would then be not merely nearly-decomposable, but completely decomposable, and this would result in enormous simplifications in analyzing it. (See Wimsatt, 1980b, pp. 222–226.)

On the classical ("beads-on-a-string") model of the gene, fitness does at first appear to be an aggregative property, since that model suggests that what (and when) a gene produces is invariant with respect to position, and the additivity condition says that the fitness effect of this production is also invariant under intersubstitution of different genes at different loci. (Recall that we are here assuming total additivity for the purposes of discussion, something which I have argued is rarely to be found in nature — see Wimsatt, 1981b, pp. 137–142.)

Additivity also suggests that there is no cooperative or inhibitory interaction among fitness contributions, so the fourth condition is met. If we take seriously the atomistic model of the classical gene, it seems that the third condition — allowing fusion or decomposition of parts

— is also met, as long as these operations do not change the character of the basic genetic atoms.

What about the second condition — would fitness contributions of genes be invariant under addition or subtraction of components from the system? Here the answer must be no: any biologist would say that all bets are off if chunks of the genome are deleted or added. The effects of such changes were sufficiently well known even at the time of the classical model (e.g., through Boveri's (1902) experiments involving multi-polar mitoses and consequent random sampling in the construction of genomes of different sizes and compositions) that it would be unfair to attribute this view to any geneticist. Taken to the extreme, it would imply that individual genes would make the same fitness contribution in the absence of the entire rest of the genome, and the "selfish gene" would never (in Dawkins' picture) have had any cause to seek the cooperation of other genes. To put it in another way, context-independence (or additivity, or heritability) does not imply causal independence of the genes in producing their contributions to fitness. This is a point made elegantly in another context by Richard Lewontin (1947b).

Finally, of course, if we move away from the classical model to allow for "frame-shift" mutations and transposition of elements into the domain of a different gene-control complex, even the picture of constancy of effect under different rearrangements of the genome breaks down, and with it the satisfaction of the first criterion. The additivity and aggregativity of the genome turns out to be highly conditional and is preserved, if at all, only under highly selective rearrangements. Even with total additivity, the genome would only show very restricted aggregativity, usually involving nothing more than intersubstitution of alternative alleles at the same locus, which would preserve prior translation frame and control element relations.

When the assumption of total additivity is dropped, the domain of aggregative effects is even more restricted. If the effect is totally additive for alleles determining a given trait, intersubstitutability is restricted to those alleles at loci affecting that trait. And if we have only an additive component affecting a given set of alleles, nothing whatever follows about intersubstitutability, since a non-additive component may also be found for the same alleles. We have effected here only an analytical decomposition which, useful as it may be for solving certain problems, implies nothing about causal independence.

Even if it suggests that they may be treated as units of selection (evolving under their additive component of fitness), it may still be necessary to recognize complexes of them as higher-level units of selection.

Another interesting problem is raised in this area by the suggestion of several authors (e.g., Maynard Smith (1978), p. 80, and Templeton (1979), pp. 1265–1267) that multiplicative fitness models indicate a kind of independence of selection at different loci. It might seem that this involves a different way in which one could find lower-level units of selection, but it does not. I will defer discussion of it until I have looked at some other cases of multiplicative models.

Multiplicative models arise in several contexts in population biology. The first is in the context of life-history models, where different selection forces act at different stages in the life history of the organism. Here, the selection forces are treated as independent filters or hurdles which the organism must successfully pass if it is to survive and reproduce, and its probability of passing all of them is the product of the probabilities of passing each of them. The claim of independence of components here arises from the multiplicative law of probabilities, in which the individual events must be assumed to be stochastically independent if the probability of their co-occurrence is to be calculated as their product. But this independence and the commutativity and associativity in their composition function does not translate into meaningful operations on the life-history of the organism. It is true that if one interchanged the probabilities of surviving from birth to leaving the nest, and from leaving the nest to reproduction in brown jays, one would get the same net probability of survival. But organisms cannot play their lives in different orders, and even if they could, juveniles would do poorly as nestlings and nestlings as juveniles.

A second context arises in Richard Levins' (1968) discussion of "fine" vs. "coarse" grained patterns of environmental variability. Levins contrasts two ways in which the organism may "perceive" its environment, as a function of its properties and those of the environment. If variation in the environment is perceived as "fine grained", then the fitness of the organism is the arithmetic average of its fitnesses in the various environment types weighted by their relative frequencies. This results in a formula (or "adaptive function") similar in form to that for mean Darwinian fitness, but differing in that it is different environments, rather than different genotypes which produce differences in fitness.

Levins has in mind cases like that of local temperature variation which are small in spatial extent relative to the thermal mass and rate of travel of an organism, so that it spends a relatively short time in each patch and the expected rate of heat gain or loss is the same as it would be in a homogeneous environment with a temperature equal to the average of the environment temperatures weighted by their frequencies of occurrence. He gives similar arguments for small vs. large prey (the plankton of a filterfeeding whale vs. the caribou of a wolf pack), where for the small prey, it is reasonable to talk about a homogeneous rate of capture per unit time, and for large prey it is not.

For a coarse-grained environment (the large temperature patches or the large prey), the environments are "perceived" as alternatives (high vs. low rather than average temperature, and prey vs. none, rather than mean resource input). Here, the organism must get through a series of environments which are treated as "alternatives" (Levins, 1968, p. 18) or as independent hurdles to survival (as in the life-history models), in which the fitness function has a multiplicative form in which the frequencies occur as exponents of the fitnesses in the different environments. These concepts are discussed in more detail in Wimsatt (1980a), pp. 292–297, as well as in Levins (1968) Strobeck (1975).

While the mathematical forms of both fitness functions are associative and commutative, the more paradigmatically additive one, the "fine-grained" adaptive function, arises through interactions among successive sub-environments which are not assumed to be present in the coarse-grained case. In the derivation of the "fine-grained" function, either the temporal duration of the sub-environments is assumed to go to zero (Levins, 1968, p. 18) or, equivalently, the number of sub-environments experienced in a finite life-cycle is assumed to go to infinity (Strobeck, 1975, p. 421.) In the case of both functions, the life-history is characterized as a sequence of environments of the various types. If we assume that the fine-grained fitness function is commutative with respect to the order of its sub-environments, then rearrangement of these sub-environments such that all of the environments of a given type are temporally continuous should not matter. But if we do so, we have converted a fine-grained situation into a coarse-grained one, and the fitness should be characterized by the coarse-grained function. Thus, the terms in the fine-grained function cannot be regarded as commutative if they are to describe properly the fitness of the organism!

This is because the "averaging" implicit in the fine-grained function assumes that the sub-environments will be homogeneously distributed in time. It applies in the case of the temperature patches because the thermal mass or inertia of the organism prevent it from immediately assuming the temperature of its environment and the consequences which would then ensue. If a representative variety of patches of other temperatures follows, its temperature (ignoring the effects of metabolic, locomotory, and thermoregulatory activity) deviates only negligibly from the mean temperature of the patches it has experienced, weighted by their frequency. Thus it is this mean temperature of the environmental patches, rather than some weighting of its fitnesses in the patches of different temperatures which is important in assessing its fitness. If the patches are reordered, then this assumption no longer applies, and the properties of the different individual environments are the relevant ones to use in assessing its fitness. A similar explanation applies in the case of fine vs. coarse-grained prey, with the "inertial" term being the energy reserves and food storage capacity of the organism. It is important to note that in this case, unlike the preceding life-history case, no use has been made of the non-interchangeability of life stages for reasons of developmental constraints or functional inappropriateness. Nonetheless, the fitness response of an organism to a given sub-environment is a function of the sub-environments it has recently experienced, so they cannot be permuted without changing the fitness value of the sequence.

Here we have a case in which the formal analysis of aggregativity (in terms of formal properties of the composition function) will not do: the fine and coarse-grained adaptive functions are associative, commutative, and relevantly qualitatively similar under addition or subtraction of parts, but the physical operation associated with commutativity changes the form of the function which applies. A similar argument could be made for commutations equivalent to inter-locus intersubstitution of alleles with totally additive variance in fitness. What has gone wrong is that idealizations made in order to derive the given composition function are violated if the parts are intersubstituted. This kind of failure seems likely to be sufficiently general as to render formal attempts at characterizing aggregativity just in terms of properties of the composition function impossible. This does not render such an analysis useless however, for the formally stated conditions have been a useful guide in analyzing why the system property is not aggregative, and in localizing the relationships responsible for this failure. *This is indeed probably one of the most pervasive functions of idealizations in science — their use as a heuristic template to organize information for further analysis, including analysis of when or why the idealization is correct or incorrect.*

Multiplicative fitness models in population genetics are models in which the net fitness contributions of a set of alleles are the product of their fitness contributions considered individually. (See, e.g., Ewens (1962) pp. 97–100 and Maynard-Smith (1978), p. 80.) They share some properties of intersubstitutability with the additive models, in that it does not matter which other (multiplicative) alleles are present in determining the fitness of the genotype, but only how many of them. This produces a kind of context-independence of the fitness contributions of multiplicative alleles *within classes of genotypes with equivalent numbers of multiplicative alleles.* But the relevant equivalence classes for fitness contributions of additive alleles are far broader — the same allele makes the same contribution in the presence of *any* number of additive alleles! This difference means that in the comparable reference class for multiplicative alleles, the fitness contribution of a multiplicative allele is context-dependent – on the number of other multiplicative alleles present. The presence of more multiplicative alleles in effect enhances the fitness contribution of a given allele. This means that in contexts in which Mendelian segregation and recombination can change the number of multiplicative alleles present in a genotype (as it can, in these models), the fitness effect of an allele is not completely heritable or context-independent. (The sign of its contribution will not change, but its magnitude will.)

Why does this case differ from the superficially similar one of the linear amplifiers discussed earlier? The main reason appears to be that we are interested in different properties of the two systems. The motivating question for any population geneticist working at the level of individual genes is: "What properties of this gene affect its relative rate of reproduction and probability of survival?" In this case, the presence of a multiplicative interaction in fitness between it and other alleles indicates a context-dependent interaction which can enhance its relative probability of survival in some contexts — those with a higher number of multiplicative alleles in the genome. The context of differential selection and replication of multiplicatively interacting units is not present in the linear amplifier case. Each linear amplifier is a device which multiplies any input signal (within the range in which it is linear) by a scalar quantity. While it gets these input values from any preceding amplifiers and passes its output to any succeeding ones, it is not seen as interacting with them in producing its particular increment in the signal. The evolutionary context forces us to look at a class of cases parallel to that of sets of amplifiers with different net amplification ratios, basically because the whole organism is the unit of selection in this case. The importance of the heritability of fitness of the phenotype introduces a background against which the multiplicative component (or actually, the difference between it and a corresponding additive contribution) is seen as an interaction or context-dependent term, so this case is not viewed in the same way, even though the form of the composition functions in the two cases are similar. This provides another sort of case of which a purely formal analysis of aggregativity seems problematic: here it is the theoretical context and use of the composition function which seems to affect whether we regard it as aggregative.

It is hard to know why biologists might be tempted to regard multiplicative fitness models as aggregative, or to make claims which seem to imply this. Maynard-Smith suggests (1978, p.80) that "Multiplicative fitness models are what we would expect if selection acts independently on the two loci." This much is true, as we have seen in the life-history models, and may provide some of the temptation for this conclusion. But it does not, of course, follow that if we have a multiplication model that selection forces are acting independently.

Furthermore, even if selection forces were acting independently, this would not be enough. A case discussed by Falconer (1977) demonstrates that multiplicative interactions between loci due to independent selection forces can convert variance in fitness which is additive within loci into variance in fitness which is non-additive or epistatic between loci. Thus independence of selection forces on different loci in even the most favorable case (additivity within loci) does not yield the conclusion that the genes or the loci can be treated as independent units of selection.

Templeton (1979) suggests that in both additive and multiplicative models, individual genes can be treated as units of selection. Subsequent clarification (Templeton, personal communication) however, revealed that he was merely talking about conditions under which simple composition functions could be constructed relating individual gene effects to total genomic or phenotopic effects. As such, this is what I have elsewhere (Wimsatt, 1980b, 1981b) called "genetic bookkeeping," and has no implications for this aggregativity of these effects.

9. AGGREGATIVITY, EMERGENCE, AND REDUCIBILITY

It was common during the dominance of logical positivism to treat reducibility as the universal solvent for conceptual problems in the sciences. Perhaps for this reason, there arose the view that for a property to be emergent was simply for it not to be reducible to the properties and laws at lower levels. This discussion of aggregativity and non-aggregativity provides a quite different answer to the nature of emergence. We have seen a variety of ways in which the properties of the whole may be "more" than the "sum" of its parts if by this we mean that properties of the whole depend not only upon the presence of the parts but also upon how they are arranged and how they interact. Probably the majority of interesting properties of complex systems are of this sort. If we may choose to regard non-aggregativity as a criterion for emergence, then we get not only a non-trivial and intuitive concept of emergence, but a classification of different kinds of emergence as a function of which conditions fail to be met. I suggest that this concept of emergence comes closer to capturing what most people have meant by that term than an analysis which ties emergence to non-reducibility.

In all of the cases discussed, I have supposed the existence of a "composition function" relating the system property to a decomposition of the system into parts, which possess properties whose relationships constitute or generate the system property. But to support the existence of such a composition function is to suppose a relationship which would count as a reduction of the system property to properties of the parts and their relations. This severs the connection between emergence and irreducibility. Both emergent and non-emergent properties may be reducible. Conversely, if a property is non-reducible (or not yet reduced) we do not have a composition function for it, so the question of whether it is emergent is open. Emergence and aggregativity are features of the world, not of our knowledge of it.

It is worth noting that in the progress of a reductionistic research program earlier simpler models which tend to treat parts as isolated and as characterized in terms of context-independent monadic properties are replaced by later more complex models in which relational properties of the parts enter and their behavior is thus rendered increasingly context-dependent. We saw that this arose in a variety of ways in the move from the classical beads-on-a-string model of the genome to more complexly interacting models in molecular genetics. (See also, Wimsatt, 1979, for more examples.) If this picture is correct then we have, if anything, the opposite of the picture painted by the positivists. Rather than emergence disappearing with the progress of reductionistic theories, we have it growing in demonstrated scope and importance as we move from the earliest and simplest models to more complex and realistic ones. Thus some holists, at least, can draw only comfort from the continued successes of reductionistic approaches.

10. CONCLUSION

I have provided 4 conditions which seem to be centrally involved in judging whether a system property is an aggregate of the properties of its parts, and discussed a variety of cases which fail to meet one or more of these conditions. In addition, I have characterized a number of more restricted forms of these conditions. Taken together, they provide a variety of tools for describing different forms of organization. Where possible (as with the first two and possibly the third condition) I have described formal conditions on the form of the composition function relating system and parts' properties, but in some cases (notably for the last condition) there seemed to be no unambiguous and adequate way to do so. Extended discussions of cases in population biology raised questions as to whether any analysis which looked only at the form of the composition function, and not for example at the idealizations involved in the derivation of the composition function, or the theoretical context of the use of that function, could determine whether the system property was aggregative or not. This and the complexities of interpreting a variety of other cases suggest that the analysis proposed should be viewed not as a definition or complete characterization of aggregativity, but as providing a set of heuristic tools for the description of systems and the modes of interaction of their parts.

University of Chicago

ACKNOWLEDGEMENTS

The majority of work on this analysis was done between 1970 and 1973 and laid aside until my interest in it was rekindled by the units of selection controversy, and the apparent aggregative picture of gene action presupposed by the more reductionistically inclined biologists. As much of my other work, it owes a great deal to the views of Richard Levins — in particular in this case to his view that statistical properties of a system are "aggregates" and to his distinction (Levins, 1970) between "aggregates," "engineering systems," and "evolved systems". I would like to thank the students in my classes at Chicago and the faculty (particularly Ron Laymon) and students at Ohio State who gave me useful feedback and criticism during its long gestation period, and Jim Griesemer and Leigh Star who read this last version. Richardson (1982) has since proposed a somewhat different but not unrelated analysis of some of these same ideas which I recommend to the interested reader.

Historically, probably the best discussion of many of these issues (but with a more negative conclusion) is Ernest Nagel's essay, "Wholes, Sums, and Organic Unities," reprinted later as a section of his classic *The Structure of Science*. When I gave an earlier version of this paper to Ken Schaffner in 1973, he suggested that I look at Nagel's piece. To my surprise, when I read it, I discovered that I had read it before in 1963, and underlined just those passages where Nagel's views are most similar to those expressed here. I have consequently never been able to determine how much of this I owe to Nagel, but I would be surprised if his views had no influence on me at the time — in my first philosophy of science course.

BIBLIOGRAPHY

Almli, C. and Finger, S. (eds.): (1984), *Early Brain Damage, Volume 1: Research Orientations and Clinical Observations*, Academic Press, New York.

Asquith, P. and Giere, R. (eds.): (1981), *Proceedings of the Philosophy of Science Association — 1980*, vol. 2, The Philosophy of Science Association, East Lansing, Michigan.

Asquith, P. and Kyburg, H., Jr. (eds.): (1979), *Current Research in Philosophy of Science*, The Philosophy of Science Association, East Lansing, Michigan.

Asquith, P. and Nickles, T. (eds.): (1982), *Proceedings of the Philosophy of Science Association — 1982*, vol. 1, The Philosophy of Science Association, East Lansing, Michigan.

Boveri, T.: (1902), 'On Multipolar Mitoses as a Means of Analysis of the Cell Nucleus', translated and reprinted in Willier and Oppenheim, 1974, pp. 74–97.

Brewer, M. and Collins, B. (eds.): (1981), *Scientific Inquiry and the Social Sciences*, Jossey-Bass, San Francisco.

Buck, R. and Cohen, R. (eds.): (1971), *PSA 1970. In Memory of Rudolf Carnap*, Boston Studies in the Philosophy of Science, vol. 8, D. Reidel, Dordrecht, Holland.

Carlson, E.: (1967), *The Gene: A Critical History*, Saunders, Philadelphia.

Edsall, J.: (1972), 'Blood and Hemoglobin: The Evolution of Knowledge of Functional Adaptation in a Biochemical System. Part I: The Adaptation of Chemical Structure to Function in Hemoglobin', *Journal for the History of Biology* **5**, no. 2, pp. 205–257.

Ewens, W.: (1962), *Population Genetics*, Methuen, London.

Falconer, D.: (1977), 'Why Are Mice the Size They Are?' in Pollack, Kempthorne, and Bailey, 1977, pp. 19–21.

Glassman, R. and Wimsatt, W.: (1984), 'Evolutionary Advantages and Limitations of Early Plasticity', in Almli and Finger, 1984, pp. 35–58.

Kauffman, S.: (1971), 'Articulation of Parts Explanations in Biology and the Rational Search for Them', in Buck and Cohen, 1971, pp. 257–272.

Levins, R.: (1968), *Evolution in Changing Environments*, Princeton University Press, Princeton.

Levins, R.: (1970), 'Complex Systems', in Waddington, 1970, pp. 73–88.

Lewontin, R.: (1970), 'The Units of Selection', *Annual Review of Ecology and Systematics* **1**, pp. 1–18.

Lewontin, R.: (1974a), *The Genetic Basis of Evolutionary Change*, Columbia University Press, New York.

Lewontin, R.: (1974b), 'Analysis of Variance and Analysis of Causes', *American Journal of Human Genetics* **26**, pp. 400–411.

Lotka, A.: (1924), *Elements of Physical Biology*, reprinted (1956) as *Elements of Mathematical Biology*, Dover Publications, Inc., New York.

Maynard Smith, J.: (1978), *The Evolution of Sex*, Cambridge University Press, London.

Milkman, R. (ed.): (1982), *Perspectives on Evolution*, Sinauer, Sunderland, Massachusetts.

Nagel, E.: (1961), *The Structure of Science*, Harcourt, Brace, and World, New York.

Nickles, T. (ed.): (1980), *Scientific Discovery: Case Studies*, D. Reidel, Dordrecht, Holland.

Pollack, E., Kempthorne, D., and Bailey, T. (eds.): (1977), *Proceedings of the International Congress on Quantative Genetics*, The Iowa State University Press, Ames, Iowa.

Richardson, R.: (1982), 'Grades of Organization', in Asquith and Nickles, 1982, pp. 324–340.

Schaffner, K. and Cohen, R. (eds.): (1974), *Proceedings of the 1972 Biennial Meeting, Philosophy of Science Association*, Boston Studies in the Philosophy of Science, vol. 20, D. Reidel, Dordrecht, Holland.

Simon, H.: (1962), 'The Architecture of Complexity', reprinted as chapter 7, of his (1982), *The Sciences of the Artificial*, 2nd ed., MIT Press, Cambridge, Massachusetts, pp. 192–229.

Sober, E.: (1984), *Conceptual Problems in Evolutionary Biology*, Bradford Books – MIT Press, Cambridge, Mass.

Strobeck, C.: (1975), 'Selection in a Fine-grained Environment', *The American Naturalist* **109**, no. 968, pp. 419–426.

Templeton, A.: (1979), 'The Unit of Selection in Drosophila mercatorm-2', *Genetics* **92**, pp. 1265–1282.

Templeton, A.: (1982), 'Adaptation and the Integration of Evolutionary Forces', in Milkman, 1982, pp. 15–31.

Waddington, C. (ed.): (1970), *Towards a Theoretical Biology*, vol. 3, Edinburgh University Press, Edinburgh.

Willier, B. and Oppenheim, J. (eds.): (1974), *Foundations of Experimental Embryology*, Hafner, New York.

Wimsatt, W.: (1972), 'Teleology and the Logical Structure of Function Statements', *Journal for the History and Philosophy of Science* **3**, no. 1, pp. 1–80.

Wimsatt, W.: (1974), 'Complexity and Organization', in Schaffner and Cohen, 1974, pp. 67–86.

Wimsatt, W.: (1979), 'Reduction and Reductionism', in Asquith and Kyburg, 1979, pp. 352–377.

Wimsatt, W.: (1980a) 'Randomness and Perceived Randomness in Evolutionary Biology', *Synthese* **43**, pp. 287–329.

Wimsatt, W.: (1980b), 'Reductionistic Research Strategies and Their Biases in the Units of Selection Controversy', in Nickles, 1980, pp. 213–259.

Wimsatt, W.: (1981a), 'Robustness, Reliability and Overdetermination', in Brewer and Collins, 1981, pp. 124–163.

Wimsatt, W.: (1981b), 'Units of Selection and the Structure of the Multi-Level Genome', in Asquith and Giere, 1981, pp. 122–183.

Wimsatt, W.: (1984), 'Von Baer's Law of Development, Generative Entrenchment, And Scientific Change', *Philosophy of Science* (in press).

PART IV

RENCONTRE

LESLEY COHEN

DESCARTES AND MERLEAU-PONTY ON THE *COGITO* AS THE FOUNDATION OF PHILOSOPHY*

When one refers to René Descartes' dualism one usually refers to his metaphysical dualism of mind and body. However, there is also an epistemological dualism which is fundamental to his philosophy. Not only are the mind and body distinct substances, but one can know that one's body and the rest of the material world exist only after one knows that one exists as a mind or thinking thing. Although Descartes never calls himself a dualist in either of these senses, these expressions accurately describe his view. Maurice Merleau-Ponty rejects Descartes' epistemological project as a whole and, in particular, rejects the *cogito*, as understood by Descartes, as the foundation of knowledge. In effect, he denies that what I call "epistemological dualism" is legitimate and questions the conclusions Descartes draws from it.

In this paper I will discuss Descartes' doctrine of how one knows that one exists and also the claim he bases on that doctrine, *viz.*, that one's knowledge of the existence of one's mind is prior to all other knowledge. I will argue that although Merleau-Ponty provides an insightful analysis of the *cogito*, he does not succeed in defeating Descartes' position because he does not address directly the issues Descartes considers. I conclude that an adequate treatment of these issues entails epistemological dualism. In addition, I will indicate that an examination of the relationship between epistemological and metaphysical dualism reveals that it is possible to accept Descartes' epistemology without accepting his metaphysics.

At the beginning of the *Meditations* Descartes explains that it is his intention to re-examine his former opinions in order to discover what can be known with complete certainty. To this end he doubts everything he previously believed to be true including the things which appear to be most certain — for he may be deceived by illusions, he may be dreaming when he believes that he is awake, or there may be an evil genius who is able to deceive him about many things. As a result, he withholds judgment concerning whether his senses, his body, and the external world exist or are as they appear to be. It is Descartes' hope that if he can find one thing which is certain in spite

A. Donagan, A. N. Perovich, Jr., and M. V. Wedin (eds.), Human Nature and Natural Knowledge, 295–312.

of his extreme doubts, it may be able to serve as a foundation for knowledge.

In *Meditation* II, Descartes recognizes that no matter how much he doubts and no matter how much he is deceived he can find one thing which is indubitable, namely, he can know that he exists whenever he thinks he exists. "*I am, I exist,* is necessarily true each time I pronounce it, or that I mentally conceive it."[1] What is unique about the *cogito* is that it is self-evident and self-guaranteeing. Whenever I decide that I am thinking I cannot be mistaken that I am thinking and, thus, I cannot be mistaken that I exist as a thing which thinks.

At this point in the *Meditations* Descartes is not yet claiming that he can know his own nature. He remains in doubt as to whether the mind and body are the same thing or even whether anything bodily exists. However, he does know that even if he has no body and the external world does not exist, he exists as a thing which thinks, i.e., "A thing which doubts, understands, conceives, affirms, denies, wills, refuses which also imagines and feels,"[2] whenever he thinks.

It is important to distinguish what Descartes says about the order of *knowing* from what he says about the nature of his *experience* of what is known. Descartes attempts to prove that his knowledge of the existence of his mind is prior to his knowledge of the existence of his body and the rest of the external world. However, he does not claim that his experience of himself as a thinking thing is temporally prior to or more certain than his experience of himself as having a body and being in a world. What he does claim is that his experiences are not sufficient to prove that he has a body and is in a world of material objects. It is possible to have the experiences that a human being has and yet be a mind without a body. Some evidence in addition to my thinking that I perceive my body and objects in the world is needed to prove that they exist. Descartes believed that he found this evidence, but not until *Meditation* VI after he proves that his mind and God exist. Thus, before *Meditation* VI, he does allow for the possibility that he exists only as a thing which thinks and that his body and the objects which he perceives, or seems to perceive, do not exist — but without asserting that he experiences himself as a mind without a body.

In *Phenomenology of Perception* Merleau-Ponty argues that the Cartesian position is not tenable:

> Perception and the percept necessarily have the same existential modality, since perception is inseparable from the consciousness which it has, or rather is, of reaching the thing itself.[3]

To know that I perceive is to know that I perceive *something*. If what I think I perceive is merely an illusion, then I have not really perceived. Thus, if I see a chair in what Merleau-Ponty calls "the full sense of the word 'see'," then the chair actually exists. If perception

> . . . presupposes our opening on to a reality or a world, how can we possibly dissociate the certainty of our perceptual existence from that of its external counterpart? It is of the essence of my vision to refer not only to an alleged visible entity, but also to a being actually seen.[4]

On the basis of Merleau-Ponty's position we cannot accept the Cartesian dichotomy between one's knowledge of the existence of oneself as a thinking thing and that of oneself as an extended thing in the material world.

By insisting that perception and the percept have the same existential modality, Merleau-Ponty is not eliminating all doubt concerning the existence of perceived objects. Rather, he is extending doubt to the level of perception. Illusions with respect to any act of thought and illusions with respect to the objects thought about involve one another:

> It is no less difficult for me to know whether or not I have felt something than it is to know whether there is really something there, for the victim of hysteria feels yet does not know what it is that he feels, as he perceives external objects without being aware of that perception.[5]

There are illusory or imaginary perceptions, feelings, and thoughts as well as true ones. Thus, we must distinguish between illusory or mistaken love and true love, and between the pain of a victim of hysteria and real pain.

Perceptions, feelings, and thoughts which are illusory or imaginary actually are experienced but, unlike true perceptions, feelings, and thoughts, they are "on the outer fringes of ourselves" and are performed inauthentically or artificially.

> Thus we are not perpetually in possession of ourselves in our whole reality, and we are justified in speaking of an inner perception, of an inward sense, an 'analyser' working from us to ourselves which, ceaselessly, goes some, but not all, the way in providing knowledge of our life and our being.[6]

We may eventually discover that our pasts acts of consciousness were false and inauthentic, but prior to this discovery our inauthentic acts of consciousness are indistinguishable from our authentic ones. As a result there are times when we can be mistaken about the existence and nature of our perceptions, feelings, and thoughts. During those times everything is present within us about which we may later learn and of which we may become aware. We simply were not aware of it all in the past due to our being blinded by particular aspects of the situation which we are in and being unaware of other aspects. Because I always view it from a particular perspective, my situation is never totally transparent to me. This is not to say that it is possible that all consciousness is illusory. Although "there are imaginary sentiments to which we are committed sufficiently for them to be experienced, but insufficiently for them to be authentic,"[7] Merleau-Ponty states that absolute commitment is possible when one takes a "blind plunge into 'doing'."[8] At those times I can be certain that I actually have the perceptions, feelings, and thoughts I think I do. In this way I

> . . . fulfill my own existence. If this were not so, an ineradicable doubt would spread over the world, and equally over my own thoughts. I should be forever wondering whether my 'tastes', 'volitions', 'desires' and 'ventures' were really mine, for they would always seem artificial, unreal and unfulfilled.[9]

Merleau-Ponty stresses that he is not certain of his existence because he thinks; he is certain of his thoughts because of their genuine existence. The existence of my thoughts is not assured as mere thoughts about something. Their certainty, "is owed to that of the acts of love, hatred or will of which I am quite sure because I *perform* them."[10]

It is important to note here that it is problematic for Merleau-Ponty to treat perceptions and feelings in the same manner. Whereas I can be mistaken about whether I perceive an object, I cannot be mistaken about whether I have a particular feeling. When a person with an amputated limb feels pain in what he believes to be his limb he is not mistaken that he feels pain. What he is mistaken about is the cause of the pain. The patient who claims he feels a cold, clammy hand on his neck whenever he is about to fall asleep is not mistaken about what he feels, but about whether there is a hand on his neck. There is nothing inauthentic about these feelings. They are no less feelings than the pain one feels in an actual limb or the sensation one has when an actual cold, clammy hand is on one's neck.

Merleau-Ponty claims that his thoughts or inner perceptions cannot provide him with the certainty of his own existence. It is only his relationship to the objects in the world which gives him this certainty. Whenever I doubt, I must doubt something; there must be a relationship between my doubting and the world. I could not doubt (or have any mode of thought), if my doubt were not related to an object of doubt (of thought). If Merleau-Ponty is correct, neither the existence of objects in the world nor of my own thought can be known in the manner Descartes claims occurs. Merleau-Ponty argues that certainty of the reality of any act of consciousness, such as doubting or perceiving, does not derive from my thoughts about my doubting or perceiving but from the doubting or perceiving itself, as an act. There can be illusions both with respect to the world and with respect to consciousness. But there can be knowledge. We can know that there are objects and inner perceptions. However, these two types of knowledge occur together; neither is epistemologically prior to the other. Merleau-Ponty agrees with Descartes that "I think" and "I am" are equivalent, and that it is this which makes the *cogito* possible. However,

> . . . it is not the 'I am' which is pre-eminently contained in the 'I think,' not my existence which is brought down to the consciousness which I have of it, but conversely the 'I think,' which is re-integrated into the transcending process of the 'I am', and consciousness into existence.[11]

As required by his claim that one cannot separate the knowledge of one's own existence as a thing that thinks from the knowledge of one's body and the world, Merleau-Ponty presents a new interpretation of the *cogito*. One's own existence, according to Merleau-Ponty, is disclosed with that of the world. Thus, one's existence as a thing which thinks is indubitable only when the existence of the world is indubitable. Merleau-Ponty does not merely challenge Descartes' interpretation of the *cogito* and its place in epistemology as the foundation of all knowledge, he challenges *any* attempt to find a self-evident, presuppositionless truth and claims that there is *no* knowledge which is totally explicit. There is always some aspect of one's knowing which remains unknown and unexplained, yet is required for understanding what is known. What many previous philosophers have called "self-evident truths" can occur only within a situation, they cannot be totally presuppositionless. They are tied to our nature and are "implied in any conception that we may form of being and the possible."[12] All of our knowledge and all of our truths are of a psycho-physical being who is in the world and has a perspective and a history.

> Once launched, and committed to a certain set of thoughts, Euclidean space, for example, or the conditions governing the existence of a certain society, I discover evident truths; but these are not unchallengeable, since perhaps this space or this society are not the only ones possible.[13]

According to Merleau-Ponty, Descartes is correct in asserting that "I think" is the "primary truth," but only if "we understand thereby 'I belong to myself' while belonging to the world."[14] The *cogito* does not reveal a body-independent mind, alone and separate from the world. Rather, it reveals "the simultaneous contact with my own being and with the world's being."[15] If I separate my existence from that of the world by accepting the certainty of my own existence while calling into doubt that of the world, I am no longer talking about a human ground of experience. I am always an embodied subject in a situation in the world. For man, inside and outside cannot be separated:

> The world is wholly inside and I am wholly outside myself. . . . In so far as, when I reflect on the essence of subjectivity, I find it bound up with that of the body and that of the world, this is because my existence as subjectivity is merely one with my existence as a body and with the existence of the world, and because the subject that I am, when taken concretely, is inseparable from this body and this world.[16]

In his analysis, Merleau-Ponty attempts to restore "temporal thickness" to the *cogito.* By acknowledging my existence as a thing which thinks, he argues, I must acknowledge more than the timeless, worldless thought of Descartes' meditator of *Meditation* II. It is not that the *cogito* does not reveal something significant about the self and self-knowledge. However, in order to understand properly what it reveals it must not be considered in isolation from perception, the body, and the world.

Contrary to what Merleau-Ponty indicates, the difference between his and Descartes' view stems not from a disagreement about the nature of one's experience but rather from a disagreement about the status of that experience with regard to philosophy. Descartes doubts whether he is in a world or has (is) a body until the point in *Meditation* VI when he proves that he is the union of mind and body, but he never doubts that his thoughts are thoughts of a world. Thus, although Descartes allows for the possibility that he is only a mind, he never claims that he experiences himself as being a body-independent mind separated from the world. The issue with which Descartes is concerned is whether he can *know* that he is an embodied subject. If there were an evil genius who makes me believe that I have (am) a body when I do not and that there is a world when there is not, it nevertheless would be true that I seem to have perceptions of the sort that an embodied subject in the world has, even if these seeming perceptions are not enough to guarantee that there is a material world. What Descartes and Merleau-Ponty dispute is whether one ought to begin philosophy with perceptual experience or with pure thought.

In opposing Descartes' epistemological dualism and the quest for a presuppositionless foundation for knowledge, Merleau-Ponty begins philosophy with the natural faith in the world of *Meditation* VI. Thus, an examination of being-in-the-world and a phenomenological description of experience as it is lived serve as his starting point. On the basis of the descriptions he presents, Merleau-Ponty claims not only that all acts of consciousness are acts of an embodied being in the world, but that these acts, themselves, are bodily. He argues that one could not be a body-independent and worldless mind and have the experience a human being has. From Merleau-Ponty's point of view experience indicates that there is no pure thought – all thoughts are to some extent perceptual – and therefore there is no choice between starting with perception or pure thought. I will argue below that if we are to begin philosophy with what is first in the order of knowing, we must begin with the *cogito*. But this is not to say that we ought entirely to dismiss the importance of the phenomenological descriptions of the sort Merleau-Ponty provides. Descartes, himself, employs such descriptions in *Meditation* VI to learn about the relationship between the mind and body.

Central to Merleau-Ponty's criticisms of Descartes' view is his claim that perception and the percept have the same existential modality. Does it make any sense to say that I have perceived when I doubt whether the object I perceive exists? Merleau-Ponty is correct in answering this question negatively in the sense that part of what we mean when we state that we have seen or heard something is that the perceived object is not illusory. If we are in some doubt as to whether the object is illusory, we say that we *think* that we see or hear something. However, Descartes' argument in *Meditation* II does not rest upon whether he can know whether he has perceived an object; it rests upon his being correct in thinking that he has performed a *cogitatio.* His being correct about what sort of thought he actually performed is unimportant. The certainty that what I take to be a perception is, in fact, a perception does occur with the certainty that the object which is perceived exists. Descartes argues that if he is to have further knowledge about perception and the world he must eventually have a criterion for distinguishing true from illusory perception and be able to know when the object of apparent perception is an actual object. Nevertheless, whether I have a true or illusory perception, I think, and whenever I think my existence as a thing which thinks is indubitable.

It is unproblematic for Descartes to assert that he can be certain that he exists as a thing which thinks while he doubts the existence of objects. Even if my perceptions and, hence, my percepts are both dubitable, my existence as one who is doubtful about the existence of my perceptions and percepts remains indubitable. The hallucinator does not exist any the less because he hallucinates. His hallucination is an act of thought even if it is not true perception. My perception may be called into doubt when my percepts are called into doubt, but this is not to say at the same time that my being correct in thinking that I have performed some act of thought can be called into doubt nor that my existence as a thinker can be called into doubt. It is this that assures the possibility of my doubting whether the material world exists while being certain that I exist as a thinking thing.

Marjorie Grene, in her book *The Knower and the Known,* in a fashion similar to Merleau-Ponty's, argues against Descartes' position, stating that we can never have knowledge which is self-evident and indubitable.

She argues that for Descartes

> . . . it was the total luminousness of the act of understanding which was to guarantee the truth of its content. But there are no such acts. On the contrary, it is because *no* knowledge is wholly focal, wholly present all in one to the knowing mind, that it is always capable of reassessment.[17]

A subsidiary or non-thetic aspect of knowledge which cannot be totally eliminated always underlies what is focally known. If this aspect becomes focal, there will be other aspects which will be subsidiary. There is always room for further reflection. Professor Grene presents a convincing case that this is true with respect to science.[18] But is it true that *no* knowledge is wholly focal? What about the first thing which Descartes considers to be certain and indubitable, the knowledge that "I am, I exist" is true whenever I pronounce it or mentally conceive it? Can this too be reassessed? It may be true that there are aspects to my knowing that I exist which are not focal and are necessary for my knowing that I exist, but the *cogito* is unique in that no matter what these aspects are I cannot be mistaken about my existing whenever I think that I exist. The fact that knowledge may not be wholly focal does not mean that one can never be certain that what is non-focal will not affect what is focal. With respect to the *cogito*, if what was tacit becomes focal I may revise my views concerning what it is that I claim I am, or what is involved in thinking, or makes my thinking possible; but that I am a thing that thinks remains indubitable as long as I think that I am. Professor Grene is correct in asserting that "no subject matter, no concept, however precise, can be, as Descartes thought it, wholly at the centre of attention. . . . " However, it is not true that " . . . no subject, no concept is ever self-evident and indubitable."[19] This, at least, is not the same as the preceding and does not follow from it. Therefore, the *cogito*, even if it has tacit aspects, can still be certain – as, indeed, it appears to be. It is not that for Descartes thinking is totally luminous but that what is in the center is.

Professor Grene goes on to talk about self-knowledge in particular.[20] She asserts that Descartes claims that " . . . my *own* mind is what I know directly, luminously and without difficulty. . . . I know who I am, but . . . I must reason somehow by analogy to the existence of others."[21] Professor Grene denies this and claims that we know ourselves indirectly by looking at ourselves as if we were another and examining the impact we make. However, others are known directly from what they do, what they say, and their facial expressions, bodily movements, and overall behavior. Thus she states: "other selves, even sticks and stones, are known to me before and better than I myself am known.[22] She admits that one directly experiences one's own pains, whereas one does not directly experience those of another. It is in this sense that one is "within" one's own experience and not "within" that of anyone else. But Professor Grene claims that this is not knowledge;

. . . it *is* simply. . . . To *know* is to interpret clues of which I am only subsidiarily aware, perhaps even wholly unaware, in terms of the entity or event upon which they bear – to see *through* them what they mean. Only because of this indirectness can such insights be conveyed, indirectly, to others, be publicly substantiated, become knowledge rather than a blind given. But this sort of mediation is just what I cannot perform upon myself, precisely because what should be focussed on, attended to, is already subsidiary, part of what I dwell in, of my own reality, and the proper clues to knowledge of myself are accessible in countenance and tone of voice, in turn of phrase or speed of reaction, to others more readily than to me. They are part of my bodily existence, so interiorized already that I can project from them only outward, not inward.[23]

According to this view, the knowledge I acquire of other people and things in the world is both prior to and more direct than my knowledge of myself. I must, for the most part, always remain unknown to myself.

This is a significant analysis of knowledge and must be taken into account. Professor Grene is correct in asserting that in an important sense self-knowledge is more difficult to acquire than knowledge of people and things external to oneself. It is not clear, however, that Professor Grene's position opposes that of Descartes. *Meditation* II is entitled "Of the Nature of the Human Mind; and that it is more easily known than the Body." Despite this title Descartes does not claim that every aspect of the mind can be known before and better than every aspect of the body. What he does claim is that the *existence* of one's mind, *as a thinking thing*, is more easily known than the *existence* of the body (or bodies in general), *as an extended thing*. When he states that he knows himself before and better than he knows everything else, Descartes is asserting that he knows *that* he exists, *qua* mind, and not that he knows *what* he is completely, before and better than he knows that everything else exists. Descartes never claims that whenever he thinks about his thinking he cannot be mistaken about the nature of his thought or that all of his psychological states are transparent to him. He merely states that his thinking about his thinking is sufficient for his knowledge that he thinks and exists, whereas his perception of bodies, including his own, is not sufficient for his knowing that they exist. It is in this sense and only in this sense that the knowledge of one's mind is more certain than and prior to the knowledge of one's own body, others, and the rest of the material world.

Once again, it is important to point out that Descartes does not deny that one's *experience* of people and things in the world may be temporally prior to one's becoming aware of oneself. Self-awareness comes only with reflection. It is with Descartes' decision to doubt all that can be doubted and his attempt to determine which of his former beliefs are true that he realizes that the *knowledge* that he exists, as a thinking thing, is prior to the knowledge that other people and matter exist. Subsequently, more may be known about the various properties of these people and material objects than can be known about oneself. But Descartes' thesis, that the knowledge of one's own existence as a thinking thing is prior to all other knowledge, remains unaffected.

An important problem remains. If it is true that my experience of my body and the world is not enough to enable me to know that they exist, is there any sort of evidence that can give me this knowledge with the degree of certainty which Descartes demands? There are only two possible means. The first of these involves my stepping outside my situation and looking at myself perceiving. If I could see myself perceiving and see the object I am perceiving at the same time, I would be certain that I am having a true perception. Clearly this is impossible. But even if it were not impossible, this problem would not be solved. If I can doubt whether the objects I appear to perceive exist, then I can doubt whether my perception of myself perceiving an object is veridical. I would then have to seek some guarantee that this second-level perception is not illusory. Hence, there would be an infinite regress of my having to step outside my present situation and look at myself perceiving my perceiving, etc.

The second alternative is to employ some infallible guarantee which enables me to know that the objects I seem to perceive actually exist. Descartes employs this alternative when he uses God to prove that the external world exists. If I can know that a non-deceiving God exists and that to allow me to believe that there is a material world when there is not, is a deception, then I can prove that I actually perceive existing objects. The *only* way I can prove that I perceive is to prove first that a non-deceiving God or some other infallible guarantee exists. If this cannot be done on the basis of what is known in *Meditation* II, then I can never be certain that my body and the rest of the material world exist.

Assuming for the sake of argument that such a proof cannot be provided, what would it mean to deny or even doubt the existence of my body and the external world? Illusion and reality can have meaning only within the context of my own situation because I can never separate myself from it. Hence in this sense Merleau-Ponty's analysis is correct. It is not because we simply have not been able to find the evidence that is needed to satisfy Descartes' goal that we accept less than he asked for with respect to our knowledge of material objects. We must accept less because Descartes asked for more than anyone (human) could ever achieve. Descartes is correct when he claims that knowledge of the existence of one's mind requires different evidence than knowledge of the existence of one's body. Hence, it *is* intelligible to doubt the existence of material objects. But one soon discovers that there is no possible evidence that would confirm or deny a claim concerning their existence apart from one's situation in the world.

I have explained why I think that Merleau-Ponty is wrong in claiming that one's knowledge of his existence as a thinker must come with his knowledge of the external world. I can withhold judgment about whether anything outside my mind exists and yet at the same time be certain that I exist as a thing which thinks. However, in an important sense, Merleau-Ponty is correct in claiming that he cannot separate himself from the world. If I am to go beyond the fact that I exist and learn more about my nature, others, and the world, I must examine the experience revealed by the *cogito* which is never of a mind without a body but always of an embodied subject in a situation within the world.

Descartes teaches us that unless we accept *Meditation* III or find some other kind of guarantee for our experiences, claims about the world always will be uncertain to some degree. Whereas it is true that I will never be able to advance beyond *Meditation* II if I continue to separate myself as a thinker from the world which my thoughts reveal, in advancing beyond this *Meditation* (without *Meditation* III) the ideal of certainty must be relinquished. Beginning with epistemological dualism enables us to recognize the limits of what can be known with certainty; Merleau-Ponty's phenomenology provides a method for proceeding once those limits have been discovered.

Why should we begin with the Cartesian *cogito* if it gives us so little? First, it enables us to recognize the difference between what is required to know that one exists as a thinking thing and what is required to know that there is an external world or that one's perceptions are veridical. And second, there would be no reason to begin with Merleau-Ponty and a rejection of the quest for philosophical certainty if we had not first discovered the limits of that certainty.

According to the argument I have presented, we can accept

Descartes' epistemological starting point, i.e., begin philosophy with the *cogito*, and yet continue with Merleau-Ponty's position and the claim that knowledge can be derived from one's experience which reveals one as an embodied subject in the world. I tried to show that this is the way one ought to proceed. In order to determine whether this is a tenable position it is important to examine the relationship between epistemology and metaphysics. Many claim that one cannot be an epistemological dualist without being a metaphysical dualist, but I will argue that metaphysical monism is consistent with epistemological dualism.[24]

In her book, *Sartre*, Professor Grene denies the possibility of beginning philosophy with the *cogito* and not being a metaphysical dualist.[25] According to Professor Grene, if we accept the *cogito* as the foundation of knowledge and claim that there is an epistemological priority to consciousness, then we must accept a metaphysical dualism of mind and body. She states that the *cogito* separates reflecting consciousness from the world and explains that this does not mean that there must be a dualism of substance but it does mean we cannot avoid

> . . . a dualism of consciousness against what is rejected; the out-there, the extended and external, the other-than-myself-here-now-aware-or-myself-here-now.[26]

Beginning with the *cogito*, Professor Grene maintains, results in the opposition of self and other, and act and object. "The man of the *cogito*" must be a man "of detached self-consciousness."[27] Hence, if we accept Descartes' epistemology we can never accept Merleau-Ponty's position, even partially, because we can never get consciousness back together with the world.

It is not clear how recognizing, as Grene and Merleau-Ponty do, that human beings are embodied subjects in the world completely eliminates the oppositions they both try to avoid. Although there is opposition of mind and body with respect to man, a dualism of man as a mind-body against the world remains. Even if one rejects Descartes' *cogito* as the starting point of philosophy, as soon as one talks about man's being *in* the world it is difficult to avoid some distinction between man and the world. The fact that Merleau-Ponty differentiates between the structure of behavior of the human order and that of the physical order indicates that he is committed to a dualism of human beings, as subjects, and the world.[28] As Professor Grene points out, this is not necessarily a dualism of substance but it is, nevertheless, a dualism.

Professor Grene is correct when she says that the *cogito* separates reflecting consciousness from the world, but it does this in one sense only. The *cogito* enables us to recognize the fundamental difference between what is required to prove that one exists as a thinker and that one's body, whether it is a material object as Descartes claims or a subject as Merleau-Ponty claims, and the world exist. I have argued above that this is a legitimate distinction. However, there is another sense in which it does not separate consciousness from the world: this has to do with one's experience of consciousness and the world. When one examines this experience one recognizes that consciousness cannot be separated from the world because consciousness is always of an object; consciousness always reveals a thinker in a world. This is something which Descartes never denies. He would agree that consciousness is intentional and thus that the *act* of thinking cannot be separated from the *object of thought.* Whenever I am certain that I think, I am certain that I am thinking about something. Thus one can begin with the *cogito* and agree with Grene and Merleau-Ponty that one's experience is always that of an embodied being in the world. Descartes explicitly endorses this in *Meditation* VI. However, because Merleau-Ponty uses this experience as the foundation of philosophy, he claims that Descartes' position concerning the *cogito* is untenable. He states that Descartes' methodical doubt does not accomplish what it was supposed to " . . . since the whole world, at least in so far as we experience it, is reinstated in the *Cogito,* enjoying equal certainty, and simply labelled 'thought about . . . '."[29] Merleau-Ponty's position is even stronger than Grene's as he is claiming that we cannot, in principle, begin with the *cogito* because we cannot dissociate thought from the object of which thought is.

When Merleau-Ponty denies that we can separate thought and the world, and Grene states that we are making a mistake if we begin epistemology by separating them, they both assume that the object of thought (what I think about) is identical with the object itself. However, whereas we cannot dissociate the *cogito* from the object of thought, we can separate it from the actual object we are thinking about. When I perceive a table, based upon my experience alone, I can be certain that I am thinking, and that I am thinking about (apparently perceiving) a table, but I cannot be certain that the table exists.

Accepting the latter doubt does not force us to deny the former certainty. Although it is legitimate to deny that thought and object can be separated, it is only in the first sense that this is true. If there really were an evil genius who caused me to believe that the world exists when it does not, I could still not deny that thought and object cannot be separated with regard to my experience. Whether or not the material world exists, my experience of the inseparability of the act and object of thought is unchanged. But because the existence of thinking is self-guaranteeing and that of objects is not, there is always a legitimate separation of thought and the world in that sense. This is a gap that can never be bridged.

The distinction I am making between one's experience of objects or objects of thought and the objects themselves is the distinction Descartes makes in his *Reply to Objections* I to Caterus. He explains that the thing itself, which exists outside the understanding, is one thing whereas the idea of the thing, which does not exist outside of the mind, is quite another.

Hence the idea of the sun will be the sun itself existing in the mind, not indeed formally, as it exists in the sky, but objectively, i.e., in the way in which objects are wont to exist in the mind.[30]

Beginning philosophy with the *cogito* in no way denies that one's experience is always that of an embodied subject in the world, or that man is part of the world, or that mind is bodily or part of the body. It simply denies that we can have a secure foundation for knowledge if we accept experience unanalyzed by philosophic reflection. I have tried to argue that once we begin an analysis of experience, the truth of Descartes' claims about the *cogito* becomes clear. However, that analysis also reveals how little we can add to the foundation built in *Meditation* II. In addition, it should be emphasized that there is nothing inherent in this position which promotes metaphysical dualism. Descartes begins with the *cogito* and eventually affirms a metaphysical dualism which includes the claim that there is a substantial union of mind and body and that this relationship is not like that of a pilot in a ship. But one could begin with the *cogito* and develop a metaphysics which is monistic with respect to man's nature (e.g., materialism, idealism, some species of neutral monism, or Merleau-Ponty's view which describes man as a unity of mind and body).[31] It is possible to claim that knowledge of the mind and knowledge of the body are mutually independent and yet at the same time claim that the mind is metaphysically part of the body, for example, the central nervous system. Descartes allows for this possibility in *Meditation* II when he says:

> But perhaps it is true that these same things which I supposed were non-existent because they are unknown to me, are really not different from the self which I know. I am not sure about this, I shall not dispute about it now; I can only give judgment on things that are known to me.[32]

It is important to make clear that if I am wrong about epistemology, and starting with the *cogito* is untenable, this does not mean that metaphysical dualism has been refuted. If my knowledge of the existence of myself as a thinking thing and my knowledge of the existence of myself as an extended thing are acquired together, this does not prove that the mind and body are identical or inseparable and that some form of monism is true. Metaphysical dualists and monists can agree with Merleau-Ponty and Grene with respect to epistemology and their claim that experience is primary, yet disagree as to how one goes on from there to metaphysics. Clearly there must be a connection between the two, but epistemological dualism (or monism) does not lead inevitably to metaphysical dualism (or monism). I have tried to argue, with Descartes, that in order to begin epistemology with a secure foundation we must begin with the *cogito.* In doing so we maintain an epistemological dualism of mind and body, but we are not at the same time forced to deny the unity of mind and body which experience reveals.

University of Illinois at Chicago

NOTES

* I am grateful to Alan Donagan, Paul Ricoeur, Daniel Garber, John Bannan, and Thomas Short for their helpful comments on various versions of this paper. When I was a freshman at the University of California at Davis, Marjorie Grene introduced me to the pursuit of wisdom begun in ancient Greece. At least as important, by her example she introduced me to the love of that wisdom. Part of what I learned from her is that the respect we give to philosophers is to continue their pursuit, not by admiring their works as if they were in a museum, but by critically analyzing the questions and answers they offered. I consider myself fortunate to have had the opportunity to have studied with Marjorie Grene and hope that she will forgive me for my interpretation of and views about Descartes, Merleau-Ponty, the *cogito*, and herself.

[1] Descartes (1897–1913), vol. 7, p. 25; Descartes (1972), vol. 1, p. 150.
[2] Descartes (1897–1913), vol. 7, p. 28; Descartes (1972), vol. 1, p. 153.
[3] Merleau-Ponty (1945), p. 429; Merleau-Ponty (1962), p. 374.
[4] Merleau-Ponty (1945), p. 429; Merleau-Ponty (1962), p. 375.
[5] Merleau-Ponty (1945), p. 431; Merleau-Ponty (1962), p. 376.
[6] Merleau-Ponty (1945), pp. 435–436; Merleau-Ponty (1962), p. 380.
[7] Merleau-Ponty (1945), p. 437; Merleau-Ponty (1962), p. 382.
[8] Merleau-Ponty (1945), p. 438; Merleau-Ponty (1962), p. 382.
[9] Merleau-Ponty (1945), p. 438; Merleau-Ponty (1962), p. 382.
[10] Merleau-Ponty (1945), p. 438; Merleau-Ponty (1962), pp. 382–383.
[11] Merleau-Ponty (1945), p. 439; Merleau-Ponty (1962), p. 383.
[12] Merleau-Ponty (1945), p. 455; Merleau-Ponty (1962), p. 397.
[13] Merleau-Ponty (1945), p. 454; Merleau-Ponty (1962), p. 396.
[14] Merleau-Ponty (1945), p. 466; Merleau-Ponty (1962), p. 407.
[15] Merleau-Ponty (1945), p. 432; Merleau-Ponty (1962), p. 377.
[16] Merleau-Ponty (1945), p. 467; Merleau-Ponty (1962), p. 407–408.
[17] Grene (1974), p. 84.
[18] See *ibid.*, pp. 74 ff.
[19] *Ibid.*, p. 86.
[20] See *ibid.*, pp. 86–87.
[21] *Ibid.*, p. 86.
[22] *Ibid.*
[23] *Ibid.*, p. 87.
[24] When I refer to metaphysical dualism and monism I am referring to positions with respect to the sort of entity man is. In this sense Merleau-Ponty is a metaphysical monist in spite of the fact that he distinguishes among the physical, vital, and human orders. This monism, however, is radically different from the materialist and idealist metaphysics which Merleau-Ponty attacks.
[25] See Grene (1973), pp. 36–38.
[26] *Ibid.*, p. 46.
[27] *Ibid.*
[28] See Merleau-Ponty (1942), p. 255; Merleau-Ponty (1963). As indicated in note 24, Merleau-Ponty actually makes a three-fold distinction among three orders, viz. the physical, vital, and human.
[29] Merleau-Ponty (1945), p. iii; Merleau-Ponty (1962), p. ix.
[30] Descartes (1897–1913), vol. 7, pp. 102–103; Descartes (1972), vol. 2, p. 10.
[31] See Armstrong (1968).
[32] Descartes (1897–1913), vol. 7, p. 27; Descartes (1972), vol. 1, p. 152.

BIBLIOGRAPHY

Amstrong, D.: (1968), *A Materialist Theory of the Mind*, Routledge and Kegan Paul, London.

Descartes, R.: (1897–1913), *Oeuvres*, ed. C. Adam and P. Tannery, 11 vols., Cerf, Paris.

Descartes, R.: (1972), *Philosophical Works*, ed. and trans. E. Haldane and G. Ross, 2 vols., Cambridge University Press, Cambridge, England.
Grene, M.: (1973), *Sartre*, New Viewpoints, New York.
Grene, M.: (1974), *The Knower and the Known*, 2nd ed., University of California Press, Berkeley.
Merleau-Ponty, M.: (1942), *La Structure du Comportement*, Presses Universitaires de France, Paris.
Merleau-Ponty, M.: (1945), *Phénoménologie de la perception*, Gallimard, Paris.
Merleau-Ponty, M.: (1962), *Phenomenology of Perception*, trans. C. Smith, Routledge and Kegan Paul, London.
Merleau-Ponty, M.: (1963), *The Structure of Behavior*, trans. A. Fisher, Beacon Press, Boston.

ALAN DONAGAN

THE WORST EXCESS OF CARTESIAN DUALISM

Since I cannot separate the influence of Marjorie Grene's writings and conversation on me from her influence as a human being, I cannot write a paper to be presented to her, even one in which she is referred to in the third person, without addressing it to her. And to address her in the common philosophical mode, even on topics of shared philosophical interest, would be like putting on academic dress to lunch with her at the Algonquin.

A theme in her work that never fails to enlist my agreement is her ridicule of science-fiction philosophizing about human beings — as if there were no more in principle to them than there is to the fabrications of technology. But that theme seldom occurs in her writings or conversation except in conjunction with another: denunciation of Cartesian dualism. Not unintelligibly. Since the first theorists of *l'homme machine* traced their ancestry directly to Descartes, it is not surprising that somebody to whom mechanistic reductions of living things are philosophical death should identify the theory that gave birth to those reductions with philosophical sin.

What is surprising is that reductive materialists today should (with a few distinguished exceptions like David Armstrong) have abandoned their predecessors' ancestral piety. However appalled Marjorie Grene's materialistic contemporaries may be by her antireductionism, they rejoice to join her in denouncing the errors of Descartes. And with what fervor they do it! Invited to name the recent article from whose introductory paragragh the following is plucked,

> In spite of dark warnings from Wittgensteinians, many philosophers have veered back to a Cartesian view of the mind. I do not mean Cartesian *dualism* (only a crank would wish to revive that) . . .

even sedulous readers of journals would probably be at a loss: it would be in place in so many. And having been informed that the following four chapter-titles of Marjorie Grene's *The Knower and the Known* each have the name of a different philosopher in the blank: "-----ian Certainty," "The Errors of -----," "-----'s Premises," "-----: the

A. Donagan, A. N. Perovich, Jr., and M. V. Wedin (eds.), Human Nature and Natural Knowledge, 313–325.

Knower as Agent," who would fail to guess that 'Descartes' goes in the second? The point is not that she thinks that Descartes made more errors than Aristotle, Hume, or Kant (the other three), but that she endorses the consensus that his errors are paradigmatic — and his dualism above all. That is the error not to be mentioned without appropriate abhorrence. The tone of that abhorrence in papers I read as I began this essay seemed to echo something in literature, which I half-remembered and which Anthony N. Perovich finally found for me. It was Lady Bracknell in *The Importance of Being Earnest* comparing the circumstances of Ernest's birth to the ultimate social horror:

> To be born, or at any rate bred, in a hand-bag, whether it had handles or not, seems to me to display a contempt for the ordinary decencies of family life that reminds one of the worst excesses of the French Revolution.

Here care must be taken not to simplify. The dualism of the ghost in the machine, ritually flagellated every quarter in English language journals — the dualism of Descartes *au goût britannique* — while it makes an appearance in *The Knower and the Known*, "which was written during the years 1961–1963,"[1] is absent from Marjorie Grene's recent essays. And the blame for that early appearance must go largely to Merleau-Ponty, whom she followed, and whose Descartes is not readily distinguished from the Descartes of Gilbert Ryle.

"The errors of Descartes," as exposed in *The Knower and the Known*, are at bottom four: "the principles of (1) indubitability, (2) self-evidence or total explicitness, (3) the unity of science, (4) the duality of man;" and they form a structure. Scientific knowledge is indubitable because it is self-evident; self-evident knowledge is a unity because it is the work of the mind alone; and it could not be the work of the mind alone "were the mind not cleanly and essentially separate from the body in nature and function."[2] The duality of man, so understood, is the real distinction in man of mind and body. All four principles are of course rejected; but the fourth, the duality of man, is denied with an echo of Ryle, "We are not pure cognitions absurdly attached to a machine," and a quotation from Merleau-Ponty: "we are not in some incomprehensible way an activity joined to a passivity, an automatism surmounted by a will, a perception surmounted by a judgment."[3]

The criticism in *The Knower and the Known* of Descartes's

principles of indubitability and self-evidence are decisive; those of the principles of the unity of science and the duality of man disappear from her later work, not because they are intrinsically unsound, but because further study of Descartes, under the influence of Martial Gueroult rather than of Merleau-Ponty, showed them to be misplaced: Cartesian science is not a unity; Cartesian man is not simply a duality, but a union of two substances; and Descartes considered the phenomena distinctive of that union to be intelligible, but not by any Cartesian science.

The Knower and the Known went astray partly because in it the source of Descartes's thought was taken to be the fragmentary *Regulae ad Directionem Mentis*, composed between 1622 and 1629, and not published during his life-time. Gueroult, however, has shown that for understanding Descartes's mature thought, *Meditationes de Prima Philosophia*, which he published in Paris in 1641, is canonical. And, as Gueroult has put it, the six Meditations composing that work divide into two parts, like the panels of a diptych. In the first three, the meditator "goes against the bent of his natural inclinations by a strong effort, and abstracts himself from what constitutes him specifically as a man . . . in order to attach himself exclusively to the understanding by itself, the extended by itself, regions which the pure natural light illumines."[4] This picture of metaphysics and the exact sciences as the end of philosophy, which has "traditionally passed as summing up the whole of the Cartesian enterprise,"[5] is no more than the first panel, which the second, composed of the last three Meditations, reverses. In them, the meditator comes to perceive that, as "author of the composite substance, of the qualitative, of the obscure, and of the confused, God must guarantee its truth in a measure as he grounds its reality. . . . The opposition of the clear and distinct to obscurity and the confusion of the senses can therefore no more appear as that of the true and the false, of being and not being, but . . . as that of two heterogeneous ways of being."[6]

To the mature Descartes of the culminating Meditations, human beings are not intelligible by way of the two exact sciences of Cartesian physics and (for want of a better name) Cartesian *scientia cogitationis*. Paradoxically, as Spinoza saw, Descartes dismissed in advance the "dualist" scientific project of treating human bodies as independently intelligible in terms of physics, and human minds as independently intelligible in terms of *scientia cogitationis*; and he dismissed it on the ground that it consigns to oblivion what is distinctively human about human life: namely, sensation, passion, and voluntary bodily action.[7] As he was to write in *Principia Philosophiae* I, 48:

we experience certain other things in us, which can be referred neither to mind alone, nor at all (*etiam*) to body alone, and which . . . originate from the close and intimate union of our mind with a body: namely the appetites of hunger, thirst, etc.; and also the emotions (*commotiones*) or *animi pathemata* that do not consist in thought alone, such as being moved (*commotio*) to anger, to mirth, to sadness, to love, etc., and finally all sensations (*sensus*), such as of pain, of pleasure (*titillationis*), of light, and colors, of sounds, smells, tastes, of heat, hardness and other tactile qualities.[8]

Here we encounter what to materialists and behaviorists today is the worst excess of Cartesian dualism. It is not asserting the existence of *res cogitans* as well as of *res extensa*: bad though that is, it can do little harm in its parallelist and epiphenomenalist forms. Rather, it is asserting their union — their interaction. Yet the standard reason for dismissing interaction as impossible, even ludicrous, is astonishing: namely, that it is scientifically unintelligible. For, by the standards of non-Cartesian science, both Aristotelian and Newtonian, the action of non-mechanical forces to produce mechanical effects is not unintelligible at all. Descartes pointed this out, with respect to Aristotelian gravitational theory, in a letter to Arnauld.[9] By the standards of *Cartesian* science it is indeed unintelligible: but how came it about that philosophers who deny that science is what Descartes took it to be are not laughed at when, against Descartes's thesis that certain phenomena originating in mind-body unions are not intelligible in terms of Cartesian science, they object that — they are not intelligible in terms of Cartesian science? With dry finality, Robert C. Richardson has said what needs saying:

The problem of how two such heterogeneous things as mind and body could interact is, as Descartes so clearly saw, no problem at all. The illusion of a problem has two sources: the presumption that psychophysical interaction must be similar in kind to physical interaction, and the conviction that psychophysical dualism must lead to a naive attempt to attain an exhaustive segregation of properties into those that are attributable to the purely physical and those that are attributable to the purely mental.[10]

In Marjorie Grene's later writings on Cartesian dualism, although the theme that mind-body interaction is not to be thought of unless it can in principle be scientifically explained is an *obligato*, her fundamental objection is different and independent: namely, that because Descartes neglected to develop a coherent conception of what animality is, he not only degraded brute animals into machines, but also implicity denied the animality of human nature. This objection is both sound and fruitful. It draws attention to a disabling omission in the foundations of Descartes's theory of the nature of human beings, and identifies an ill consequence of it that may fairly be set down as an excess. However, as I shall try to show, it is an excess deriving from a defect in his concept of *cogitatio*, not from his dualism.

It is not accidental that the paper in which this objection is most fully developed, "People and Other Animals,"[11] while formally directed against Cartesian dualism, is largely about a lecture by the most distinguished living Cartesian dualist, Sir John Eccles, in which he advocates the three-world theory of his friend Sir Karl Popper.[12] According to Popper's theory, reality is divided into three "worlds": World (1), consisting of physical objects and states; World (2), consisting of states of consciousness; and World (3), consisting of knowledge in the objective sense, which, by a *principium divisionis* I have failed to grasp, is divided into records of intellectual efforts stored in libraries and archives, and intellectual problems, arguments, and systems.[13] Worlds (1) and (2) are identified with Descartes's *res extensa* and *res cogitans;* and World (3) is added as a portion of reality which Descartes overlooked.

The thesis of "People and Other Animals" is that, while Popper and Eccles are right in asking what "worlds" reality is composed of, they have answered their question wrongly. World (2), identified with *res cogitans,* an "inner life of consciousness" which "Descartes split apart . . . from the external, material world," is a myth.[14] Reality is the realm of nature. And the realm of nature is divided hierarchically, like Aristotle's realm of *psuche*, into (*a*) merely physical sysems, (*b*) physical systems that are living, (*c*) living physical systems that learn, and (*d*) living learning physical system that think, and hence have language.[15] World (1) is therefore confused in its very conception: it is the realm of nature falsely identified with that portion of nature that consists of merely physical systems. And World (3) is, if possible, even more confused: it is that part of the natural world that has been transformed by the manipulative social activity of thinking animals into cultural products.

The hierarchy of "People and Other Animals" is one of levels of explanation. Merely physical processes — or processes occurring in objects considered merely with respect to their physical properties — are explicable in terms of merely physical models, that is, of processes describable wholly in terms accepted as proper to the science of physics, and occurring according to what are accepted as laws of that science. Processes belonging to the higher level of living things, while physical, and hence both describable in physical terms and occurring according to physical laws, are also describable in biological terms that are not definable in terms of physical ones, and occur according to biological laws that are not analysable as special cases of physical laws. The biological level is higher than the physical, and less fundamental, because biological processes are, in the sense explained, physical but not merely physical. Far from being "exempt" from laws of lower levels, it is essential to processes of a higher level that they also belong to those lower levels and "obey" their laws.[16]

One of Marjorie Grene's most effective illustrations of a higher-level concept is found, not in "People and other Animals," but in her paper on Helmuth Plessner's concept of "positionality," where she draws attention to Sir Macfarlane Burnet's argument that a deep-seated property of living bodies is their ability to "know" their own cells from invading material. [17] Although this immunological property is not definable in physico-chemical terms, it is by physico-chemical mechanisms that physiologists will seek to explain the operation of the semi-permeable membrances in virtue of which living bodies have that property. Or, to use another of her illustrations: Fraenkel and Gunn point out that "A traffic jam . . . is not explained through given drivers, or any group of drivers, trying to get to work; it is the overall laws of the collective that explain the situation, despite not because of, the behavior of each individual."[18] From the higher level point of view of traffic engineering certain patterns of movement of vehicles on a road system are classified as "traffic jam"; but that such a pattern will occur at certain regions of a certain road system, given that more than a certain number of vehicles follow certain routes within a certain period, is a matter of lower level physical causation.

Before foisting on Descartes anticipations of twentieth century reductionism, let us review the evidence. *Prima facie,* not only *Le Monde,* which he decided to suppress, but also the collection of scientific *opuscula* which he published instead, contain unreduced descriptions of processes of various levels, which are speculatively explained as the functioning of natural mechanisms described in terms of Cartesian physics. Digestion is the first of the "functions which can be imagined to proceed from matter and to depend on nothing but the arrangement (*disposition*) of the organs" which he discusses in the *Traité de l'Homme*.[19] But how can it be maintained that the concept of digestion is less irreducibly biological than the immunological property discussed by Macfarlane Burnet? It is true that Descartes does not assert its irreducibility, but neither does he show the slightest interest in reducing it. What he does try to do is to explain the mechanism by which that irreducibly biological process occurs. And why should that be reductionist when Descartes does it, but not when Macfarlane Burnet does?

Marjorie Grene correctly maintains that there are sciences of nature of a higher level than physics, in that they employ concepts not definable in terms of physics and offer explanations (for example, functional ones) not reducible to physical ones. At the same time, she acknowledges that the processes of which a higher level science treats may be causally explained by mechanisms describable in terms of a lower-level science. It follows that Descartes's doctrine that *all* "functions which can be imagined to proceed from matter" can be causally explained in terms of physics is *consistent* with admitting that there may be natural sciences of a higher level than physics. Descartes himself expresses no opinion one way or the other, because the issue of reductionism in the twentieth century sense had not arisen. But Cartesian biology and physiology are no more branches of Cartesian physics (in the sense in which the theory of light is a branch of it) than twentieth-century biology and physiology are branches of twentieth-century physics.

Cartesian dualism is *not* a poverty-striken doctrine about the hierarchy of sciences, which is to be corrected by a richer theory. Taken generally, that is, without any commitment to Cartesian physics as distinct from late twentieth-century physics, Cartesian dualism is compatible with ordering the realm of nature into the first three of Marjorie Grene's levels (the merely physical, the living physical, and learning and living physical) each being the subject of a distinct level in the hierarchy of sciences (physics, biology, ethology); but it is not compatible with recognizing her fourth level as unqualifiedly a part of the realm of nature, or as the proper field of any science of nature whatever, no matter how exalted its level. Why?

In dissociating her hierarchy theory from vitalism, the doctrine of biologists like Driesch that "there is some mysterious something about living things exempt from physical laws," she repeatedly insisted that higher order processes are not exempt from lower-level laws, and may in fact be explained in terms of lower-lever mechanisms (like the immunological powers of living beings). This can be so because the same processes can belong to two levels: a living being is a physical being that, in virtue of its special physical character, has properties of a higher level. But it is Descartes's contention that the attribute of *cogitatio* cannot belong to anything in virtue of its special physical character. Of course a certain complex sound produced by a human voice may, by virtue of human linguistic practice, express a certain thought in a certain language: but it does not do so in virtue of its physical character as such; nor (despite some Wittgensteinians) does it do so in virtue of its setting in the context of its speaker's and others' physical behavior. It is a property *qui ne peut être imaginée procéder de la matière.*

Descartes was not the first to describe the feature of *cogitatio* in virtue of which he maintained its radical distinction from anything material — and hence from anything within the realm of nature. Medieval Aristotelians in expounding their teacher's *De Anima* had noticed that thinking is characteristically thinking *that something is so*, and that the thought that *p* can take place whether or not *p* is true, and whether or not the concepts composing *p* apply to anything in reality or not. The objects of thought have being (*esse*) as such, but that being must be distinguished from independent or "natural" being — *esse naturale.* As Wittgenstein was to remark, we can think of a man who does not exist, but we cannot hang a man who does not exist. An object of thought was therefore described as having the being of something-to-which-a-thought-is-directed, or *esse intentionale*, even when it did not have *esse naturale.* Descartes expressed this in terms of a later scholasticism when he maintained that every thought whatever has as a component something he called an *idea*, or an image of a thing (*rei imago*), and pointed out that that of which an *idea* is the image must have being as an imaged object, *esse objectivum*, but not necessarily being as an independent thing, or *esse formale.*[20] Here we have what Descartes took to be the essence of a *res cogitans*, a thinking thing: it is a thing in which occur processes involving *ideae*, processes in virtue of which the thinker represents to himself objects that may have only *esse objectivum* or *esse intentionale.* The foundation of Cartesian dualism is the conviction that no process involves such representation — exhibits what Brentano was to teach us to call 'intentionality' — by virtue of its physical character, no matter how special that physical character.[21] A living body's ability to "know" its cell from invading material is indeed an ability it has in virtue of its special physical character, but that ability is not a genuine case of intentionality. It is convenient for human beings to use the language of intentionality in describing a variety of functional processes in both animals and machines, even though there is no reason to believe that either animal or machine represents anything to itself.

Descartes seems to have been confident that nobody who clearly and distinctly grasped what a *cogitatio* is and what *res extensa* is could entertain the notion that a *cogitatio* could be a mode of any *res extensa.* Certainly he did not argue it.[22] His reticence was well-judged: not because his case was weak; but because, if true, it is a matter of principle, and matters of principle cannot be deductively demonstrated. What was left to do, having explained what his dualism amounts to, was to demolish attempts to show that whatever has a certain special physical character must, by virtue of the character, exhibit intentionality. For those who incline to such attempts, Quine's *Word and Object*, pp. 216–221, is salutary reading. The least unpromising line for an anti-dualist to take seems to be to deny the existence of any process involving intentionality, while confessing that, for the present, such processes are indispensable in our thinking.

One excess of Cartesian dualism, however, remains: the denial of consciousness to brute animals. Practically, as Majorie Grene rejoices to point out, it is incredible. Hence it appears that if dualism in any form is true, the line between living beings that possess consciousness and those that do not cannot be drawn where Descartes draws it. Yet how can it be drawn anywhere else except arbitrarily?

Descartes's reasons for drawing it where he did can be reconstructed with fair probability. As we have seen, he took *cogitatio* to be a process in which the *cogitans* stands in a relation to an intentional object, that is, an object having *esse objectivum* but not necessarily *esse formale.* And since he took it for granted that linguistic capacity is decisive evidence that something is a *cogitans,* and lack of it decisive evidence that it is not, he may be presumed to have presupposed that a *cogitatio* is characteristically linguistically expressible. Yet there are processes in living beings that, while not *cogitationes* by this criterion, are intimately related to them: namely, sensations, feelings of pleasure and pain, and emotions. When a *cogitans* feels a hard cold surface, feels it as painful and is frightened by it, he normally is "impelled" or "taught" by nature to think that there is a hard cold surface which he is touching, which it is not good for him to touch and the presence of which bodes him no good.[23] That is, a *cogitans* cannot have sensations, feelings of pleasure and pain, and emotions without being impelled by nature to have certain thoughts — some true and some false. The linkage between sensation and thought sufficiently justifies his extension of the term '*idea*' to what is "properly and immediately" sensed, and of the term '*cogitatio*' to the sensing of it.[24] As he assumed in the *Meditationes*, and explicitly pointed out in *Principia Philosophiae*, sensations "when they are viewed only as sensations" do not take us beyond themselves.[25] They lack intentionality or *esse objectivum*. (Here, of course, he is referring not to animal sensation — events in the nervous systems and brains of animals — but to events in the human mind caused by those bodily events.) Sensations acquire their intentionality by their natural connection with thought — with sensory judgements. But if mental sensation (and with it, feeling and emotion) affects those that have it solely by its natural connection with thought, there is no reason to ascribe it to non-rational animals. In them it would be an idle epiphenomenon, having no effect on their lives. So, I think, Descartes reasoned, and not implausibly.

This position is formidable, however hard it may be to believe its implications. Incidentally, it explains why Descartes showed little interest in consciousness in general, as opposed to the consciousness exemplified by thinking that something is the case.[26] The only serious internal strain to which it is subject is its identification of intentionality, in its primary application, with linguistically expressible thought. The medieval Aristotelians had attributed intentionality to the higher brute animals, although they are incapable of such thought, on the ground that their behavior is explained by desire and sense-perception. Descartes of course dismissed this, on the ground that desire, itself an intentional state, is a *cogitatio*; and, in addition, one that can affect behavior only by way of further *cogitationes* — beliefs about what behavior is in the thinker's power.

The medieval Aristotelians had unwittingly anticipated objections like Descartes's by including in the sensory powers with which they credited the higher brute animals the power to form sensory images of behaviors open to them, and by conceiving animal desire simply as a non-rational impulse or tendency to do certain of the things the animal imagines as possible. They described the imaginative power of such animals as "estimative," because the imagined behaviors which their desires can move them to initiate may or may not be within their power.[27] A cat, having seen a mouse, having imagined itself pouncing on it, and having desired what it imagined, as a result pounces; but whether it lands on the mouse or not depends on how good it is at imagining what it can do — how good its estimative power is. Plainly, in describing the higher animals as imagining themselves as doing what they may or may not be able to do, as desiring to do what they imagine, and as succeeding in doing it or failing, the medieval Aristotelians have developed a non-materialist theory of what it is to be a higher animal, but a theory that is also non-Cartesian: one by which there occur in animals genuinely intentional processes that are not processes of linguistically expressible thought. Is it a true theory? That depends on how good an explanation of higher animal behavior it is. Those who deal practically with animals seem mostly to endorse it, as well as observers of wild animals.

For my own part, I accept the common opinion that the higher animals are capable of behavior that is intentional, although not intellectual, and therefore conclude that, in implicitly denying it, Cartesian dualism has committed an excess of simplification. But what follows? Not reductive materialism. And not a naturalistic hierarchy theory that fails to recognize the unique status of intentionality. Nor should Cartesian dualism be replaced by a three-substance theory. What is needed is a return to nineteenth-century common sense: the realm of the mental is the realm of actual and potential consciousness, and not all that is actually or potentially an act of consciousness is an intellectual act. But, as Brentano saw, it must be a return to nineteenth-century common sense with a Cartesian (and medieval) difference. The essence of actual and potential consciousness is *intentionality*, not privileged access or infallible knowability. Hence, since there is consciousness, nature is not everything. There is also Spirit.

California Institute of Technology

NOTES

[1] Grene (1974a), p. 7.
[2] *Ibid.*, p. 79.
[3] *Ibid.*, p. 91. The quotation from Merleau-Ponty is from Merleau-Ponty (1962), p. 428.
[4] Gueroult (1968), vol. 2, p. 274. (This and subsequent translations from the same work are mine.)
[5] *Ibid.*
[6] *Ibid.*, pp. 274–275.
[7] "Quid, quaeso, per Mentis, et Corporis unionem intelligit? Quem, inquam, clarum et distinctum conceptum habet cogitationis arctissime unitae cuidam quantitatis portiunculae?" Spinoza, *Ethica*, Part V, Praefatio (Spinoza (1925), vol. 2, p. 279). Spinoza's rhetoric is that of a true ghost in the machine dualist, scandalized by heresy, and unable to believe that the heretic can really mean what he has plainly said: that the union of mind and body cannot by understood in terms of the exact sciences, and that there is no clear and distinct conception of a thought closely united to a certain particle of quantity.
[8] Descartes (1964–1975), vol. 8-1, p. 23.
[9] 29 July, 1648. In Descartes (1964–1975), vol. 5, pp. 222–223 (translated by Kenny in Descartes (1970), pp. 235–236). The passage is discussed by Robert C. Richardson in Richardson (1982).
[10] Richardson (1982), p. 36.
[11] Grene (1974b), pp. 346–360; cf. also pp. 242–253 and 320–345.
[12] 'The Brain and the Soul', in Eccles (1970), pp. 151–175. The definitive statement of Popper's theory is Popper (1972). Cf. also Popper and Eccles (1977).
[13] Grene (1974b), pp. 349–350.
[14] *Ibid.*, p. 346.
[15] *Ibid.*, pp. 352–354.
[16] *Ibid.*, pp. 53–54.
[17] *Ibid.*, p. 234.
[18] *Ibid.*, p. 67.
[19] Descartes, *Traité de l'Homme*, in Descartes (1964–1975), vol. 11, p. 120. Cf. Descartes (1972), p. 4 and p. 2 of facsimile text.
[20] Descartes, *Meditationes* III, in Descartes (1964–1975), vol. 7, pp. 37, 40–41.
[21] Brentano (1973), pp. 88–89. Cf. Brentano (1969), pp. 50–54.
[221] Descartes, *Meditationes*, in Descartes (1964–1975), vol. 7, p. 78, lines 8–20 (*Med.* VI).
[23] *Ibid.*, pp. 38, 77 (*Med.* III, VI).
[24] *Ibid.*, p. 75 (*Med.* VI).
[25] Cf. Gewirth (1968), pp. 257–258. Cf. *Meditationes* III and *Principia Philosophiae* I, 68 (in Descartes (1964–1975), vol. 7, p. 35, (1964–1975), vol. 8-1, p. 33.)
[26] It should be evident that I cannot accept G. E. M. Anscombe's view that, in Descartes's work, "the notion of *cogitatio* . . . has turned into that of *consciousness* or *experience*" (Anscombe (1981), vol. 2, p. 60). On the contrary, he held all states of consciousness to be *cogitationes* because he held that a state of a substantial union of mind and body could be a state of consciousness only if it naturally impelled its possessor to make certain intellectual judgments. In the absence of such a connection

with judgments and their contents (*ideae*, in the having of which *cogitationes* consist), no sensation would be a state of consciousness.

[27] For a detailed references, see Mahoney (1982), pp. 606–607.

BIBLIOGRAPHY

Anscombe, E.: (1981), *Collected Papers*, 3 vols., University of Minnesota Press, Minneapolis.

Brentano, F.: (1969), *The Origin of Our Knowledge of Right and Wrong*, ed. O. Kraus and R. Chisholm, trans. R. Chisholm and E. Schneewind, Routledge and Kegan Paul, London.

Brentano, F.: (1973), *Psychology from an Empirical Standpoint*, ed. O. Kraus and L. McAlister, trans. A. Rancurello, D. Terrell, and L. McAlister, Routledge and Kegan Paul, London.

Descartes, R.: (1964–1975), *Oeuvres*, ed. C. Adam and P. Tannery, new ed., 11 vols., Vrin, Paris.

Descartes, R.: (1970), *Philosophical Letters*, trans. A. Kenny, Clarendon Press, Oxford.

Descartes, R.: (1972), *Treatise of Man*, ed. and trans. T. Hall, Harvard University Press, Cambridge, Massachusetts.

Doney, W. (ed.): (1968), *Descartes: A Collection of Critical Essays*, University of Notre Dame, Indiana.

Eccles, J.: (1970), *Facing Reality*, Springer, New York.

Gewirth, Alan: (1968), 'Clearness and Distinctness in Descartes', in Doney, 1968, pp. 250–277.

Grene, M.: (1974a), *The Knower and the Known*, 2nd ed., University of California Press, Berkeley.

Grene, M.: (1974b), *The Understanding of Nature: Essays in the Philosophy of Biology*, D. Reidel, Dordrecht, Holland.

Gueroult, M.: (1968), *Descartes selon L'Ordre des Raisons*, 2nd ed., 2 vols., Aubier, Paris.

Kretzmann, N., Kenny, A., and Pinborg, J. (ed.): (1982), *The Cambridge History of Later Medieval Philosophy*, Cambridge University Press, Cambridge.

Mahoney, E.: (1982), 'Sense, Intellect and Imagination in Albert, Thomas, and Siger', in Kretzmann, Kenny, and Pinborg, 1982, pp. 602–622.

Merleau-Ponty, M.: (1962), *The Phenomenology of Perception*, trans. C. Smith, Routledge and Kegan Paul, London.

Popper, K.: (1972), *Objective Knowledge: An Evolutionary Approach*, Clarendon Press, Oxford.

Popper, K., and Eccles, J.: (1977), *The Self and Its Brain*, Springer, New York.

Quine, W.: (1960), *Word and Object*, MIT Press, Cambridge, Massachusetts.

Richardson, R.: (1982), 'The "Scandal" of Cartesian Interactionism', *Mind* **91**, pp. 20–37.

Spinoza, B.: (1925), *Opera*, 4 vols., Carl Winter, Heidelberg.

ANTHONY N. PEROVICH, JR.

GENIUS, SCIENTIFIC METHOD, AND THE STABILITY OF SYNTHETIC *A PRIORI* PRINCIPLES

It is no secret that history has been a source of trouble for Kant. Not only have the favored theories which his critical philosophy in some sense aimed to support been superseded by others less congenial to his system, but also the historical perspective deemed necessary for appreciating the import of such developments has been denied to him. Marjorie Grene tells us these two points are related: the latter failing, an inability to recognize "the historicity of knowledge," is the source of Kant's attachment to certain scientific theories of his day as permanent achievements, to be added to but never replaced. "Human reason is, for him, essentially unhistorical. Once it finds its way, the road, as Kant says, can never more be missed. That is what we must deny. Science is no more exempted from historical development, from historical change — from temporality — than is any other human activity."[1] There can be no doubt that Kant saw his transcendental inquiries as in some way grounding the physical theory of his day; the consequences of his failure to appreciate the tentative status of that theory are less clear. Some modern writers, among them Professor Grene herself, have suggested that to appreciate that status is to recognize that "synthetic *a prioris* change."[2] On the contrary, I shall try to show first, that the route from stable synthetic *a priori* principles to uniformly progressive scientific theory is, as Kant conceives it, almost astonishingly straight-forward, second, that Kant's own disscussion of genius contains passages with sufficient parallels to accounts of science that assign a prominent role to change of theories to show how dependent Kant's theory of rectilinear scientific progress is, not on the stability of synthetic *a priori* principles, but on that unsatisfyingly straightforward picture of scientific activity which he combines with them, and finally, that there is no incompatibility between a doctrine of unchanging principles and an acceptance of the fact that history displays to the spectator a large number of competing and successive conceptual schemes.

The certainty and finality of science do appear to follow from the certainty and finality of the system of principles Kant believes himself

A. Donagan, A. N. Perovich, Jr., and M. V. Wedin (eds.), Human Nature and Natural Knowledge, 327–339.

to have established in the *Analytic*. Not, of course, that Kant believes that Newtonian physics is a deductive consequence of the synthetic *a priori* principles of the *Critique of Pure Reason*.[3] Nevertheless, he does maintain that, "although we learn many laws through experience, they are only special determinations of still higher laws, and the highest of these, under which the others all stand, issue *a priori* from the understanding itself."[4] This determination occurs only through the addition of empirical elements; for example, to derive Newton's first law Kant must appeal to characteristics of "the empirical concept of a matter"[5] as well as the second analogy: "In universal metaphysics there is laid down the proposition that every change has a cause; here [in mechanics] there is only to be proved of matter that its change must always have an external cause."[6] He establishes this latter proposition by pointing out that matter has only external relations; thus the second analogy shows that its changes must have causes, its nature as matter shows that these causes must be external.

There is consequently a gap between the synthetic *a priori* principles of the first critique and the laws of motion, a gap that can be bridged only empirically. As Professor Grene notes, "we have to add empirical content, even though not empirical procedures, to our original principles in order to derive such laws from them."[7] A further look at the remarks Kant makes regarding this movement from synthetic *a priori* principles to empirical laws of nature confirms the view that for him the scientific enterprise of bridging this gap is relatively straightforward once the proper approach has been discovered.

According to Kant the human mind begins its scientific inquiries in possession of a number of certain, necessary, and universally valid principles which also serve to delineate the general structure of the natural world. Only when consciously armed with these synthetic *a priori* principles do we succeed in entering on the highway of science which leads to the discovery of natural laws, but we learn these laws from nature herself, not from mere reflection on the principles. "Reason, holding in one hand its principles, according to which alone concordant appearances can be admitted as equivalent to laws, and in the other hand the experiment which it has devised in conformity with these principles, must approach nature in order to be taught by it."[8] While reason cannot prescribe the results of scientific inquiry, "the sure path of a science" is entered on rather easily once we learn to model those results on reason's own principles, as Newton's first law was seen to be framed in accordance with the second analogy.

Even physics, therefore, owes the beneficent revolution in its point of view entirely to the happy thought, that while reason must seek in nature, not fictitiously ascribe to it, whatever as not being knowable through reason's own resources has to be learnt, if learnt at all, only from nature, it must adopt as its guide, in so seeking, that which it has itself put into nature.[9]

The discoveries for which the scientist searches are thus found to "lie in the natural path of him who investigates and reflects according to rules,"[10] in particular according to those rules by which the understanding has given the law (or, at any rate, the general lawful structure) to nature.

Kant's attempt to illustrate such surefooted scientific practice turns out embarrassingly, however: as Professor Grene points out, "one of the three instances he refers to serves for modern historians of science precisely as the arch-example of the ways in which scientific conceptions may *mis*lead and may even lead to new truths *by* misleading."[11] Not merely the sure path of science but also one's faith in the sureness of that philosophy of science which seeks to explain and justify it face serious doubts once one encounters the claim that with the advent of the phlogiston theory "a light broke upon all students of nature."[12]

Where does the blame lie for such an untoward outcome? Historically minded thinkers have tended to locate the trouble in Kant's search for permanently valid synthetic *a priori* principles. Collingwood claims that the heritage of the Enlightenment kept Kant from appreciating the historically relative character of such principles.

The short historical perspective which Kant inherited from Voltaire was at this point his undoing, and made it possible for him to write what was in reality neither more nor less than a history of the absolute presuppositions of physics from Galileo's time to his own, without being aware that this was what he was doing, and in the mistaken belief that he was writing an account of the absolute presuppositions of any possible physics.[13]

Kant failed to understand that "absolute presuppositions change."[14] Professor Grene agrees. To draw the appropriate lesson from Kant's unfortunate appeal to phlogiston theory she appeals to Thomas Kuhn's *The Structure of Scientific Revolutions*[15] and its familiar notion of a paradigm. "But no theory in any field is ultimate. Paradigms, though they begin, as Kuhn has shown, in revolution, become the repositories of tradition, and they grow and change and decay as all traditions do. Or, in Kantian terms, synthetic *a prioris* change."[16]

Now Kuhn differs from Kant not merely in that he fails to see the uniform advance in the history of science that the latter describes. He also has an entirely different view of scientific practice. One may suspect that with a more Kuhnian view of scientific activity Kant might well have been led to abandon his vision of a rectilinearly progressive science. Might he have abandoned this view while retaining his notion that there are stable synthetic *a priori* principles which constrain all scientific theorizing? After all, the upshot of the discussion to this stage is that for Kant bridging the gap between such principles and the laws of nature is a remarkably straightforward and almost mechanical affair. Should this practice be replaced in Kant's eyes with one less straightforward and mechanical and more Kuhnian, that alone might have been enough to temper his claims for scientific progress without necessarily upsetting his view that synthetic *a priori* propositions do not change. These suspicions are confirmed by a glance at Kant's discussion of genius, whose activities bear considerable resemblance to those of Kuhnian science, whose productions are nonprogressive, and whose background includes the stable principles of the Kantian understanding.

"*Genius* is the talent (or natural gift) which gives the rule to art"[17] through the production of a work of art that is capable of serving as a model; thus according to Kant genius "is attributed to this artist only when his creation is exemplary, that is, when it is worthy to be imitated as an example (*exemplar*)."[18] There seem to be three different ways in which the work of the genius serves as a model or exemplar. First, his work may merely be copied [Nachmachen], deformities and all; this is mere aping.[19] Second, his product may inspire gifted pupils to imitation [Nachahmung]; this results in the founding of a school.[20] Third, because "genius is entirely opposed to the *spirit of imitation*",[21] the product of genius "is an example, not to be imitated . . . , but to be followed [Nachfolge] by another genius, whom it awakens to a feeling of his own originality and whom it stirs so to exercise his art in freedom from the constraint of rules, that thereby a new rule is gained for art. . . ."[22]

The idea of a paradigm or exemplar[23] is of course quite prominent in Kuhn's philosophy of science, and the parallels between Kuhn's and Kant's idea are numerous. There is little place in either art or science for mere copying, but the difference between the second and third ways in which the artist's product may serve as a model corresponds in several respects to Kuhn's distinction between "normal" and "extraordinary" science.

The function of a paradigm is clearly similar for both the Kuhnian normal scientific tradition and the Kantian artistic school. Kant of course does not claim that the pupil of the genius is engaged in "puzzle-solving" (which is how Kuhn describes the activity of the practitioner of normal science), but puzzle-solving for Kuhn consists largely in the not un-Kantian "ability to see a variety of situations as like each other".[24] Moreover, the Kuhnian scientist very frequently manifests this ability merely by directly "imitating" (to use the Kantian term) the paradigm: "Scientists solve puzzles by modeling them on previous puzzle-solutions, often with only minimal recourse to symbolic generalizations."[25] Reference to rules (which include "symbolic generalizations") may be helpful, then, but it is not necessary. "Normal science can be determined in part by the direct inspection of paradigms, a process that is often aided by but does not depend upon the formulation of rules and assumptions."[26]

As suggested, the story is much the same in the aesthetic realm. The pupil in the Kantian school likewise comes to see the world through the spectacles provided by the work of genius. Although the school may seek to derive rules from the artistic paradigm,[27] Kant recognizes that appealing to rules is an inadequate way to describe the activity of the successful pupil. As in Kuhn, the follower displays his talent by modeling his work on the paradigm, as something to be imitated without mere copying. Moreover, Kant admits that it is difficult to explain such imitation, precisely because (as in Kuhn) the influence of the work of genius is direct, not mediated through rules. Somehow the "ideas of the artist excite like ideas in his pupils"[28] just as paradigms inspire the normal scientist in the activity of puzzle-solving.

Ideas excite the pupil's mind, and ideas are what the genius produces; in Kant's remarks regarding them a number of parallels to Kuhnian extraordinary science emerge. Ideas are either rational or aesthetical. The rational ideas are a product of our reason and receive treatment in the *Transcendental Dialectic.* The aesthetical ideas are produced by the imagination. Rational ideas are concepts to which no intuition can be adequate; aesthetical ideas are intuitive representations of the imagination to which no concept can be adequate.[29] Both go beyond the cognitive limits (one set on the theoretical, the other on the empirical side) imposed on us by the understanding in its activity of gaining knowledge of the world (an activity which culminates in its trek down the sure paths of science). Rational ideas transcend the limits of all experience; they provide language which adequately describes no object of experience; conversely, aesthetical ideas offer sensible representations that "cannot be completely compassed and made intelligible by language."[30] These aesthetical ideas of the genius strive to represent another world beyond that conceived through the understanding. "The imagination . . . is very powerful in creating another nature, as it were, out of the material that actual nature gives it."[31]

How another nature comes to be envisioned is not something that can be explained, as Kant remarks in an extremely important passage that highlights what he sees to be the differences between scientific and artistic endeavors:

> Even if a man thinks or composes for himself and does not merely take in what others have taught, even if he discovers many things in art and science, this is not the right ground for calling such a (perhaps great) *head* a genius (as opposed to him who, because he can only learn and imitate, is called a *shallowpate*). For even these things could be learned; they lie in the natural path of him who investigates and reflects according to rules, and they do not differ specifically from what can be acquired by industry through imitation. Thus we can readily learn all that Newton has set forth in his immortal work on the *Principles of Natural Philosophy*, however great a head was required to discover it, but we cannot learn to write spirited poetry, however express may be the precepts of the art and however excellent its models. The reason is that Newton could make all his steps, from the first elements of geometry to his own great and profound discoveries, intuitively plain and definite as regards consequence, not only to himself but to everyone else. But a Homer or a Wieland cannot show how his ideas, so rich in fancy and yet so full of thought, come together in his head, simply because he does not know and therefore cannot teach others.[32]

The mathematician and the natural scientist have the advantage of recourse not only to intuition but also to definite rules to justify their results, whereas the artist "gives the rule to art" through the production of ideas that strive to represent what transcends the world of our understanding and the experience constituted by it.

There is much in Kuhn's account of the introduction of a new paradigm that is reminiscent of Kant's discussion of the aesthetical ideas. Just as the genius creates a new nature through his ideas, so through the introduction of a new paradigm the revolutionary scientist comes to practice his trade in "a different world".[33] And just as the genius creates the new nature out of the old, the flashes of intuition by which a revolutionary scientist's new paradigm comes to birth "gather up large portions of that experience [gained with the old paradigm] and transform them to the rather different bundle of experience that will thereafter be linked piecemeal to the new paradigm but not to the old."[34] Again, just as Kantian rational and aesthetical ideas elude capture in the experience and language dominated by the conceptual framework of the understanding, so the concepts and experience of postrevolutionary science are incommensurable with those of the prerevolutionary period.[35] Finally, just as the genius is unable to show that the connection of ideas in his artistic vision is in any way necessitated (while the scientist can make his discoveries "intuitively plain and definite as regards consequence, not only to himself but to everyone else"), so in Kuhn we find that "a law that cannot even be demonstrated to one group of scientists may occasionally seem intuitively obvious to another. . . . [T]he transition between competing paradigms cannot be made a step at a time, forced by logic and neutral experience."[36]

Given these parallel pictures of Kuhnian scientific activity and Kantian artistic activity, it is hardly surprising that neither is represented as progressive in the ordinary sense. Each in some way is engaged in the project of envisioning a new world, a vision which may prove enormously attractive to followers but which can never be shown to be necessitated. Moreover, each vision is destined to fall from favor through the production of new works of genius and new paradigms, with the consequent formation of new schools and research traditions, respectively.[37] Thus for Kant, new and different works of genius may appear, but they do not surpass the older masterworks: "art stands still at a certain point; a boundary is set to it beyond which it cannot go. . . ."[38] According to Kuhn the "[c]umulative acquisition of unanticipated novelties proves to be an almost non-existent exception to the rule of scientific development",[39] and over the long course of scientific development he sees innovations, but not progress, at least in ontological terms.[40]

The non-progressive artistic project is carried on against the background of unchanging synthetic *a priori* principles. Kant claims that the artistic imagination, powerful though it is in creating another nature, is nonetheless subject to a certain sort of constraint by the understanding, for in imaginative creation "we remold experience, always . . . in accordance with analogical laws. . . . "[41] This remark recalls Kant's description in the first critique of the fanciful activity of seeking "to frame quite new concepts of substances, forces, reciprocal actions, from the material which perception presents to us, without experience itself yielding the example of their connection. . . ."[42] These references to analogical laws, to new concepts of substances, forces, and reciprocal actions suggest that while the imagination is not bound by, say, the Analogies, it is constrained in its activity to produce analogical substitutes for them. Kuhn somewhat similarly remarks that, "[e]ffective research scarcely begins before a scientific community thinks it has acquired firm answers to questions like the following: what are the fundamental entities of which the universe is composed? How do these interact with each other and with the senses?"[43] Yet Kuhn's remark is even more suggestive in so far as it implies that there are higher level constraints operating on all conceptual framworks, a point to be returned to below.

Synthetic *a priori* principles thus make their presence felt in imaginative activity, but in a different way than in scientific activity. In the latter such principles serve as a particular conceptual scheme, to be specified and articulated into scientific theory in the manner indicated earlier. In the former they serve a double role: they are still the necessary conceptual scheme for the actual world, but they also serve more generally as embodiments of conditions which must be satisfied if any imaginatively constructed world is to be a world at all. Kant's position seems to be that only, say, the Analogies apply to the world of experience, but when we seek to go beyond that world in imaginative creation we are still limited to analogies of Analogies.

Thus Kant has no difficulty in reconciling the stability of synthetic *a priori* judgments with the existence of change of and variety in the worlds created in the imaginative realm because these latter are not the relatively direct consequences of the former (as is the scientific theory of the actual world, according to Kant's view of science) but rather alternative realizations of general conditions which they incorporate. But it is his view of artistic rather than scientific practice that more closely approximates to a position such as Kuhn's that recognizes the "historicity of knowledge" in scientific endeavors. Hence, there is no difficulty in concluding that with a rather different and more plausible view of scientific activity (one which in his way he did a good deal to anticipate) Kant could subscribe both to a doctrine of stable *a priori* principles and to a view in which revolutionary changes of theory and non-uniform progress are prominent features of the history of science.

Substituting theses regarding science for corresponding claims he makes about art produces a demand for further modification in Kant's position, for what in art are merely imaginative creations analogous to the genuine world of experience turn out in science to be contending accounts of reality. This means that the double role of Kant's principles noted above must be separated: stable synthetic *a priori* propositions may set out conditions to be satisfied by any conceptual scheme at all, but the claim must thereby be relinquished that they themselves represent in addition a particular, privileged conceptual scheme. If the Kantian doctrine of unchanging synthetic *a priori* principles is to be maintained along with a recognition of the sometimes radical changes of theory witnessed in the history of science, it must also be compatible with the existence of quite different, specific conceptual schemes underlying those different theories. This sort of Kantianism has sympathies both with the point of view of those seeking out the one, necessarily unique conceptual scheme (i.e., stable principles)[44] and with the point of view of those who find history filled with alternative conceptual schemes (i.e., varying specifications of the conditions laid down by the stable principles),[45] by drawing a distinction between particular conceptual frameworks and the more general structural principles requisite for anything to count as such a framework at all. (Indeed, one would think that for there to be alternative conceptual schemes there would also have to be some common conceptual constraints on what counts as a conceptual scheme in general.)

That Kant did not himself clearly draw such a distinction is one of the main reasons why his project is often seen as so parochial by members of the second, historicist group mentioned above.[46] Yet thinkers clearly inspired by Kant, such as Stephen Körner (whose work on categorial frameworks represents a notable attempt to spell out the general conditions to be met by any conceptual scheme[47]), have recognized this distinction and traced it, at least implicitly, to Kant himself: "Kant's transcendental deduction in particular not only sharpens some of the points made in his metaphysical exposition of his categorial framework, but provides together with its preliminaries many of the notions needed for a logico-philosophical inquiry into the structure of categorial frameworks."[48] That the foundation for drawing such a distinction and thereby reconciling the project of discovering stable synthetic *a priori* propositions with a recognition of the tentative and changing nature of our scientific theories can be found in Kant (though in his discussion of the artistic rather than the scientific endeavor) has been the concern here, and the identification and justification of such unchanging principles as Kant sought, and an examination of the theoretical constraints imposed by them, remain untreated.[49]

Why then, to return to our initial concerns, are changes in theories thought to undermine the stability of synthetic *a priori* principles? There are two reasons: first, Kant's remarks about scientific activity suggest a straightforward and tight connection between his principles and the particular physical theory of his day; second, some of his putative principles seem historically conditioned and parochial. It can now be seen that neither of these reasons is satisfactory. The same account of science (viz. Kuhn's) that so greatly emphasizes changing theories also presents an alternative picture of scientific activity which even when conjoined with the idea of unchanging synthetic *a priori* propositions need not lead to rectilinear progress, as Kant's comparable account of artistic activity seems to suggest. And the distinction between unchanging principles and their alternative historical instantiations, while blurred in Kant, permits even the admission of alternative conceptual frameworks along with the stability in principles that Kant sought.

Hope College

NOTES

1 Grene (1966), p. 145.

2 *Ibid.*

3 Although some have suggested as much. See, e.g., Popper (1968), p. 192: "Since Kant believed that it was our task to explain the uniqueness and the truth of Newton's theory, he was led to the belief that this theory followed inescapably and with logical necessity from the laws of our understanding."

4 Kant (1968c), pp. 147–148, A 126. (References to Kant will follow the page number of the English translation with the volume and page numbers of Kant (1968a), or in the case of the *Critique of Pure Reason* with the page number of the first or second edition.)

[5] Kant (1970), p. 6; vol. IV, p. 470.
[6] *Ibid.*, pp. 104–105; vol. IV, p. 543.
[7] Grene (1966), p. 126.
[8] Kant (1968c), p. 20; B xiii.
[9] *Ibid.*, B xiii–xiv.
[10] Kant (1968b), p. 151; vol. V, p. 308.
[11] Grene (1966), p. 144.
[12] Kant (1968c), p. 20; B xiii.
[13] Collingwood (1972), p. 249.
[14] *Ibid.*, p. 48.
[15] Kuhn (1970).
[16] Grene (1966), p. 145.
[17] Kant (1968b), p. 150; vol. V, p. 307.
[18] Kant (1978), p. 123; vol. VII, p. 224.
[19] Kant (1968b), p. 162; vol. V, p. 318.
[20] *Ibid.*
[21] *Ibid.*, p. 151; vol. V, p. 308.
[22] *Ibid.*, p. 162; vol. V, p. 318.
[23] Kuhn introduces the term ‘exemplar’ in the “Postscript” to (1970) to distinguish paradigms in the sense of “concrete problem-solutions” (what corresponds in Kant to “the product of a genius”) from the other components belonging to what Kuhn calls the “disciplinary matrix”. See p. 187. The discussion here shows, I think, that Kuhn’s claim to novelty in regard to the idea of an exemplar (p. 208) is incorrect.
[24] Kuhn (1970), p. 189.
[25] *Ibid.*, pp. 189–190.
[26] *Ibid.*, p. 44.
[27] Kant (1968b), p. 162; vol. V, p. 318.
[28] *Ibid.*, p. 152; vol. V, p. 309.
[29] *Ibid.*, p. 157; vol. V, p. 314
[30] *Ibid.*
[31] *Ibid.*
[32] *Ibid.*, pp. 151–152; vol. V, pp. 308–309.
[33] Kuhn (1970), e.g., pp. 111, 121, 150.
[34] *Ibid.*, p. 123.
[35] *Ibid.*, pp. 141, 158ff. Of course, the regulative function of the rational ideas and the necessary inadequacy of any sensible representations to the latter and of any concept to the aesthetical ideas offer important respects in which Kantian ideas are not merely incommensurable with the concepts and experience of the Newtonian understanding. The favored status of the latter keeps Kant from seeing metaphysics as the adumbration of new scientific theory. (Cf. Feyerabend (1968), p. 39.) An attempt to link the regulative function of the ideas with their potential for serving as an alternative model for the construction of cosmological theories is made in Schäfer (1972).
[36] Kuhn (1970), p. 150.
[37] Of course, the sources of this fall from favor will be rather different in the two cases, showing that the similarities between the two must not be overestimated.
[38] Kant (1968b), p. 152; vol. V, p. 309.

39 Kuhn (1970), p. 96.
40 *Ibid.*, p. 206.
41 Kant (1968b), p. 157; vol. V, p. 314.
42 Kant (1968c), p. 241; A 222, B 269.
43 Kuhn (1970), pp. 4–5.
44 E.g., Strawson (1963) and Harrison (1974).
45 E.g., Professor Grene, Collingwood, Kuhn, and Feyerabend (1975).
46 See the remarks of Collingwood above and of Feyerabend (1962), p. 29.
47 Körner (1974). Körner, however, sees no necessity even in the demand that some categorial framework or other be employed (see p. 13); still, the evidence is not entirely unambiguous (see also his remarks regarding the necessity of classificiation for thought, p. 1).
48 *Ibid.*, p. 72.
49 For some preliminary attempts see Körner (1974) and Perovich (1978).

BIBLIOGRAPHY

Beck, L. (ed.): (1972), *Proceedings of the Third International Kant Congress*, D. Reidel, Dordrecht, Holland.

Collingwood, R.: (1972), *An Essay on Metaphysics*, Henry Regnery Co., Chicago.

Feigl, H. and Maxwell, G. (eds.): (1962), *Scientific Explanation, Space, and Time*, Minnesota Studies in the Philosophy of Science, vol. 3. University of Minnesota Press, Minneapolis.

Feyerabend, P.: (1962), 'Explanation, Reduction, and Empiricism', in Feigl and Maxwell, 1962, pp. 28–97.

Feyerabend, P.: (1968), 'How To Be a Good Empiricist — A Plea for Tolerance in Matters Epistemological', in Nidditch, 1968, pp. 12–39.

Feyerabend, P.: (1975), *Against Method: Outline of an Anarchistic Theory of Knowledge*, New Left Books, London.

Grene, M.: (1966), *The Knower and the Known*, Basic Books, New York.

Harrison, R.: (1974), *On What There Must Be*, Clarendon Press, Oxford.

Kant, I.: (1968a), *Kants Werke*, 9 vols., Walter de Gruyter, Berlin.

Kant, I.: (1968b), *Critique of Judgment*, trans. J. Bernard, Hafner Publishing Co., New York and London.

Kant, I.: (1968c), *Critique of Pure Reason*, trans. N. Kemp Smith, St. Martin's Press, New York.

Kant, I.: (1970), *Metaphysical Foundations of Natural Science*, trans. James Ellington, The Bobbs-Merrill Co., Inc., Indianapolis and New York.

Kant, I.: (1978), *Anthropology from a Pragmatic Point of View*, trans. V. Dowdell, rev. and ed. H. Rudnick, Southern Illinois University Press, Carbondale and Edwardsville, Illinois.

Körner, S.: (1974), *Categorial Frameworks*, Basil Blackwell, Oxford.

Kuhn, T.: (1970), *The Structure of Scientific Revolutions*, International Encyclopedia of Unified Science, vol. 2, no. 2, 2nd enl. ed., University of Chicago Press, Chicago.

Nidditch, P. (ed.): (1968), *The Philosophy of Science*, Oxford Readings in Philosophy, Oxford University Press, Oxford.

Perovich, A.: (1978), 'Conceptual Schemes, Incommensurability, and the Concept of an Object', Ph.D. Dissertation, University of Chicago.
Popper, K.: (1968), *Conjectures and Refutations: The Growth of Scientific Knowledge*, Harper and Row, Publishers, Inc., New York and Evanston.
Schäfer, L.: (1972), 'The Restrictive and Proliferative Function of Kant's Regulative Ideas', in Beck, 1972, pp. 480–485.
Strawson, P.: (1963), *Individuals: An Essay in Descriptive Metaphysics*, Doubleday and Co., Inc., Garden City, New York.

RICHARD RORTY

SHOULD HUME BE ANSWERED OR BYPASSED?

Marjorie Grene is among the most vigorous contemporary exponents of what she has called "the philosophical reform central to the thought of Bergson, Whitehead, Collingwood, Heidegger, and Merleau-Ponty. That is the reform which would make *time* the central category of metaphysics."[1] The line of thought which leads to this reform goes roughly as follows:

(1) Hume is a reductio ad absurdum of a traditional, but inadequate, picture of the relation between the knower and the known.
(2) The trouble with Hume's picture is that it is an attempt to use Newtonian spatial relationships as the grid upon which to construct a model of knowing; it is an attempt to create an epistemology on the model of mechanics.
(3) Such an epistemology presupposes a reductionist account of the known — e.g., a reduction of the living to the non-living.
(4) So we need a new epistemology based upon a non-Humean "theory of the person" — one which emphasizes the *activity* and the *personal* character of the knower rather than reducing the knower, as Hume did, to an assemblage of interacting sense-contents.
(5) The way to get such a theory is to look to biological phenomena rather than non-living phenomena.
(6) But such a theory of the person must be embedded within a larger metaphysical view so that "the structure of tacit knowing" may be "mirrored in the structure of comprehensive entities because they both mirror the metaphysical structure of time".[2]

Since writing *The Knower and the Known*, Grene has altered some emphases and scaled down some demands, but I think it fair to say that the general line of thought presented in that book, the line I have just attempted to summarize, still dictates her approach to philosophical problems. My own approach to these problems differs radically

A. Donagan, A. N. Perovich, Jr., and M. V. Wedin (eds.), Human Nature and Natural Knowledge, 341–352.

from hers, and in this essay I should like to exhibit the similarities and differences between our approaches by focusing on Hume.

Grene thinks that Hume's attempt to formulate a theory about the relation between Knower and Known was a good thing to do. She criticizes neo-Thomists who invoke *noûs* to dismiss Hume's problems, Hegelians who invoke *Geist* for the same purpose, and Popperians who "dismiss the whole inductive context" by saying "None of these systems bears any relation to Hume's problem, and however one answers Hume, one cannot, or ought not to, ignore him."[3] She applauds Kant's insistence that "we see every act of intelligent apprehension including even perception, as an achievement which has always already transformed what it has received."[4] She thus joins a long line of writers who agree with Kant that we need the sort of account of knowledge which Hume attempted, who think that Kant explained why Hume failed, and who want to follow Kant's lead in doing Hume's job better. Like Whitehead and the others whom she cites, she criticizes Kant for an inadequate account of the knower. We need, she says, to replace Kant's "pin-point 'I' of the transcendental analytic" with "a fully structured, fully functioning, human person."[5] The theory of such a person will be made possible, she thinks, by a metaphysics which takes time as central.

Contrast this view with another — one which takes the definitive criticism of Hume to have been Reid's rather than Kant's. If one begins with the so-called "Scottish answer to Hume" rather than the "German answer," one will be dubious about the very idea of a "science of man" of the sort which Hume envisaged. One will see that idea as made possible by the Cartesian "theory of ideas," the positing of inner, psychical, equivalents for Boylean corpuscles. Once one drops the notion of "ideas" one will not be tempted by Kant's addition of an active faculty of spontaneity to Hume's assemblage of associating sense-date. One will not be tempted by the idea of a theory of the relation between Knower and Known, nor by a "theory of the person" which will shore up such an account, nor by a metaphysics which will shore up that theory of the person. One will do just what Grene deplores: ignore Hume rather than answering him.

Grene criticizes Ryleans like Passmore for "pretending that there is no problem at all" about the nature of the mind, and for relying on "ordinary language" with no "metaphysical pangs."[6] By contrast, I sympathize with Ryle's feeling that we need less metaphysics rather than a new and better metaphysics. I think that the idea that we should have a theory that shows how "our minds are once more at home in a world ontologically suited to contain them"[7] is an unfortunate remnant of the same hope which motivated Hume — that we can somehow see human knowledge as itself "natural," as integrated within some large context. Following Sellars, I see this hope as a confusion of explanation with justification. It is an attempt to transcend the pragmatic day-to-day problems of revising our beliefs, in the hope that by rising above details we can see the process of revision *en bloc*. On my view, this is an attempt to jump outside our skins, or, more precisely, outside our historical situation. So whereas Grene turns from physics to biology in the hope of a biologistic metaphysics, I am inclined to turn from the sciences to history and the arts for an understanding of our existence as historical and creative beings. As I see it, the great split within contemporary philosophy is not between those who want the temporalizing "reform" of which Grene speaks and neo-Hobbesians who want a "naturalized epistemology" which is consciously and deliberately reductive, but between both and those who want no metaphysics because they see the metaphysical urge itself as the cause of Hume's problems — problems which should be dissolved rather than solved.

Before attempting to lay out this large issue, however, I shall try to fill in some details of how one will view Hume if one prefers Reid to Kant. Consider first a criticism of Hume (and of British Empiricism generally) which Reid and Kant share — namely, that Hume runs together the propositional and non-propositional, the mere occurrence of a sensation and the acquisition of knowledge. The criticism is very widespread. Thus Jonathan Bennett says of Locke that his "use of 'idea' to cover both sense-data and concepts" "embodies his substantive mistake, shared with Berkeley and Hume and others in the empiricist tradition, of assimilating the sensory far too closely to the intellectual."[8] Wilfrid Sellars says of Locke, Berkeley, and Hume that they all "take for granted that the human mind has an innate ability to be aware of certain determinate sorts — indeed, that we are aware of them simply by virtue of having sensations and images."[9] Sellers notes that "the idea that sense is already a cognitive faculty, acts of which belong to the intentional order, the order of signification" helps bring about what he elsewhere calls "that cancer at the heart of modern philosophy": the Cartesian doctrine that the contents of mental acts are "the immediate and primary objects of knowledge."[10] About a hundred years earlier, T. H. Green claims that Locke effectuated

> the fundamental confusion on which all empirical psychology rests, between two essentially distinct questions — one metaphysical, What is the simplest element of knowledge? the other physiological, What are the conditions in the individual human organism in virtue of which it becomes a vehicle of knowledge?[11]

Only through this confusion, Green says, can "any idea . . . be described as an 'impression' at all . . . A metaphor, interpreted as a fact, becomes the basis of his [Locke's] philosophical system."[12]

Still a hundred years further back, this same criticism is made, for the first time, by Reid. The confusion involved in thinking that an immediate "presentation to sense" is somehow also the knowledge of the truth of a singular proposition is described by Reid as the confusion between feeling and perceiving. Reid says

> When I am pained, I do not say that I perceive pain, but that I feel it, or that I am conscious of it. Thus *perception* is distinguished from consciousness . . . .
>
> The observations made on this word would have been unnecessary, if it had not been so much abused in philosophical writings upon the mind. Although this offense is not chargeable on Mr. Hume only, yet I think he has carried it to its highest pitch . . . under the name of impressions, he comprehends all our sensations, passions and emotions. . . . I believe no English writer before him ever gave the name of a perception to any passion or emotion.[13]

So much, then, for the criticism of Hume on which Kant and Reid can agree. In Reid this criticism takes the form of saying that we should not try to get back behind the propositional to the non-propositional. In Kant, however, it takes the form of offering a more complex account of the genesis of the propositional out of the non-propositional. Kant thinks that we should, in Grene's words, "see every act of intelligent apprehension, including even perception, as an achievement which has already transformed what it has received," and that we can do this by explaining the genesis of conscious judgments out of an unconscious manifold of non-physical entities. For Reid the notion of such entities is just the original Cartesian sin all over again. Reid abjured the whole idea of explaining the nature of knowledge. He breaks decisively with Hume on the question of whether the "presence to mind" of our ideas enables us to develop a science of man without violating Newton's injunction to eschew hypotheses — an injunction which both Hume and Reid cite with approval. He thinks that the attempt to explain how we are conscious is based on a mistaken analogy between the intellectual and the physical worlds, an analogy which tempts us to postulate Rube Goldberg mechanisms. Thus he says:

All the systems of perception I have mentioned . . . [are] all hypothesis, made up of conjectures and suppositions, without proof. The Peripatetics supposed sensible species to be sent forth by the objects of sense. The moderns suppose ideas in the brain, or in the mind. Malebranche supposed, that we perceive the ideas of the Divine mind. Leibnitz supposed monads and a pre-established harmony. . . . In like manner, the Indian philosopher supposed that the earth is supported by a huge elephant, and that the elephant stands on the back of a huge tortoise. . . . Men then only begin to have a true test in philosophy, when they have learned to hold hypotheses in just contempt . . .

I believe no man is able to explain how we perceive external objects, any more than how we are conscious of those that are internal. Perception, consciousness, memory and imagination are all simple and original powers of the mind, and parts of its constitution. For this reason, though I have endeavored to show that the theories of philosophers on this subject are ill-grounded and insufficient, I do not attempt to substitute any other theory in their place.[14]

To justify the claims that we should hold hypotheses in contempt, Reid uses the standard example of Newton being content to state the laws of gravitation without attempting to discover a medium for gravitational force. To justify his claim that we have as little idea how we perceive inner objects as external, he anticipates Ryle and Wittgenstein in saying "if we should admit an image in the mind, or contiguous to it, we know as little how perception may be produced by this image as by the most distant object."[15] The latter point, that inner representations are of no use in explaining *consciousness* of themselves or of anything else, is essential to the appreciation of the former. Only the Cartesian confusion between occasions of perception and objects of perception will explain why such explanations were thought possible, and why men should have thought themselves capable of forming hypotheses on the matter.

From Reid's point of view, the Kantian attempt to explain knowledge by assuming the transcendental standpoint would have seemed just one more attempt to do the impossible by one more Rube Goldberg device — the interactions of transcendental faculties. Reid would happily appropriate all of Strawson's polemic against "the imaginary subject of transcendental psychology." Sellars is, in this respect, Reid's most faithful follower. Sellers makes clear that there is no room for a middle ground between neurology (or a place-holder for neurology such as Fodorian cognitive psychology), the discipline which will tell us as much as we shall ever know about "how consciousness is possible" and "how the mind works", and the sociology of knowledge (the Reid- and Wittgenstein-like reminders of our ordinary ways of phrasing and judging knowledge claims). Whereas Reid and Sellars, like Grene, counter Hume by discounting the whole idea of "what the mind has received" and rejecting a primordial non-physical relation between knower and known, Green, like Kant and Russell, keeps this notion and worries about how to describe its "transformation." The study of that transformation provides the occasion and the terrain for the construction of an epistemology, a "theory of the person," and a project of metaphysical reform.

On the view I am expounding there is a hidden link between, on the on hand, such metaphysical reformers as Bergson, Whitehead, (early) Heidegger, Merleau-Ponty and Grene and, on the other, proponents (such as Fodor) of "cognitive psychology" as the legitimate heir of philosophical empiricism. This latter school of thought is often denounced by Grene and her companions-in-arms (such as Hubert Dreyfus and Charles Taylor) as "reductive." But both sides presuppose just what Reid and Sellars reject — the notion that there really are introspectible "inner states" of the sort Descartes and Hume tried to isolate, and that we can do well what Hume did badly by describing these states better.

Grene thinks that Fodor-like projects — which take nonliving computers as models for living knowers — are the result of an inadequate, reductive, "spatial" metaphysics. This should be replaced by a more adequate and less reductive temporalistic metaphysics, one which will do more justice to what these inner states are really like. On my view, this issue between reductionism and anti-reductionism is artificial. What is wrong with Fodorian cognitive psychology is not that it is inadequate to explain knowledge. As a place-holder for neurology, cognitive psychology may do all that an ideal neurology could do by way of explaining information processing. But explaining information processing is not what people hoped to get out of a "science of man." For that science should be relevant to the justification of our beliefs and not just to their acquisition. Fodor and Grene share what Sellers calls the urge "to analyze epistemic facts without remainder into non-epistemic facts — a mistake of a piece with the so-called 'naturalistic fallacy' in ethics."

For Sellars, confusions between sense and intellect, particulars and propositions, physiological mechanism and rational inference, are simply instances of the general Cartesian confusion between description and explanation on the one hand and justification on the other. In a much-quoted passage, Sellars says "In characterizing an episode or a state as that of *knowing*, we are placing it in the logical space of reasons, of justifying and being able to justify what one says."[16] The common fallacy in Grene and her "reductionist" enemies, on this view, is the attempt to reduce justification to description by assuming that the ways in which we justify our fellows' utterances and other actions will somehow be better understood by gaining a deeper understanding of how human beings work — a metaphysical understanding in Grene's case and an empirical understanding in Fodor's.

Grene and Fodor both hope to get a science of man by positing a special sort of "non-epistemic fact," one which will explain the "logical space of giving reasons" in terms of something more primordial. Fodor prides himself on the large ontological tolerance which permits him, despite the strictures of Skinner and Wittgenstein, to countenance "inner representations" and a "language of thought." He condemns the inadequacies of attempts to do psychology with more limited ontological resources. Similarly, Grene thinks that one must eschew Hume's Newtonian paradigm and start from biology in order to get a theory of the person which will give us "the mediation we are seeking between knower and known."

Grene implicitly agrees with Fodor that empiricism posed the right problem when it asked how we get from impoverished and unimaginative internal representations to rich and interesting ones. Fodor expresses this as follows:

> . . . the core of the empiricist theory of perception is inevitable. In particular, the following claims about the psychology of perception seem to me almost entirely true
>
> 1. Perception typically involves hypothesis formation and confirmation.
> 2. The sensory data which confirm a given perceptual hypothesis are typically internally represented in a vocabulary that is impoverished compared to the vocabulary in which the hypotheses themselves are couched.[17]

In other words, there must be room — if we are to have an account of human beings which lies between neurology and sociology — for an analysis of the process which gets us from bits of impoverished data, like red and rectangular sense-impressions, to daring hypotheses like Sellars's parody: "By the methods of Mill, that must be what Mother calls 'red'!"

Fodor's (1) is not itself an empirical result of psychological inquiry; rather it plays the same constitutive role for cognitive psychology as "motion typically involves interaction between bodies" did for mechanics. For Fodor, the reason why people who — like Skinner — refuse to press on from Hume to Kant have gotten nowhere is that, in Grene's words. "Hume's genetic principle [that ideas are merely copies of impressions] . . . overlooks the validating and organizing contribution of mind to knowledge".[18] But Fodor's way of updating Kant — finding a discipline which applies equally to people and machines — seems to Grene just the wrong way of taking account of this contribution. For machines are not proper models of the "fully structured, fully functioning, historical human person, with all its levels of significance, and its ineluctable contingency."[19]

Grene's criticism of quasi-mechanical accounts of human knowledge is, I think, misplaced. She thinks of them as failing to give an account which ought, in principle, to be available, and which will become available once our philosophical thinking is sufficiently de-spatialized. On my view, however, the mistake is in asking for an *account*, or a *model*, of historicality and contingency in the first place. Phenomena such as these (or "creativity", or "evaluative character") are not the sort of thing one ought to ask for an account of. No new vocabulary will make intelligible what had previously been puzzling. Nothing could satisfy the demand which Grene is making. The jargon of *Process and Reality* or *Being and Time* will fare just as badly as that of MIT-style psycholinguistics. For any such jargon — metaphysical or empirical — will be "reductive", in the sense that it will make the historical and the contingent a special case of the universal and necessary. Any *model* of the historical and contingent will have failure built in — for the urge which Grene shares with the other temporalizing metaphysicians she mentions is an urge which model-building cannot satisfy. One will only get, as Heidegger sneeringly said about the notion of "value" as a ontological category, the self-destructive attempt to "round out" the ontology of "things" by sticking in extra quasi-things — Rube Goldberg devices such as transcendental synthesis, prehensions, the lived body, the structures of pre-linguistic consciousness which constitute the *Lebenswelt*, etc., etc.

If one wants to preserve the historicality and contingency of human existence in one's philosophical reflections, one should not try for a better, more comprehensive, richer, "theory of the human person." Rather, one should forget "the human person" as such, and reflect on, e.g., us late-twentieth-century inheritors of the European intellectual tradition, on the differences between us and human persons of other places and times, real (historical) or imaginary (dreamed up by novelists). Among the temporalizing reformers Grene mentions, I think Collingwood comes closest to doing what she wants done. For reflection on historiography is better calculated to produce a grasp of our historicality and contingency, and of our fullness of structure, than reflection on biology. Further — and this brings me back to Sellars and Reid — such reflection is what takes one from explanation to justification, from a grasp of the processes involved in the acquisition of beliefs to the "logical space of reasons." The historicality of our existence is best expressed by the fact that we take the assertion of p to justify that of q because we are creatures of our time. We take a given vocabulary as a sensible one in which to formulate our philosophical reflections because we have read certain books which our ancestors had not, and have not read certain books which our descendants will read. We inhabit the logical space of reasons we do because we are not just any old information-processing mechanisms, but mechanisms programmed by a given culture. From this historicist point of view, the difference between a human being so programmed and a computer so programmed is of no interest. We and our computers share precisely the same historicality and contingency. The computer can be programmed to be creative, as we program our children, but the limits of the creativity of either is a function of the bounds of sense which the culture imposes on both.

Hume's *épatant* assimilation of our information-processing abilities to "the reason of animals" lies in the background of Bergson's and Grene's biologizing — the idea that if we could recapture what we share with the lower orders we would be in a better position to understand our own humanity, less in danger of confusing ourselves with machines. The same idea lies behind Nagel's and Searle's attempt to find something we share with bats but not with computers.[20] This proto-Darwinian, naturalizing, vitalistic side of Hume is as important to Grene as the Newtonian, mechanistic, Rube Goldberg, side is to Fodor. But neither brutes nor computers will help us get a sense of a "fully structured human person" — both attempts at analogies with our own condition are equally "reductive." The sense of "something missing" when we try to think of ourselves as animals or as machines is, I think, correctly identified by Grene as the inability to preserve a sense of historicality and contingency — where "contingency" is just the realization that we might have been in a different logical space of reasons than we are, that there is no unwobbling pivot of rationality which links our justifications of belief and action to the atemporal. But this sense cannot be overcome by software breakthroughs or by further information on phylogenetic or ontogentic development. It is simply the metaphysical pathos which comes from watching our language go inauthentic on us — watching our brave new *Existentiale* freeze into one more set of ontic categories, our latest attempt at temporalizing our metaphysics reveal itself as one more set of impotent metaphors drawn from (where else?) the *spatio*-temporal world.

If we fellow Reid in saying that we cannot get back behind the common sense principles which make our logical space of reasons the space it is, then we can draw a line from Reid to Hegel, bypassing Kant. We can view Hegelian historicism as the idea that logical spaces are themselves historical products. All that then remains to arrive at the position common to Derrida and the later Wittgenstein — the one which Grene describes in "Life, Death, and Language"[21] — is to give up the notion of *Seinsgeschick* common to Hegel and Heidegger, the notion that there is some larger "totalizing" context beyond the causal contingencies of history. The Humean notion of a science of man, extended down through systems of naturalistic metaphysics such as those of Dewey and Whitehead, is the distinctively Anglo-Saxon form of the quest of such a context. "Historicism," in the objectionable sense which Karl Popper has given the term — as the quest for an immanent teleology in history — has been the distinctively German form. The moral of Reid is that we should stop thinking of knowing as a relation between two sorts of things — the knower and known — and thus give up both sorts of attempt to ground our logical space of reasons in a metaphysics. Only then, it seems to me, will we be able to do what Grene wants done: to treat "history as our basic phenomenon and draw the world of science, as a limiting case, out of historical reality."[22] This aim, which I fully share, seems to me to conflict with her project of following up on Hume and Kant by discovering "the structure of tacit knowing [which] is mirrored in the structure of comprehensive entities because they both mirror the metaphysical structure of time."[23]

I have not argued for the crucial premise of this paper, viz.: that the development of a new metaphysical vocabulary will always leave unsatisfied just those motives which drive us from Hume to Kant to Grene. Without such an argument, my claim that we would do better to follow Reid with regard to knowledge, Ryle and Dennett in regard to minds and computers, and Collingwood in regard to history, remains unsupported. In the space at hand, all I can do is gesture in the direction of Heidegger's progress. He started by developing an ontological jargen which would attempt to get Being and Time back in touch, but ended by writing a History of Being which helps illustrate the historicity of our own notions of knowledge, mind, and history. I think that Heidegger moved in the right direction, but I know no way to show this save by detailed examination of Bergson's, Whitehead's, the early Heidegger's, and Grene's own efforts to capture "the metaphyscial structure of time."

It is a trifle ironic that, having first heard Heidegger's name from Grene — thirty years ago — I should now find myself adopting a more-Heideggerian-than-thou stance in regard to her work. The Heidegger whom Grene told me of then (at a time when I was a starry-eyed young Whiteheadian) looked good to me for quite a while. Indeed, I still think that *if* one is going in for metaphysics at all, then something like the temporalizing project which Grene outlined in *The Knower and the Known* is the sort of metaphysics to try for. Certainly if one thinks that Hume asked the right questions then one will want to go on to Kant, and past Kant to the sort of more "fully structured" analysis of the activity of the knower which Grene demands. If one wants to hook up this analysis with science (as one then should) one will go in the direction Grene has taken — towards biology. But if one thinks — for the Reidian, Sellarsian, and Wittgensteinian reasons I have outlined in this paper — that Hume asked the wrong questions, one will find congenial the later Heidegger's account of why he asked those questions. This will lead one to think that to get a grasp of what it is that makes us human persons we should turn to history and the arts, rather than to the sciences.

University of Virginia

NOTES

[1] Grene (1974), p. 243.
[2] *Ibid.*, p. 252.
[3] *Ibid.*, p. 114.
[4] *Ibid.*, p. 153.
[5] *Ibid.*, p. 157.
[6] *Ibid.*, p. 242.
[7] *Ibid.*, p. 243.
[8] Bennet (1971), p. 25.
[9] Sellars (1963), p. 160.
[10] *Ibid.*, p. 44 and 42.
[11] Green (1885), vol. 1, p. 19.
[12] *Ibid.*, p. 11.
[13] Reid (1895), vol. I, pp. 9–10 (*Essays on the Intellectual Powers of Man*, Essay I, Chapter I).
[14] *Ibid.*, p. 239 (Essay II, Chapter XV).
[15] *Ibid.*, p. 220.
[16] Sellars (1963), p. 169.
[17] Jerry Fodor (1978), p.44.
[18] Grene (1974), p. 153.
[19] *Ibid.*, p. 157.
[20] I discuss Nagel's and Searle's strategies in Rorty (1982).
[21] In her (1976).
[22] Grene (1974), p. 243.
[23] *Ibid.*, p. 252.

BIBLIOGRAPHY

Bennett, J.: (1871), *Locke, Berkeley, Hume*, Clarendon Press, Oxford.

Fodor, J.: (1978), *The Language of Thought*, Thomas Y. Crowell Co., New York.

Green, T.: (1885–1888), *Works*, ed. A. L. Nettleship, 3 vols., Longmans, London.

Grene, M.: (1974), *The Knower and the Known*, 2nd ed., University of California Press, Berkeley.

Grene, M.: (1976), *Philosophy In and Out of Europe*, University of California Press, Berkeley.

Reid, T.: (1985), *Philosophical Works*, ed. Sir William Hamilton, 2 vols., James Thin, Edinburgh.

Rorty, R.: (1982), 'Contemporary Philosophy of Mind', *Synthese* **53**, pp. 323–348.

Sellars, W.: (1963), *Science, Perception and Reality*, Routledge and Kegan Paul, London.

PART V

REFLECTIONS

MARJORIE GRENE

IN AND ON FRIENDSHIP

It is fitting to begin with the beginning of these essays. Berger writes of love and friendship in a way that sets my thoughts rambling. How his argument bears — as it does, oddly, far as my own interests have strayed from professional ethics — on my particular philosophical concerns, I shall remark in due course. Meantime, I must say thank you, first to the editors: I have done enough editing myself to know what a chore that is, and am most grateful to them, more deeply than I can say, for all the trouble they have taken in such an odd and unnecessary cause. And thank you, also, to all the friends who have contributed to this wide-ranging collection. "Thank you" is a dull, weak word, but I don't know a better. Suffice is to say that for someone who was exiled from her profession for some fifteen or twenty years (depending on whether one counts as a return to one's profession being a pariah in the U.K. — 1959–1965 — or only being a maverick in the U.S. — 1965-), it is most moving that so impressive an aggregate of colleagues should have been willing to lend their names, and thoughts, on this occasion.

When I look over the list, in fact, I find that only three of the contributors were friends, or even acquaintances, before I came home to the United States in 1965. Richard Rorty visited the farm I shared with my then husband in Cook County — yes, really, it was a farm, a shabby, back-breaking, debt-ridden farm, and it was in Cook County — probably in the spring of 1951. Although (probably because I haven't understood what he was saying, nor, perhaps, he what I was saying — see below) I have found myself in profound disagreement with him in recent years, I have always remembered and in a way fed philosophically on a remark he made on that occasion. The history of modern philosophy, he said, in effect, has consisted in knocking, one by one, the props out from under the traditional cosmology (ontology? metaphysics?). That may sound obvious enough. Yet it's not only (what I already knew, having learned it from Gilson's early work and from Cartesian texts themselves) the odd house-cleaning that starts with Descartes that's in question, but, on the positive side, for instance,

A. Donagan, A. N. Perovich, Jr., and M. V. Wedin (eds.), Human Nature and Natural Knowledge, 355–368.

the way that Hume, abandoning substance itself as well as God and mind, still clings pertinaciously to the remaining prop — looking to us perhaps more like a piece of driftwood — of a uniform (and even uniformly pretty decent) human nature. It's the same undermining of sub-structures, moreover, that necessarily brings with it the ineffectiveness and incoherence of classical philosophy of science. Not that I'm putting all those words in Rorty's mouth; we disagree about history as much as we do about contemporary philosophy. But I'm just recalling, in these rambling reflections, the beginning of a friendship and what, in the many years between its beginning and its renewal, I made of that long past conversation.

The second of my three (relatively) old friends among the contributors is Alasdair MacIntyre, whom I met first in Manchester when I was Michael Polanyi's research assistant in 1957–1958 and MacIntyre, I believe, was a lecturer in philosophy of religion. Then, when I went to Leeds, first (for my sins) as Senior Research Fellow in the Institute of Education, looking (allegedly) into presuppositions in the teaching of biology (the nadir of my strange off-and-on career), MacIntyre was lecturer in philosophy. And when the next year (in 1959) I myself fell luckily into a temporary lectureship, I remember rejoicing that I would be MacIntyre's colleague, since I was sure I would learn a lot from that experience. I did. Chiefly the impossibility of teaching in competition with so persuasive a pedagogue. I was teaching Descartes to Kant — my absolutely favorite academic occupation — and MacIntyre had drummed into the students the thesis "There is no philosophy from St. Thomas to Hegel." Every time I came close to convincing the class that something in *my* period was worth attending to, I could just see their dear little minds ticking over: "But MacIntyre says there *is* no philosophy from St. Thomas to Hegel." A labor of Sisyphus. It's strange, by the way, that while MacIntyre's recent work seems to me much more intelligible, and even convincing, than what he was publishing in those days (especially his concept of "practice," which should be fundamental, in my view, to a sound philosophy of science), he says he gets it all from Hegel, while I find in Hegel only a Great Betrayal and take my inspiration from a Kantian starting point. A colleague reading my rereading of the transcendental unity of apperception in "The Paradoxes of Historicity" remarked: "Marjorie, you know that's not Kant." Maybe not, but it's decidedly not Hegel: the World Mind sending the passions into the field to do battle for it — beastly Teutonic professorial swindle! Kant's honest effort is the place to start, I still believe, not nineteenth century preenings and pretendings. And incidentally, although I don't quite follow Perovich's attempt to import the *Third* into the *First Critique*, I am, of course, moved by any philosophizing that truly begins from Kant. My effort has been, I suppose, to bore from within from the *First Critique*, or perhaps to build extensions of and for it. These turn out to be quite irresponsible meanderings — I was told I might say what I liked! But to list the third old friend included here: I must have met Rom Harré before I came home in 1965. I don't remember where. Perhaps through our mutual friendship with Michael Polanyi, perhaps at a philosophy of science meeting. My earliest distinct recollection of an encounter with Harré — although I know we must have met earlier — is a visit to Davis when he was at Reno, possibly in '65–66. I remember dining with him at one of those dreadful darkened California restaurants, and there was Rom opposite me pronouncing with characteristic intensity, "Powers!" Only he and Whitehead could claim to derive so much insight from poor old Locke's unexplained, indeed, inexplicable concept. Since then I've met Harré at numerous places from Colorado Springs to Salzburg to Ithaca. We always share philosophical interests, but he always seems to fly off at tangents from where I think he's going — unless of course it's I who think tangentially. I'll return later, however, as in other cases, to the point at which our philosophical interests do seem to converge.

Before I settle down to at least slightly less helter-skelter jottings, however, I must acknowledge especially those who have talked here about my work, and while doing that, I'll venture some remarks, also, about some of the other essays that touch on figures and topics in the history of philosophy that I have taught, thought about and, partly, written on, even though the arguments in question do not impinge directly on any work of mine.

The most gratifying references to my work, I find, are those that mention in passing some remark of mine about a particular thinker (Perovich: Kant and Wedin: Aristotle). That's exactly the way I'd like to be remembered by colleagues and students, as someone who thought about the great thinkers of the past and encouraged others to think about them. I'm a teacher, not a philosopher. (A Berkeley colleague once remarked to me, crackling down the tie line, "Of course, you and I have upper division minds.") Accordingly, I can't take my own philosophizing very seriously.

Besides, when other people take it seriously, they appear (to judge from the present sample) to mistake my meaning. Thus, both Donagan's and Rorty's essays confirm my view that I am not a philosopher, but a teacher of the history of philosophy. People tell me I "explain things clearly," and I do hope I have made some clear, if elementary, statements about Aristotle, Descartes, Kant, Hume, Sartre, Heidegger, even possibly, Merleau-Ponty. (The last is much the hardest, since, as Rorty would put it, Merleau is an *edifying* thinker: If one is edified, so one is; but to write about the source of such edification is tough. One is transformed by the experience — so? On the contrary, it's so much easier to write about philosophers one doesn't believe!) But obviously when I mix in too much of my own opinions, as in *The Knower and the Known*, I do it badly. I don't make myself clear at all. Rorty says I want to found philosophy on biology, somewhat as his scientistic heroes (who rank high among my anti-heroes) would reduce philosophical reflection to acquiescence in some empirical discipline, whether physics or naive behaviorism. Horrors. What should be, must be, fundamental to any true counter-Cartesian reform in thinking is the recognition of the primacy of *life* — not in the spirit of Dilthey's secret inner somethings, but in something like the way Plessner, for instance, works through from positionality to eccentric positionality or Merleau from the lived body (a *live* body, not a piece of Cartesian extension) to his very de-Cartesianized *cogito*. Granted, he denied even that (tacit) *cogito* in *The Visible and the Invisible*. Maybe that was right, but I'm just trying to suggest some sources for the kind of other-than-linguistic, other-than-scientistic turn I have in mind. The point is: This has nothing special to do with the biological sciences but with the simple fact that we are *alive*. Polanyi's *Personal Knowledge*, of course, was an effort in the same direction, but I'm afraid that, as Körner once remarked, his rhetoric was hopeless, and in the end Polanyi, too, was too perversely subjectivistic. Anyone who defended "consciousness" (a concept Heidegger rightly banned from philosophical discourse) was Good and anyone else was a Wicked Materialist. But Polanyi did try to ground knowledge in life, in human existence as one form of animal existence, and certainly with more subtlety and insight than selfish-gene reductivists have shown. And it's that kind of grounding that's needed — an *archē* of thought that rejects *both* Cartesian separate, mathematicizing mind *and* Cartesian mathematicized matter as the double, and doubly barren, beginning of ontology and that reflects about fundamental problems of knowledge, of action, of being, *within* our place in nature and, within *that* place, in culture. Nor is it a question, as Donagan believes, of softening Cartesian austerity to admit that animals suffer and even, in a sort of way, think. No. What is at fault is the *radical* incapacity of the Cartesian method, and the Cartesian ontology, to interpret the very first beginnings of our being-in-the-world. There is simply no way to patch up that disastrous metaphysical methodology and produce anything but a *Nullpunktsexistenz,* whether Sartrean or Quinine. As far as intentionality goes, I would have thought (and that seems to be the upshot of Donagan's APA presidential address) that St. Thomas had got that pretty straight — why is the Cartesian twist necessary? The trouble is, though, that we can't any longer start, most of us, either, with a Christianized Aristotle. We need to start, within the tradition, indeed, but now: in our place in history, where *we* stand in the conversation that began with Thales. But not, please, both Alan Donagan and Lesley Cohen, with the *cogito*! Surely, Sartre has shown definitively by his own horrid example where that takes us. Indeed, I may remark along this partly-personally-reflective rambling path, it was the *cogito*, in a way, that led me as an undergraduate to abandon philosophy and major in science. I was told I *had* to believe that argument, and I couldn't. Four-year olds still believe they're the center of the universe; by the time one's seventeen, surely one knows better. (Now of course I know it's not an argument, but I still can't accept it.) When I came, then, three years later, to read the *Meditations* in Werner Brock's seminar in Freiburg, although I never dared raise my voice in the official meetings of the group (German exegesis of a Latin text was a bit frightening), I did remark, as some of us strolled away after a session (I can hear my own voice, a cry on a dark street): "Was bin ich ohne meine Umwelt?" That's still my leading thought. I am, each of us is, *one* expression of a world that is, in infinite dimensions, infinitely more than ourselves. It's that awareness in Spinoza, by the way, that makes him so great a thinker. If he was a reformed rake, or would-be rake, as Frankfurt seems to be arguing, so was St. Francis. It's the transcendence of self, not the self-centered starting point that matters. It is not Descartes-Roo: "Look at me thinking," but the giving of oneself to what is more than oneself that constitutes humanity: Heidegger's *Sich-vorweg-Sein,* Polanyi's "commitment".

Nor, by the way, Alan Donagan, other things being equal, would I have started any argument, ever, from that ridiculous dualist-trialist

Eccles account, or have fudged up "three worlds" of my own — really a presentation of Plessner, in any case, had I not been asked to give the Research Lecture at Davis, an occasion on which one wants some gimmicks, especially visual aids, to entertain a very general (if not largely agriculturalist) audience. And the Eccles approach was so fatuously impossible, that I thought: if he can diagram "the human condition," so can I (with much help from my then colleague Mel Beal, who, good analytical University of Rochester thinker that he is, found my argument as funny as Eccles's, but who likes to draw diagrams, bless him. In color, they were smashing!).

Indeed, much of my writing has consisted of occasional pieces, even books, that I have written because I was asked to. Why, apart from mere lightmindedness? Let me "explain" this — I can't excuse it — by a little longer narrative digression. In 1931 I was sent to Freiburg as a German-American Exchange Fellow and sat at the feet of Heidegger. That chance, plus my own laziness and cowardice, moved me from science to philosophy. Then in 1932, when Heidegger was on leave, I went, as seemed inevitable, to Heidelberg to Jaspers, and in 1933 (after witnessing the so-called *Nationale Erhebung*) I returned home, as I had planned to do, to study for a doctorate at Radcliffe. In '35 I composed a dreadful slapdash dissertation on "Existenz," because, though by then I disliked intensely all German *tiefere Bedeutungen*, it was still a new topic in this country, and I wanted to get out fast and start teaching — in the middle of the Depression! And of course when I finished they said, "Goodbye, you're a bright little girl, but nobody gives jobs to women in philosophy." So there I was, trying to fend for myself. The first job I landed, in 1936, after writing to 129 colleges and universities (there were no "Jobs for Philosophers" in those days, only an old-boy network, and I mean "boy"), was in a prep-school-turned-junior-college in Illinois, where the new president, a Harvard education M.A., hired me because after asking a Boston friend to interview me to make sure I wasn't Jewish (my name was Marjorie Glicksman in those days), he received in one and the same mail my dossier from Wellesley or Radcliffe saying I was Jewish, and a letter from his friend, saying, "Don't be silly, she's a Nordic type." So he hired me, for $50 a month and room and board, to see what I looked like. At Harvard, in reaction against existentialism, I had been attracted by the Vienna Circle, and in 1937–1938 I managed to get to Chicago as a half-time assistant in order to participate in Carnap's research seminar (this time because I *was* a woman: they said, "We're supposed to be a liberal department, and we've never hired a woman, so let's try this one." Then I was an instructor for six years, and then zero.) As someone who knew a tiny bit of zoology, however, I couldn't take the pure extensionalism of early Carnap (or Hempel, who was his assistant). When I tried to explain to Hempel that one couldn't handle biological properties adequately with extensional logic, he said in his firm but gentle way: "But we only explain what we can say *pre-cise-ly*." So I married an Irish classicist as a refutation of positivism — not an adequate philosophical ploy, I'm afraid. And many decades later, Elliott Sober has at last said something about the place of intensional discourse in biology. It was about time. (Again, you see, I'm not a philosopher; I didn't do it, I just knew it had to be done. And in fact I may have known it partly becaue of having studied epistemology with C. I. Lewis!) Meantime, though, by the time I had (I thought) settled into teaching, I had found historical questions, and specially questions about the seventeenth and eighteenth centuries, sufficiently entrancing to want to continue with them. I had given a Descartes seminar in my last year of teaching and had made a first stab at going through Adam and Tannery. (All my notes were lost by the time we sold out in 1952 and moved to an Irish farm. I've always suspected they were used for a Muscovy duck with 120 ducklings that we housed upstairs before the sale because rats were destroying them — the ducklings, not the notes — under the barn. Come to think of it, though, they — the notes, not the ducklings — were on 5 × 8 cards, not very good bedding. But if my dreadful Kierkegaard manuscript, written during a post-doc in 1935–1936, went for that end, it was probably good riddance). When I was forcibly retired in '44, then, I toyed with the idea of working on Descartes and, of all things, seventeenth century theology. But of course I didn't do it. Farming and trying to raise a family and generally struggling with assorted crises, I hadn't the gumption. (On my own definition of humanity, it seems, I turn out to be rather less than human. Still, I suppose one can be committed to one's family, and, alas, one has to be committed to the land. Philosophy came fourth in my hierarchy of obligations in those days.)

But if I hadn't the wherewithal to go on with Descartes, where, given that I was sick of existentialism, did all that continental stuff come from? It's simple, really. After I lost my job, I thought I should do anything more or less respectable I was given a chance to do. I loathed existentialism, but since I'd studied with two of the supposed representatives of that "school," I was asked to write about it when it came into fashion after the war and did so, the first time, with little understanding. Every decade or so I was asked to do something in that field and every time I finished by declaring: "Never again!" It wasn't until I read Merleau-Ponty for the first time – in 1961, after his death – that I realized I *had* learned something from that tradition. Then, of course, later, when I taught Continental German or Continental French Philosophy in Davis and, especially when I had the good fortune to direct summer seminars on Heidegger and Merleau in 1975, '76 and '79, I did come to understand them better. As most of us know, teaching texts or working through them with one's peers is the only way to learn. Further, my work in philosophy of biology, which did indeed bring me back at long last to the philosophy of science, was occasioned at the beginning by Michael Polanyi's request that I look up "heresies" in evolutionary theory. Then in Davis I worked up a course in philosophy of biology and a number of my better papers, as I think, have sprung directly from that source. But all in all, I can say with Mr. Polly, "I never planned my life. I happened, things happened to me. It's so with (almost) every one."

Now, however, in my liberated old age, I can turn back to something like my (relatively) early preoccupations, and think about the seventeenth century. So I'm glad to find some Cartesian, as well as Spinozistic, problems in the preceding pages. Janet Broughton's argument I find wholly convincing; it echoes my own uneasiness about Cartesian causality. Harry Frankfurt's Descartes, like his Spinoza, looks very different from mine. While of course acknowledging the importance of Frankfurt's work on the eternal truths, I can't believe that that thesis, constant as it was, held an *absolutely* primary place in Descartes' doctine. It must bear *some* subordinate relation to the overall Cartesian project of initiating mathematical physics. Gary Hatfield suggests it was a spinoff from the move against final causality. That sounds plausible, but I wish I could put the connection more satisfactorily. Perhaps further study of the first part of Marion's *Sur l'Ontologie blanche* . . . may shed some light on this. It's a nice problem. So is the issue raised by Ruth Marcus about Spinoza's "ontological" argument. In fact her essay raises a more fundamental question: Is the "ontological" argument epistemological? Kant surely didn't take it that way, or he wouldn't have christened it "ontological."

The question seems to be, as in Descartes' *Fifth Meditation*, does God's essence entail His existence? And the proof of *Meditation Three*, of course, is *not* of that character, nor does it belong to the class of "ontological" arguments just because it *is* clearly epistemological. The problem is: Does Anselm's "that than which no greater can be conceived" entail an epistemological moment, or doesn't it? "To be conceived" is to be grasped, if not by a knower, at least by a possible knower. True. But isn't that phrase a stand-in for "ens realisimum", the Being with the greatest possible reality? How central is the "conceptual" component? On the other hand, if the "ontological" argument is in fact "epistemological," isn't Spinoza also moving from the concept of God or substance to His (its) existence? And, indeed, his argument clearly needs the concept of attribute, and that *is* epistemological — that which the intellect perceives of substance as constituting its essence. So it seems to me, either the onological argument is not epistemological, and Spinoza's argument is as ontological as the others, or it is epistemological, and his is, too.

So much is said, as a few glancing remarks at a few of these papers, in friendship. Now *about* friendship. (Love I leave, as I have done for a good half-century, to the *Symposium* and the *Phaedrus*.) Both Fred Berger's argument and the existence of this collection are the occasion for the brief reflections that follow. Berger, like a good professional, constructs a couple of examples, not, thank heaven, science fiction examples, but still rather strained constructs, not quite human, such as might occur in a Sartrean short story, and from these he proceeds to show how utilitarianism would have to be modified so as to account for their oddity. Better, perhaps: they're good utilitarian stories; but they make us uneasy: that's not how we think of friendship or love. So the theory has to be patched up to take account, somehow, of what we do mean by these concepts. Now this seems to me a strangely abstract and devious way to think about these problems. Such a subject-matter, like every other philosophical subject-matter that concerns some aspect of human experience, needs for its starting point the context of human *being*: being-in-a-world, what it is to become a person ("person" being understood, to quote my own shorthand, as a personification of nature through participation in a culture). Only *within* an understanding of what-it-is-to-be-human (the τί ἦν εἶναι of humanity, which is, of course, the product of evolution and is itself always, fundamentally and pervasively, a history), only within the frame of such understanding, could one reflect fruitfully about the special mode of being-human that is friendship. Rules for maximizing benefits are all very well in their place. J. R. Commons, I'm told, initiated the Progressive tradition in Wisconsin on a positively Benthamite basis. But friendship (like love) is a way of being, and no head-counting individualism is going to get near it. Nor am I, of course. Like the Luneburg swine (as Kierkegaard put it), I can only throw problems over my nose for other people to solve. Briefly, though, I'll try, or try to make a start. Every human being is, has his (her) being in the world — in the wholeness of a culture built in, and by means of, nature. But no human world is entirely monolithic. Steve Gould, for instance, is a paleontologist, a scientific journalist, a Marxist and a baseball fan. He participates, more and less actively, in a number of (MacIntyrean) practices. These are all subworlds, small dwelling places within which, as our lives take place (the pun is intentional), we become the persons we are. Assimilating the standards, the excellences, of such practices, we are formed and transformed as scientists, writers, political agents, sports fans, and so on. But we also exist, *are* what and who we are, in other kinds of subworlds that are not structured as practices, and one of these is friendship. What kind of structure is this? Plainly it's a form of being-with, but so are practices. Is it the aim that's different? In a way, since friendship does not have an aim (I have always found repellent Aristotle's account of the utility of friendship). To be in friendship is to *be* more fully than one is alone, an enrichment of one's human reality (of one's freedom, in some version of the tradition of positive freedom). Thus friendship is an intrinsic good, an end in itself. It is not, like love (which I am not talking about, but see the sacred texts already referred to), a completion of a lack in one's being; it is not ec-static. It's warming, like having on one's sheepskin coat and Canadian boots in sub-zero weather. But of course such comfort can be self-deceptive. In cases of chatty acquaintanceship, even in many seeming friendships, one may be, as Sartre would say, just setting up guardrails against anguish. There is such a phenomenon; I know it well. But I also know Sartre is wrong in believeing that only the single, isolated for-itself can (if it can, and in Sartrean terms, it can't) achieve freedom. One becomes, and is, a person, responsible, and in that sense a free agent, only in a world, in a complex tissue of worlds, and more fully so in love and friendship than in isolation, however splendid.

That's a bare beginning. Maybe this is the kind of juncture at which, as Rorty would like, philosophy should give way to literature, or take on the look of literature. Maybe we need a *Symposium* or a *Phaedrus* of *philia*. Positively, there's no more I can say. But these papers have given me, when I think about them, some small clues about what friendship is *not*, although even these are not wholly clear or well thought out. For one thing, it's not a matter of communication or of agreement on issues – the latter at least not up to a point. It's clear from the very case in point, for instance, that Berger and I disagree in our fundamental approaches to philosophy. That doesn't keep us from being friends, or even, as he remarks, and I confirm, from helping one another in our philosophical reflections, divergent though they are in source, style and outcome. On the other hand, I could not, for example, I think, be a friend of a creationist. Yet superficially I could communicate with such people. I could say: "You are not fit to teach biology," and he (I hope not she) would have a pretty good idea what I meant. Does friendship then depend upon communication at some "deeper" level? Just guessing, from a long life's experience, I would say: sometimes, but sometimes not. People can be good friends, real friends, with very little genuine communication. Perhaps they have to hold some beliefs in common. I don't know (although they do, in some way, have to exist together in some shared segment of their lives). And when it's very fundamental, perhaps, as Merleau-Ponty says of perception, communication may become a kind of analogue of communion. I don't know that either. But my point is twofold: friendship does not rest in any unambiguous way either on agreement (shared beliefs) or on communication (mutual understanding of messages); and there is a complicated relation, also, between whatever role it is that each of these two factors plays. The present collection well exemplifies both these absences and these complexities.

These are slight remarks indeed, compared to what might be said. It should be kept in mind, however, whatever may be said, that, as friendship is a form of being-with, so being-with in turn is of course a form of being-in. What Polanyi called "indwelling" or Merleau-Ponty "being-in-the-world" (I know the phrase comes from Heidegger, but Heideggerian ontology is still too one-sided and, as Dorothea Frede's argument clearly shows, too ambiguous in its import, to serve for the *arche* I am trying to suggest) is, as I see it, the necessary ground, at this juncture in our intellectual history, for any fruitful reflection on any such theme as this.

I started here from Berger's essay and the difference of my philosophical perspective from his. On the other hand, there are also some essays in this collection which, although the way they put it differs in each case from the others and in all the cases from my own, yet seem to me to share in some measure the approach I have been proposing for reflection on philosophical problems, certainly in the areas (all areas, in Kant's view!) in which we are concerned with some aspect or aspects of human experience. In the cases of Burian and Gale, it is a question of shifting the focus of philosophy of science from a wholly abstract logical reconstruction, cut loose from the real family of practices that constitutes the sciences, as well as from the reality that is both the target and the milieu of scientific research, to a more careful and concrete study of special cognitive problems within the sciences. In Hacking's and MacIntyre's essays we have two beautiful studies of how particular human realities come to be. They are *philosophical* essays, however, not pieces of sociology, any more than Burian's or Gale's analyses fall into sociology of science. We need to *philosophize* about human reality from *within* human reality — since we are human and have no other place to be, but also since, being human we are moved to think about what kind of being that is (recall Socrates at the beginning of the *Phaedrus*).

To start from within human reality, however, is to start, not only within the real, natural world, but also within a culture. There is no extra-cultural way to be human. Indeed, in comparative statements like Hacking's or MacIntyre's, the analysis is doubly culture-carried. For the writer begins, as a late twentieth century western thinker, within a given culture and sets out to look, in relative detachment (a virtue acquirable, through appropriate discipline, within that culture), at an aspect of that society beside that of another time or place. I have no intention of embarking here on an analysis of the difficult question: How from within one society, with its standards, practices, aspirations, one can study another? The hermeneutical literature by now embraces numerous analyses of this special question. But I do, I fear, need to point, if very briefly, to a problem that sometimes seems to arise in the philosophical community when such accounts are presented. If human problems, problems about health and sickness, about behavior patterns, about the cognitive claims of science, are presented in their inherence in a culture, in history, rather than in the sterile milieu of logical reconstruction or of pure "argument", some people are wont to ask: Have we not succumbed to relativism? Perhaps Foucault, on whose concept of an *episteme* Hacking bases his analysis, and so Hacking himself, would say "Yes". Alas, I do not understand Foucault, and, begging his pardon, I want to take Hacking, in the present context, on my terms. *I* don't see what is "relativistic," in any sceptical or conventionalist sense, about looking at human reality where it *is* — within the unity of nature and in turn within the unity of a culture itself contained in nature. Was Aristotle a relativist because he said human beings are political animals? And he knew as well as we do that there is more than one *polis.* Yet when attempting to present a position in philosophy of science that I have tentatively labelled "comprehensive realism," I have sometimes found that my colleagues, Establishment colleagues of course, find this view relativistic if not (horror of horrors) pragmatist in its bent. I was a little shocked, here, too, to find Burian (in an earlier version of his essay) excusing himself for seeming to move "further from realism," just because he is urging us to face the *realities* of scientific life rather than to stay with the more precise abstractions of the formerly orthodox tradition, in which one had to argue, from some kind of sensationalist or observational or operational base, that one was entitled after all, *mirabile dictu,* to propose that scientific discourse refers to real objects or events — the position I like to call "thin realism". Knowing a little of Burian's academic history, I can well undestand how he comes to say that, and the remarks that follow are certainly not meant as a critique of his major argument. On the contrary, the essays of Burian himself, as of Gale, Hacking and MacIntyre, it seems to me, help us along the path to a much sturdier realism, in which we recognize that people, scientists, psychiatrists, patients, those apt for suicide in modern Europe or in traditional Japan, that all of us are real live animals seeking, within the medium of a culture and/or a subculture (a culture that itself is real, that really makes us what we really are) to orient ourselves in our world or in some aspect of that world. Reality is rich enough to permit many roads of access. If that is relativistic, so be it. But it is not a question of quasi-magical "ways of world-making," or of conventions that sharply separate the seeker from what he seeks.

Admittedly, from the point of view that demands *arguments* for realism, the position I am suggesting is a question-begging one. But, remember, logical positivism, euphemistically renamed logical empiricism, began by sweeping away what it called "pseudo-problems" from philosophy. And if there ever was a pseudo-problem, the "problem" of "scientific realism" is it. Scientists are people trained in certain practices, carried in their endeavors by a curiosity, even a passion, to understand how something in nature *really* works. (It's here, by the way, of course, that I found Harré's essay welcome. Four cheers for realism!) Thus as I (and many others) have remarked elsewhere, if scientific techniques become so refined as to produce artefact, they have failed as science. Now if logical positivism could push aside the fundamental problems of ontology as pseudo-problems, let us, please, put aside, at long last, *their* pseudo-problem. Granted, "realism" and "relativism" may not, in some sense of either term prove to be contraries; but let's begin squarely with realism as our starting point and find out, if we must, what kind of qualified "relativism" (I'd rather say, pluralism) we need within that frame. After all, communication between friends *is* a good, even if it is not an indispensable major ingredient of friendship. It would be good, therefore, to have a little more of it about this puzzle among philosophers in general, since philosophy is, on the whole, a friendly profession. So there's yet another problem I leave it to my friends to resolve. Why does more realism look like less, or the defeat of relativism appear relativistic? One is reminded of Descartes' remark that Bourdin had accused him of being a sceptic because he had refuted scepticism.

These have been very ill-organized reflections. So be it. I confess to being, in spite of myself, a "continental" writer in philosophy, and continental philosophy, as everyone knows, is poor in argument. My remarks were intended, once more, as a very patchy and roundabout way of saying: Thank you, my friends, thank you. Perhaps I may add, also , that although I am of course grateful in the first instance, on this occasion, to all those who have made the existence of this volume possible, I am grateful also to the many other friends, students, colleagues, not only in philosophy, who have so much enriched my past two decades. Bless you all.

Ithaca, New York
March, 1984

THE PROFESSIONAL ACTIVITIES OF MARJORIE GRENE

HONORS

American Academy of Arts and Sciences; Phi Beta Kappa; Sigma Xi; Pi Delta Phi
German-American Exchange Fellow (Freiburg), 1931–1932
Alice Freeman Palmer Fellow (Wellesley), 1935–1936
Lucy Martin Donnelley Fellow (Bryn Mawr), 1960–1961
Faculty Research Lecturer (University of California, Davis), 1970–1971
Inaugural Philip Merlan Lecturer (Scripps College), November 1970
President, Pacific Division, American Philosophical Association, 1971–1972
Humanities Institute Award (University of California, Davis), 1972
Arthur Skemp Memorial Lecturer (University of Bristol), 1973
Member, Council for Philosophical Studies, 1973–1978
Vice-President, Metaphysical Society of America, 1974–1975; President, 1975–1976
Fellow, American Academy of Arts and Sciences, 1976–
Fellow, American Association for the Advancement of Science, 1977–
Honorary Degree (Doctor of Humane Letters), Tulane University, May 1980

ACADEMIC EMPLOYMENT

Professor Emeritus, University of California, Davis
Professor, 1965–1978, University of California, Davis
Cowling Professor of Philosophy, 1984–1985, Carleton College, Northfield, MN
Gannett Professor of Humanities, 1983–1984, Rochester Institute of Technology, Rochester, NY
Distinguished Visiting Professor, Spring 1983, Vassar College, Poughkeepsie, NY

A. Donagan, A. N. Perovich, Jr., and M. V. Wedin (eds.), Human Nature and Natural Knowledge, 369–374.

Visiting University Professor, Spring 1982, Boston University, Boston, MA
Visiting Professor, April 1981, University of Waterloo, Ontario, Canada
Visiting Professor, Fall 1980, Yale University
Visiting Professor, Spring 1980, University of California, Berkeley
Visiting Professor, Fall 1979, Douglass College (Rutgers University)
Visiting Professor, Spring 1979, Temple University
Mellon Visiting Professor of Humanities, Fall 1978, Tulane University
Visiting Professor, Summer Semester, 1973, University of Goettingen
Visiting Professor, Fall 1972, Boston University
Visiting Professor, 1967–1968, University of Texas, Austin
Lecturer in Philosophy, 1960–1965, Queens' University, Belfast
Temporary Lecturer in Philosophy, 1959–1960, University of Leeds
Senior Research Fellow in Education, 1958–1959, University of Leeds
Research Assistant, 1957–1958, University of Manchester
Visiting Professor, Summer 1951, Northwestern University
Instructor in Philosophy, 1938–1944, University of Chicago
Assistant in Philosophy, 1937–1938, University of Chicago
Member, Teaching Staff, 1936–1937, Monticello College

OTHER PROFESSIONAL SERVICE

Director, NEH Summer Seminar, 1975, 1976,and 1979.
Director, Council for Philosophical Studies Summer Institute on Biological and Social Perspectives on Human Nature, Colorado College, Summer 1977.
Director, NEH Seminar in Residence for College Teachers, 1977–1978.
Director, Council for Philosophical Studies Summer Institute on Philosophy of Biology, Cornell, 1982.

THE PUBLICATIONS OF MARJORIE GRENE

BOOKS

Dreadful Freedom: A Critique of Existentialism (University of Chicago Press, 1948). Reissued as *Introduction to Existentialism,* 1959; Midway Paperback, 1984.
Heidegger (Hillary House, 1957).
A Portrait of Aristotle (University of Chicago Press, 1963). Reissued as Phoenix Paperback, 1967; Midway Paperback, 1979.
The Knower and the Known (New York: Basic Books, 1966; University of California Press, paperback, 1974). Reissued by University Press of America, 1983.
Approaches to a Philosophical Biology (New York: Basic Books, 1969).
Sartre (Franklin Watts, 1973). Nominated for National Book Award. Reissued University Press of America, 1983.
The Understanding of Nature: Essays in Philosophy of Biology (Reidel, 1974).
Philosophy In and Out of Europe (University of California Press, 1976)
Descartes in *Philosophers in Context,* series ed. S. Körner (Harvester Press, 1985).

BOOKS EDITED

From Descartes to Kant (with T. V. Smith) (University of Chicago Press, 1940). Reissued as two-volume paperback under the title: *Descartes to Locke; Berkeley, Hume, Kant.*
The Anatomy of Knowledge (University of Massachusetts Press, 1969).
Knowing and Being. Michael Polanyi (University of Chicago Press, 1969).
Interpretations of Life and Mind (Humanities Press, 1971).
Spinoza (Doubleday, 1973).
Topics in the Philosophy of Biology (with E. Mendelsohn) (D. Reidel, 1975).
Dimensions of Darwinism (Cambridge, 1983).
Spinoza and the Sciences (with Debra Nails) (D. Reidel, 1986).

ARTICLES

'Relativism and Philosophic Methods', *Philosophical Review,* **XLVI** (1937), pp. 649–656. (As M. Glicksman)
'A Note on the Philosophy of Heidegger', *The Journal of Philosophy,* **XXXV** (1938), pp. 93–104. (As M. Glicksman)
'Discipline and the Efficacy of Ideals', *Journal of Higher Education,* **IX** (1938), pp. 175–182. (As M. Glicksman)
'Theories of Interpretation in the Law of Contracts', *The University of Chicago Law Review* (1939), pp. 374–398.

'Hume: Sceptic and Tory?', *Journal of the History of Ideas*, **IV** (1943), pp. 333–348.
'Gerard's Essay on Taste', *Modern Philology*, **XLI** (1943), pp. 45–58.
'An Implicit Premise in Aristotle's *Ethics*', *Ethics*, **LVI** (1946), pp. 131–135.
'Kierkegaard: The Philosophy', *Kenyon Review*, **IX** (1947), pp. 48–69.
'L'Homme est une Passion Inutile: Sartre and Heidegger', *Kenyon Review*, **IX** (1947), pp. 167–185.
'Two More Existentialists: Karl Jaspers and Gabriel Marcel', *Kenyon Review*, **IX** (1947).
'On Some Distinctions Between Men and Brutes', *Ethics*, **LVII** (1947), pp. 121–127.
'Sartre's Theory of the Emotions', *Yale French Studies*, **I** (1948), pp. 97–101.
'Authenticity: An Existential Virtue', *Ethics*, **LXII** (1952), pp. 266–274.
'Men and Ideas: David Hume', *Encounter*, **IV** (1955), pp. 54–60.
'Two Evolutionary Theories', *The British Journal for the Philosophy of Science*, **IX** (1958), pp. 110–127; 185–193.
'Heidegger: Philosopher and Prophet', *Twentieth Century*, **CLXIV** (1958), pp. 545–555.
'The German Existentialists,' *The Chicago Review*, **XIII** (1959), pp. 49–58.
'The Faith of Darwinism', *Encounter*, **XIII** (1959), pp. 48–56.
'Statistics and Selection', *The British Journal for the Philosophy of Science*, **XII** (1961), pp. 25–42.
'The Logic of Biology', *The Logic of Personal Knowledge, Essays Presented to Michael Polanyi on His Seventieth Birthday, 11th March 1961* (London: Routledge and Kegan Paul, 1961), pp. 191–205.
'Leibniz's Cosmic Equation: A Reconstruction', with J. R. Ravetz, *Journal of Philosophy*, **LIX** (1962), pp. 141–146.
'Karl Jaspers: A Philosopher of Humanity', *Times Literary Supplement* (April 12, 1963), pp. 241–242.
'Causes', *Philosophy*, **XXXVIII** (1963), pp. 149–159.
'Two Evolutionary Theories: A Reply', *The British Journal of the Philosophy of Science*, **XIV** (1963), pp. 140–154.
'Beyond Darwinism: Portmann's Thought', *Commentary*, **XL** (1965), pp. 31–38.
'Descriptive and Prescriptive Statements', *Proceedings of Seminar, Institut für Wissenschaftslehre* (Salzburg, 1966).
'Positionality in the Philosophy of Helmuth Plessner', *Review of Metaphysics*, **XX** (1966), pp. 250–277
'Discussion: Mr. Manier's 'Theory of Evolution as Personal Knowledge; A Quasi-Reply', *Philosophy of Science*, **XXXIII** (March-June 1966).
'Straus's Phenomenological Psychology', *Review of Metaphysics*, **XXI** (1967), pp. 94–123.
'Martin Heidegger', *The Encyclopedia of Philosophy*, **III** (1967), pp. 459–465.
'Biology and the Problem of Levels of Reality', *New Scholasticism*, **XLI** (1967), pp. 427–449.
'Tacit Knowing and the Pre-Reflective Cogito', *Intellect and Hope* (Durham, North Carolina: Duke University Press, 1968).
'Hobbes and the Modern Mind: An Introduction', *The Anatomy of Knowledge* (London: Routledge and Kegan Paul; Amherst: University of Massachusetts Press, 1969), pp. 1–28.

'Bohm's Metaphysics and Biology', *Towards a Theoretical Biology*, **II** *Sketches*, ed., C. H. Waddington (Edinburgh: Edinburgh University Press, 1969).

'Notes on Maynard-Smith's "Status of Neo-Darwinism"', *Towards a Theoretical Biology* (Edinburgh: Edinburgh University Press, 1969).

'Sartre: A Philosophic Study', *Mind*, **LXXVIII** (1969), pp. 143–153

'Philosophical Anthropology', *Philosophy 1955–65*, ed. R. Klibansky, International Institute of Philosophy (Paris, 1969).

'Hierarchy: One Word, How Many Concepts?', *Hierarchical Structures*, ed. Lancelot Law Whyte, *et al.* (New York: American Elsevier Publishing Company, 1969).

'Die Erkenntnistheoretische Grundlage der Anthropologie', *The Human Context*, **II** (London: 1970).

'The Aesthetic Dialogue of Sartre and Merleau-Ponty', *Journal of the British Society for Phenomenology*, **I** (1970), pp. 59–72.

'Foreword', *Laughing and Crying*, Helmuth Plessner (Evanston: Northwestern University Press, 1970).

'Sense Perception: Philosophy's Step-Child?', *Aisthesis and Aesthetics*, ed. Erwin Straus (Pittsburgh: Duquesne University Press, 1970).

'Reducibility: Another Side Issue', *Interpretations of Life and Mind* (New York: Humanities Press; London: Routledge and Kegan Paul, 1971), pp. 14–37.

'Two Evolutionary Theories' and 'Two Evolutionary Theories: A Reply', reprinted in *Man and Nature* (New York: Dell Publishing Company, 1971).

'Aristotle and Modern Biology', *Journal of the History of Ideas*, **XXXIII** (July–September 1972), pp. 395–424.

'Jean-Paul Sartre: French Existentialism and Politics — The New Revolutionary', reprinted in *Political Obligation and Civil Disobedience:Readings*, eds. Michael P. Smith and Kenneth L. Deutsch (New York: Thomas Y. Crowell, 1972).

'On First Reading *L'Idiot de la Famille*', *California Quarterly* (June 1972), pp. 33–55.

'Polanyi et la Philosophie Francaise', *Archives de Philosophie*, **XXXV** (1972), pp. 3–5

'Sartre and the Other', (Presidential Address delivered before the 46th Annual Pacific Division Meeting of the American Philosophical Association, San Francisco, March 24, 1972), *Proceedings of the APA* (1973), pp. 22–41.

'People and Other Animals', *Philosophical Forum*, (1972), pp. 157–172; and *Philosophia Naturalis*, **XIV** (1973), pp. 25–38.

'Is Genus to Species as Matter to Form? Aristotle and Taxonomy', *Synthese*, **XXVIII** (1974), pp. 51–69.

'Darwin and Philosophy', *Academie Royale de Belgique: Connaissance Scientifique et Philosophie*, Colloque Organise les 16 et 17 Mai 1973.

'Three Aspects of Perception', *The Understanding of Nature* (Dordrecht: D. Reidel, 1974), pp. 13–34.

'Explanation and Evolution', *The Understanding of Nature* (Dordrecht: D. Reidel, 1974), pp. 206–227.

'Immer Noch Philosophie?', *Sachlichkeit*, Festschrift for Helmuth Plessner (1974), pp. 93–100.

'Means, Ends and Functions', Editorial, *American Review of Respiratory Disease*, **CXII** (1975), pp. 325–328.

'Life, Death and Language: Some Thoughts on Wittgenstein and Derrida', *Partisan Review*, **XLIII** (1976), pp. 265–279.

'Merleau-Ponty and the Renewal of Ontology', *Review of Metaphysics*, **XXIX** (1976), pp. 605–625.
'Mind and Brain: The Embodied Person', in *Philosophical Dimensions of the Neuro-Medical Sciences*, eds. S. Spicker and H.T. Engelhardt (Dordrecht: D.Reidel, 1976), pp. 113–129.
'To Have a Mind . . . ', *Journal of Medicine and Philosophy*, **I** (1976), pp. 177–199.
'Imre Lakatos: Some Recollections', in *Essays in Memory of Imre Lakatos*, eds. R. S. Cohen, *et al.* (Dordrecht: D. Reidel, 1976), pp. 209–212.
'Philosophy of Medicine: Prolegomena to a Philosophy of Science', *PSA 1976*, **II** (1977), pp. 77–93.
'Tacit Knowing: Grounds for a Revolution in Philosophy', *Journal of the British Society for Phenomenology*, **VIII** (October 1977), pp. 164–171.
'Individuals and Their Kinds: Aristotelian Foundations of Biology', in *Organism, Medicine, and Metaphysics*, ed. S. F. Spicker (Dordrecht: D. Reidel, 1978), pp. 121–136.
'The Paradoxes of Historicity', *Review of Metaphysics*, **XXXII** (1978), pp. 15–36.
'Knowledge, Belief and Perception: The Andrew W. Mellon Lectures, Fall 1978', (New Orleans: The Graduate School, Tulane University, 1978).
'Sociobiology and the Human Mind', *Sociobiology and Human Nature*, eds. M. S. Gregory, A. Silvers, D. Sutch (San Francisco: Jossey-Bass, 1978).
'Comments on Pellegrino's "Anatomy of Clinical Judgments"', in *Clinical Judgment: A Critical Appraisal*, eds. Engelhardt, Spicker and Towers (Dordrecht: D. Reidel, 1979), pp. 195–197
'Reply to Dr. Clarke', *Philosophy of Science*, **XLVI** (1979), pp. 326–327.
'A Note on Simberloff's "Succession of Paradigms in Ecology"', *Synthese*, **XLIII** (1980), pp. 41–45.
'The Sense of Things', *The Journal of Aesthetics and Art Criticism*, **XXXVIII** (1980), pp. 377–389.
'Changing Concepts of Darwinian Evolution', *The Monist*, **LXIV** (April 1981, pp. 195–213.
'Sartre and the Tradition', *Philosophical Quarterly* (1981).
'Empiricism and the Philosophy of Science or *n* Dogmas of Empiricism', *Epistemology, Methodology and the Social Sciences*, ed. R. S. Cohen and M. W. Wartofsky (D. Reidel, 1983).
'Philosophy of Biology, 1983: Problems and Prospects,' *Proceedings of the 7th International Congress for Logic, Methodology and Philosophy of Science* 1984.
'Idea and Judgment in the 3rd *Meditation:* An Object Lesson in Philosophical Historiography', *Independent Journal of Philosophy* (1985).
'Perception and Interpretation in the Sciences: Toward a New Philosophy of Science', in *The New Biology and the New Philosophy of Science* (Bradford Books, 1985).

TRANSLATIONS FROM THE GERMAN

C. F. v. Weizsäcker, *The World View of Physics* (University of Chicago, 1952).
Jahnheinz Jahn, *Muntu* (Faber and Faber, 1961).
Helmuth Plessner, *Laughing and Crying*, with James Churchill (Northwestern University Press, 1970).

INDEX OF NAMES

INDEX OF SUBJECTS